Science, Technology and Medicine in Modern History

General Editor: John V. Pickstone, Centre for the History of Science, Technology and Medicine, University of Manchester, UK (www.man.ac.uk/CHSTM)

One purpose of historical writing is to illuminate the present. At the start of the third millennium, science, technology and medicine are enormously important, yet their development is little studied.

The reasons for this failure are as obvious as they are regrettable. Education in many countries, not least in Britain, draws deep divisions between the sciences and the humanities. Men and women who have been trained in science have too often been trained away from history or from any sustained reflection on how societies work. Those educated in historical or social studies have usually learned so little of science that they remain thereafter suspicious, overawed or both.

Such a diagnosis is by no means novel, nor is it particularly original to suggest that good historical studies of science may be peculiarly important for understanding our present. Indeed, this series could be seen as extending research undertaken over the last half-century. But much of that work has treated science, technology and medicine separately; this series aims to draw them together, partly because the three activities have become ever more intertwined. This breadth of focus and the stress on the relationships of knowledge and practice are particularly appropriate in a series that will concentrate on modern history and on industrial societies. Furthermore, while much of the existing historical scholarship is on American topics, this series aims to be international, encouraging studies on European material. The intention is to present science, technology and medicine as aspects of modern culture, analysing their economic, social and political aspects, but not neglecting the expert content, which tends to distance them from other aspects of history. The books will investigate the uses and consequences of technical knowledge, and how it was shaped within particular economic, social and political structures.

Such analyses should contribute to discussions of present dilemmas and to assessments of policy. 'Science' no longer appears to us as a triumphant agent of Enlightenment, breaking the shackles of tradition, enabling command over nature. But neither is it to be seen as merely oppressive and dangerous. Judgement requires information and careful analysis, just as intelligent policy-making requires a community of discourse between men and women trained in technical specialities and those who are not.

This series is intended to supply analysis and to stimulate debate. Opinions will vary between authors; we claim only that the books are based on searching historical study of topics which are important, not least because they cut across conventional academic boundaries. They should appeal not just to historians, or just to scientists, engineers and doctors, but to all who share the view that science, technology and medicine are far too important to be left out of history.

Titles include:

Julie Anderson, Francis Neary and John V. Pickstone
SURGEONS, MANUFACTURERS AND PATIENTS
A Transatlantic History of Total Hip Replacement

Roberta E. Bivins
ACUPUNCTURE, EXPERTISE AND CROSS-CULTURAL MEDICINE

Linda Bryder
WOMEN'S BODIES AND MEDICAL SCIENCE
An Inquiry into Cervical Cancer

Roger Cooter
SURGERY AND SOCIETY IN PEACE AND WAR
Orthopaedics and the Organization of Modern Medicine, 1880–1948

Jean-Paul Gaudillière and Ilana Löwy (*editors*)
THE INVISIBLE INDUSTRIALIST
Manufacture and the Construction of Scientific Knowledge

Jean-Paul Gaudillière and Volker Hess (*editors*)
WAYS OF REGULATING DRUGS IN THE 19TH AND 20TH CENTURIES

Christoph Gradmann and Jonathan Simon (*editors*)
EVALUATING AND STANDARDIZING THERAPEUTIC AGENTS, 1890–1950

Sarah G. Mars
THE POLITICS OF ADDICTION
Medical Conflict and Drug Dependence in England since the 1960s

Alex Mold and Virginia Berridge
VOLUNTARY ACTION AND ILLEGAL DRUGS
Health and Society in Britain since the 1960s

Ayesha Nathoo
HEARTS EXPOSED
Transplants and the Media in 1960s Britain

Neil Pemberton and Michael Worboys
MAD DOGS AND ENGLISHMEN
Rabies in Britain, 1830–2000

Neil Pemberton and Michael Worboys
RABIES IN BRITAIN
Dogs, Disease and Culture, 1830–1900

Cay-Rüdiger Prüll, Andreas-Holger Maehle and Robert Francis Halliwell
A SHORT HISTORY OF THE DRUG RECEPTOR CONCEPT

Thomas Schlich
SURGERY, SCIENCE AND INDUSTRY
A Revolution in Fracture Care, 1950s–1990s

Eve Seguin (*editor*)
INFECTIOUS PROCESSES
Knowledge, Discourse and the Politics of Prions

Crosbie Smith and Jon Agar (*editors*)
MAKING SPACE FOR SCIENCE
Territorial Themes in the Shaping of Knowledge

Stephanie J. Snow
OPERATIONS WITHOUT PAIN
The Practice and Science of Anaesthesia in Victorian Britain

Carsten Timmermann and Julie Anderson (*editors*)
DEVICES AND DESIGNS
Medical Technologies in Historical Perspective

Carsten Timmermann and Elizabeth Toon (*editors*)
CANCER PATIENTS, CANCER PATHWAYS
Historical and Sociological Perspectives

Jonathan Toms
MENTAL HYGIENE AND PSYCHIATRY IN MODERN BRITAIN

Duncan Wilson
TISSUE CULTURE IN SCIENCE AND SOCIETY
The Public Life of a Biological Technique in Twentieth Century Britain

---

Science, Technology and Medicine in Modern History
Series Standing Order ISBN 978–0–333–71492–8 hardcover
Series Standing Order ISBN 978–0–333–80340–0 paperback
(*outside North America only*)

You can receive future titles in this series as they are published by placing a standing order. Please contact your bookseller or, in case of difficulty, write to us at the address below with your name and address, the title of the series and one of the ISBNs quoted above.

Customer Services Department, Macmillan Distribution Ltd, Houndmills, Basingstoke, Hampshire RG21 6XS, UK

Mental Hygiene and Psychiatry in Modern Britain

Jonathan Toms

Fellow of the Institute for the History and Work of Therapeutic Environments, UK

palgrave
macmillan

First published 2013 by
PALGRAVE MACMILLAN

Palgrave Macmillan in the UK is an imprint of Macmillan Publishers Limited, registered in England, company number 785998, of Houndmills, Basingstoke, Hampshire RG21 6XS.

Palgrave Macmillan in the US is a division of St Martin's Press LLC, 175 Fifth Avenue, New York, NY 10010.

Palgrave Macmillan is the global academic imprint of the above companies and has companies and representatives throughout the world.

Palgrave® and Macmillan® are registered trademarks in the United States, the United Kingdom, Europe and other countries.

ISBN 978-1-349-45808-0   ISBN 978-1-137-32001-8 (eBook)
DOI 10.1057/9781137320018

This book is printed on paper suitable for recycling and made from fully managed and sustained forest sources. Logging, pulping and manufacturing processes are expected to conform to the environmental regulations of the country of origin.

A catalogue record for this book is available from the British Library.

A catalog record for this book is available from the Library of Congress.

10   9   8   7   6   5   4   3   2   1
22 21 20 19 18 17 16 15 14 13

For Heidi, Amy and Hannah

Contents

Preface and Acknowledgements   ix

List of Abbreviations   xi

1   Moral Treatment and 'The Dialectic of the Family'   1

2   Moral Treatment for the Community at Large   7

3   The Mental Hygiene Movement's Emotional Contradictions   28

4   Dialectic Rightside Up?   53

5   Developing in the Womb of the Old?   84

6   Alternative Dialectics   104

7   Alienation Revisited   137

8   Dialectic Dismembered   172

Afterword   203

Notes   207

Bibliography   237

Index   255

Preface and Acknowledgements

This book centres on the twentieth-century mental hygiene movement in Britain and its place in the history of modern psychiatry. It is written as a counter-argument to influential Foucauldian descriptions of the history of psychiatry. It is also a reinterpretation of important elements of modern British psychiatry in its own right.

This reinterpretation, in fact, takes, as its source, Michel Foucault's description of moral treatment at the York Retreat at the turn of the nineteenth century. His appraisal of moral treatment is used to conceptually situate the mental hygiene movement within the broad sweep of modern British psychiatric history. It makes use of Foucault's assertion in *Madness and Civilization* that an ongoing 'dialectical' process was founded concurrently with the emergence of modern psychiatry and was inseparable from its development. I emphasize, however, that the structure within which this apparent dialectic was established was both political and psychological in content, and in ways that Foucault disavowed. In particular, this book offers a very different reading of the connections between notions of self-government and mental health to those that the later Foucauldian theory of power/knowledge and concept of 'governmentality' allows.

This book traces one possible historical route through which moral treatment's essential organizing principle and methodology might have been transmitted and transformed during the nineteenth century. It then examines further amendments evident with the emergence and development of the mental hygiene movement itself. Through this means the book attempts to unravel the significance and trajectory of the mental hygiene movement, from its beginnings through to the connections between its demise and the rise of a rights approach to mental health.

The book's structure and style is intended as an inherent element in its critique of Foucauldian approaches. As it progresses, the text deliberately introduces elements that challenge the habitual disavowal, in contemporary Foucauldian academic work, of any tone that can be labelled sentimental or frivolous. Elements sown into the text from the life and work of W. David Wills are an example of this. Wills was one of the early British psychiatric social workers trained in New York and, subsequently, a 'progressive educationist' working close to the mental hygiene movement. The excerpts operate as a counterpoint to the main body of the text, highlighting an important thread of the 'dialectical' story being told, and deliberately inhabiting the narrative with a more personal and less formal tone. They are not intended to constitute a biography or an appraisal of his career.

Beyond these theoretical considerations my overall intention has been to provide a sense of historical discovery and, with this in mind, I have kept signposting in the text to a minimum.

I would like to thank Professor Roger Cooter and Rhodri Hayward who supervised the PhD thesis from which a good deal of the framework of the present book was developed. I would also like to thank Mathew Thomson who supervised my subsequent and related Research Fellowship at the University of Warwick. I have benefited greatly from their support, advice and criticism. I thank also all the members of the Centre for the History of Medicine at Warwick who, during my time, there were unfailingly friendly, supportive and stimulating. I am indebted to the Wellcome Trust for funding my PhD studies and subsequent Fellowship.

I am very grateful to the mental health charity MIND who, some years ago now, allowed me access to the store cupboard that, at that time, served as their archive. Katherine Darton was especially helpful and supportive. The late Edith Morgan OBE also very kindly allowed me access to her personal papers. I would also like to express my sincere thanks to the Planned Environment Therapy Trust at Toddington in Gloucestershire. They provide a warm, engaging and stimulating environment to everyone who visits. I thank them all for their many kindnesses. John Cross was supportive (as well as giving me lifts to the station), as was Cynthia Cross, and Craig Fees, in particular, gave me much help and advice. I have benefited greatly from his knowledge. I am grateful to Liberty (formerly the National Council for Civil Liberties) for permission to access their archives at Hull Archive Centre, Hull University. Many thanks to the archivists at the University of Warwick's Modern Records Centre, Hull History Centre at Hull University Archives, the Wellcome Library and archives, and the Institute of Education Archives at the University of London.

I gratefully acknowledge Concord Media for allowing me to quote from the film *Mentally Handicapped Children Growing Up*, director Basil Wright, Realist Film Unit.

Roger Cooter, Craig Fees, Tom Harrison, Sarah Hayes, Bill Luckin and Mathew Thomson all very kindly read drafts of the book. I am very grateful for their time and advice. Responsibility for the content of the book is, of course, mine alone.

I would also like to thank Rob Kirk for his friendship, advice and support. Pam Dale and Sarah Hayes have both been kind and supportive, and Sheena Rolph gave me last-minute advice. Thanks as well to James Knight. Finally, and primarily of course, I must thank my partner Heidi for her unending patience, always-valuable suggestions and love.

List of Abbreviations

APSW	Association of Psychiatric Social Workers
BMJ	*British Medical Journal*
CAMW	Central Association for Mental Welfare
CGC	Child Guidance Council
COS	Charity Organisation Society
CRS	Civil Resettlement Scheme
GBAs	General Bugger Abouts
HSC	Home and School Council
ISTD	Institute for the Scientific Treatment of Delinquency
MHEC	Mental Health Emergency Committee
NAMH	National Association for Mental Health
NCCL	National Council for Civil Liberties
NCMH	National Council for Mental Hygiene
NEF	New Education Fellowship
PNC	Provisional National Council for Mental Health
RMPA	Royal Medico-Psychological Association
SACS	Social After-Care Scheme
WHO	World Health Organization

1
Moral Treatment and 'The Dialectic of the Family'

Could we ever truly know David Wills' thoughts and emotions as he prepared to disembark ship at New York harbour? The year was 1929 and this was Wills' first trip abroad. A few days after his arrival he began a daily journal. His first entry reflected, 'I was far too thrilled, tired and miserable (if there can exist such a combination!) to write it in the first two days'.[1]

Wills was joining the New York School of Social Work and would soon become one of the earliest English psychiatric social workers to be trained in the USA. Psychiatric social work was the new and radiant profession of the US mental hygiene movement. Full of youthful hope and ambition it was eager to apply the modern knowledge of 'the new psychology' within psychiatry, within social work, anywhere. Wills had hopes and ambitions, too. He hoped to find out more about this so-called new psychology on which the mental hygiene movement rested its claims, to apply it to his interest in rehabilitating delinquent boys and, at the same time, gain a useful professional qualification.

Wills was a Quaker. A few years after his return to England he was to have a letter published in the Quaker periodical *The Friend* appealing to the community to support a more humanitarian and radical approach to the treatment of young offenders. He soon made contact with a group of educationists and psychotherapists attempting to organize a similar experiment. Their subsequent collaboration was closely associated with the mental hygiene movement, and, though short lived, it was extremely influential, as we'll see.

There are echoes here of a much earlier Quaker venture. Back in the 1790s Quakers in York had been called upon to develop a more humanitarian approach to the treatment of people considered mad. The result was the Quaker Retreat at York and its development of a form of 'moral therapy' known as 'moral treatment'. But the links between Wills' experiment and the Retreat go beyond a shared religion. In fact, they go to the heart of this history of the English mental hygiene movement.

Moral therapy's importance in the history of psychiatry is often highlighted. There are, to begin with, obvious commonalities between moral treatment at the Retreat and contemporary psychiatric ideology. The approach emphasized that early detection and treatment of mental troubles was essential for successful recovery. Treatment should be without recourse to violence and avoid restraint wherever possible. An emphasis was placed on encouraging the 'healthy' aspects of the patient's mind. No patient was considered to be beyond all calls of reason or affection; engagement with an encouraging relationship aimed to re-connect the rational mind and promote recovery. Isolation and inactivity simply appeared to encourage mental problems, and so purposeful activity was encouraged. The actual application of these beliefs in contemporary psychiatric practice has been doubted frequently, but their rhetorical place in psychiatric ideology is not. As will become clear, the twentieth-century mental hygiene movement was a carrier of all these principles.

But, hidden at the heart of these values there lies a more fundamental issue. It runs like a thread from moral therapy, through to the mental hygiene movement and beyond. This is the issue of authority. As we will see shortly, it was crucial to moral therapists. Likewise, it was central to the concerns of mental hygienists.

During the 1960s a new academic questioning of the history of social welfare emerged. Psychiatry was an important target and moral therapy one area to come under analysis.[2] The earliest, and still the most notorious, academic critique was contained in the French philosopher Michel Foucault's book, *Madness and Civilization*.[3] A look at its essentials offers us a means to appreciate the importance of authority to our story and a starting point from which to unravel its thread.

Foucault began the section that dealt with moral therapy and the York Retreat with these words:

> We know the images. They are familiar in all histories of psychiatry, where their function is to illustrate that happy age when madness was finally recognized and treated according to a truth to which we had too long remained blind.[4]

This was the signal for an attempted demolition of the identification of medicine and psychiatry with 'progress' and 'enlightenment'. Indeed, it was an attempted demolition of the very ideas of 'progress' and 'enlightenment'.

Outright rejected by some historians, embraced by others, Foucault's analysis of moral treatment and the Retreat was part of a grander project. In its early rendition, as expressed in *Madness and Civilization*, this took the form of a radical critique of Reason itself, by means of a historical description of the forms in which it had cast 'madness'. Foucault's history couldn't help but appear romantic, the more so because of its flamboyant writing style.

But how could Foucault provide a history of 'madness' if the practice of history was itself an aspect of Reason? How, in fact, could Foucault avoid performing another manoeuvre that enclosed 'madness'? The philosopher Jacques Derrida was soon to ask the question. As Gary Gutting later put it, it seemed as if Foucault was depicting madness as an 'infrarational source of fundamental truth'.[5] The criticism hit home, and Foucault rapidly shifted his position.

Foucault's solution, in a nutshell, was to substitute the commonly accepted couplet of logic, 'knowledge/truth', with a newer one, 'knowledge/power' or 'power/knowledge', as he was to order it. Ask any theorist now the one word that immediately conjures up Foucault's thought and the answer will inevitably be 'power'. Yet, look back at his chapter on moral therapy in *Madness and Civilization* and you will see that, in fact, it is authority, just as much as power, that leaps off the page at you as a crucial component, through which the 'free terror of madness' was replaced with 'the stifling anguish of responsibility'.[6]

The York Retreat opened in 1796. It was a response to a very particular set of circumstances. A scandal at the local York asylum, in which a Quaker patient had died, precipitated calls from a group of Quakers for an asylum to cater for their brethren. The practices that they evolved there are elements of what the historian Roy Porter has called a 'psychological turn', which took place after 1750. This can't, of course, be located only in approaches to madness. But, the particular form it took in this sphere is expressed in the general activities of reform across Europe known as moral therapy, or moral treatment. To social reformers, such activities seemed a sign that newer, more rational and humane approaches to madness were at hand.

Inspired, in particular, by John Locke's sensationalist theory of the human mind, alienists began to see the ideas and associations of the mind as the main target for investigation and treatment.[7] Madness could be attributed to misplaced association of thoughts and feelings, leading to a loss of authority over one's mental state. In this sense it was a failure of 'self-government' and a form of mental 'alienation'.[8] Moral therapists aimed to reconstruct self-government in the minds of the mad through a personal relationship of authority.

Foucault didn't refer to Locke in his analysis of moral therapy. He relied, instead, on Descartes' 'cogito' to represent the transformation through which rationality placed itself in a position of supreme opposition to madness. He saw the pre-eminent issue regarding the Retreat and moral therapy in general as the introduction of a form of surveillance and judgement that was 'a mediating element between guards and patients, between reason and madness'. Before this, he claimed, there was only an 'abstract, faceless power' which kept mad people confined and didn't penetrate madness itself.[9]

Foucault emphasized that moral treatment adopted the bourgeois family as its model of care and treatment. This has since been noted by other writers. But Foucault made much more of it than most. The family was the most immediately recognizable, and concretely understood, social institution. Foucault described it as 'simultaneously imaginary landscape and real social structure' and made clear that, at the Retreat, it wasn't just a model of care—it was both an organizing principle for conceptualizing madness and a methodology for treating it. In this it accomplished a psychological task. For Foucault, the Retreat made the family 'perform a role of disalienation'.[10] This family was a hierarchical structure of authority and power. Taking the form of patriarchy it played a primary role in structuring the external conceptualization of madness and the internal experience of it. Its structure would now constitute both the 'truth' of madness, and the imperative for its treatment.

In his *Description of the Retreat* Samuel Tuke had remarked that 'There is much analogy between the judicious treatment of children, and that of insane persons'.[11] Foucault recognized that dealing with mad people by analogy with children had a much longer history, but he saw the sentiment expressed by Tuke to be of a different order. Moral treatment turned previous analogies of madness with childhood into a psychological mode of relation. For Foucault, it meant that madness was made to take on a perpetual 'minority status'.

Along with this imposition of a 'minority status' the bourgeois family also instilled its belief in the moral value of work. Tuke explained this at the Retreat in terms of the empirical discovery that it provided the patients with purpose and encouraged those elements of their minds which remained un-affected by madness. For Foucault, work had long been an element of moralizing power. At the Retreat, it took on only the guise of a therapeutic. Beneath this it was another means to structure the very experience of madness, and enclose it in order and control.[12]

So the moral therapy exemplified at the Retreat was a form of power and authority that organized the experience of madness on its own terms. Madness couldn't exist except as a form of childhood. And the frame for this childhood was a set of hierarchical relations founded on an ideal of patriarchal family relations. In Foucault's eyes these produced not only a conceptualization and treatment of madness, but a way in which madness must experience itself. What Foucault gave us was a description of the family with a capital 'F'—it was the Family. It was both organizing principle and methodology, and, in the face of it madness, was 'alienated in guilt'.[13] His analysis remains useful and illuminating.

Yet Foucault's examination of authority was truncated. He relayed a passage from Samuel Tuke's *Description of the Retreat* to depict the role of the moral therapist:

> The superintendent was one day walking in a field adjacent to the house, in company with a patient, who was apt to be vindictive on very slight occasions. An exciting circumstance occurred. The maniac retired a few

paces, and seized a large stone, which he immediately held up, as in the act of throwing at his companion. The superintendent, in no degree ruffled, fixed his eye upon the patient, and in a resolute tone of voice, at the same time advancing, commanded him to lay down the stone. As he approached, the hand of the lunatic gradually sunk from its threatening position, and permitted the stone to drop to the ground. He then submitted to be quietly led to his apartment.[14]

'Something had been born', commented Foucault, 'which was no longer repression, but authority... The space reserved by society for insanity would now be haunted by those who were "from the other side" and who represented both the prestige of the authority that confines and the rigor of the reason that judges'. But, actually, Foucault's pitting of 'Reason' against 'madness' entailed a slippage away from analysing the dynamics of authority along with power. He wrote that 'In fact, though, it is not as a concrete person that he [the superintendent] confronts madness, but as a reasonable being, invested by that very fact, and before any combat takes place, with the authority that is his for not being mad', adding that 'unreason's defeat [is] inscribed in advance' in the battle between 'madman and man of reason'.[15] So any examination of the relations of authority between moral therapist and madman was lost through a description that implied a disembodied and binary split between power as Reason, and madness. In fact, Foucault's image of madness suggested something with all the qualities of a spirit. And his 'man of reason' stalked madness like a cleric performing an exorcism—except, of course, that he, too, appeared a spook.

Samuel Tuke, however, clearly saw a dynamic relationship between authority and power in the treatment of madness. He was not, for instance, unaware that paternalistic authority might itself need restraining:

What a reflection upon human nature, that the greatest calamity to which it is incident, should have been frequently aggravated by those who had the power, and whose duty it was to employ means of mitigation. Hence, we may derive a practical comment on the observation of the wise Montesquieu, which every one interested in establishments for the insane ought constantly to remember:

"Cest une expérience éternelle, que tout homme qui a du pouvoir est porté, à en abuser; il va jusqu' à ce qu'il trouve des limites. Qui le diroit! La vertu même a besoin des limites*

L'Esprit des Loix, Liv. II. Cap. IV.

*Experience continually demonstrates, that men who possess power, are prone to abuse it: they are apt to go to the utmost limits. May it not be said, that the most virtuous require to be limited?[16]

So Foucault's analysis of the Retreat offered an illuminating description of the Family as organizing principle and methodology. But, though he cited authority as an intrinsic element, any examination of its dynamics slipped away under a rendition that emphasized a binary battle of power.

Foucault concluded that 'Henceforth, and for a period of time the end of which is not yet possible to predict, the discourse of unreason will be indissociably linked with the half-real, half-imaginary dialectic of the Family'.[17] In retrospect it's a curious statement. In Britain moral treatment died out as the nineteenth-century public asylum system rapidly expanded along with medical practitioners' dominance within it. Foucault maintained that, in fact, psychoanalysis became the exemplar of the Family principle as its 'dialectic' progressed. Yet, even if we accept this as it was baldly stated, he offered no examination of intermediary processes.

And Foucault's use of the terminology of the dialectic might also seem odd, as his entire corpus can be read as a sustained attempt to destroy any philosophy of history derived from Hegel. But this, however, is less problematic. Even here, with this terminology, he was trying to turn the tables on Hegel—historicizing the dialectic itself and, in the process, undermining *any* derivative notions of progressive historical dialectic and associated notions of human alienation. For Foucault, dialectics were injected into history and the Family was a prime location. Both alienation and the means of disalienation required to be seen through, not investigated.

But, even so, let's take this 'half-real, half-imaginary dialectic of the Family' seriously—or half-seriously anyway. Foucault focussed on how moral treatment made madness permanent childhood. But the issue from Tuke onwards became 'Of what is this 'childhood' constituted? How might it be brought to 'maturity'? What is the productive relationship between 'our' authority and 'their' growth to maturity? All aspects of a dialectical process, perhaps. And, if we trace out some of the ways in which it might seem to work itself out, we find that relationships of authority are crucial to the contradictions that give it motion.

The dialectic of the Family—let's hitch a ride and see where it takes us. Let's follow the dialectic of moral treatment and let's also follow David Wills as he travels into the heart of the mental hygiene movement.

2
Moral Treatment for the Community at Large

Maybe moral treatment didn't die a lonely death in the asylum. Maybe it shifted outwards from the asylum to the community.

The French sociologist Robert Castel claimed, in the 1980s, that the radical treatments associated with post-World War II social psychiatry held a guilty secret. Their origins were that of moral treatment. But, as their radical credentials disclaimed any association with moral agents of social control, they repressed any memory of such origins.[18] Castel was right about the continuity, but, as we'll see, he was wrong that memories were guiltily repressed. It's really that the continuities were so much more complicated and less linear than the later psychiatrists allowed. And, in fact, Castel inadvertently provides a clue to the path of this more complicated tale. According to him, moral therapists' organization of their institutions represented an attempt to produce an enclosed realm in which only their own reasonable and unyielding mind would hold sway; a realm in which the 'intrusion of history' would be negated in order to provide a clean slate on which the pure 'will' of the therapist could be inscribed.[19] But if we trace how moral treatment can be understood as having diverted outwards from the asylum into the community, then we can also see how history was re-animated in its support. This shift involved a classification and ordering of minds grounded in a particular knowledge of developmental time. History was co-opted as part of the authority of the modern moral therapist. But, at the same time, the Family remained central, as organizing principle and methodology. So, too, the accompanying presumption that reason is adulthood and madness is childhood, and the bourgeois association of rationality with the mental discipline of work.

* * *

Opportunity came, as it so often used to, in the form of a letter. Wills had been loitering with intent around the hall at Woodbrooke College waiting to pounce on afternoon tea when it arrived. The postman left the letter on a table nearby. Idle hands do the devils work, and Wills picked it up. It was addressed to his tutor J.C. Kydd, who lived in a nearby house. The frank mark showed it to be from the New York School of Social Work, and the envelope was unsealed. So Wills told the Warden's secretary that he would take it to Kydd. He did, but not before he'd slipped up to his room and found the contents to be a prospectus. All very interesting, and more interesting still was the list of 'fellowships' awarded each year. One was a Willard Straight Fellowship for a foreign student who intended to return to their own country to do social work. 'Tailor-made for W D Wills!'.[20]

* * *

Founded in 1869, the Charity Organisation Society (COS) represented the most prominent example of late Victorian philanthropic charity. From moral treatment to moral philanthropy, moral treatment lives on, I suggest, in the COS. It may have died in the asylum, but it was resurrected in the community. Like all resurrections, however, this revival of moral treatment had a countenance that was recognizably of its present time and place. It emerged in a very different social and political context to that of the earlier moral treatment.

There was a good deal of concern in nineteenth-century thought about an apparent shadow-side to the onward march of reason. Behind its leading light lay a growing spectre of social and moral estrangement. To many theorists and social reformers, the informal authority of shared values traditionally expressed through family and church, seemed undermined. Meanwhile, the state and the law, which appeared to be at least their partial successors, seemed inadequate to the task.

That was the backdrop. But the trigger for COS theorizing and activity was the perceived threat of the 'idle'. Samuel Tuke had caustically written, in his account of moral treatment at the Retreat,

> There is no doubt, that if the same exertions were used for this purpose, as are frequently employed to amuse the vain, the frivolous, and the idle, many more gleams of comfort would be shed over the unhappy existence of lunatics.

Peter Linebaugh has shown that the term 'idle' served multiple purposes and was widely used by the bourgeois class during the eighteenth century. As in the case of Jack Sheppard, for example, the notorious thief and popular working class hero, 'idleness' signified something more than simple inactivity and lack of effort. In the emerging capitalist economy it denoted refusal of the bourgeois work ethic, and its accompanying spatial and temporal control. It was a refusal of authority.[21]

Under the COS the idle were no longer to be 'amused'; they were to become the beneficiaries of an organized application of authority that would reposition them within the grain of reason and progress. During the nineteenth century this notion of 'idleness' became associated with the terms 'egoism' and 'individualism'. The word 'individualism' encompassed a cluster of concepts, including egalitarian ideas associated with 'the rights of man' and utilitarian doctrines of *laissez faire*. By the mid-nineteenth century it had become used as a derogatory label.[22] 'Egoism' became especially associated with it, and was seen as a subversive threat to the moral and political order. It was self-obsession, vanity, idleness.

But if, as I claim, this COS quest amounted to a form of moral treatment, why did it target 'the idle'? After all, there was no claim that 'the idle' were 'mad'. The question brings us back to the notion of 'mental alienation'. We've already seen its use by Tuke to denote loss of mental faculties. But there were other related meanings signifying separation and estrangement.[23] Egoism, individualism and idleness were all terms wrapped up with these notions of alienation. The late Robert Nisbet's mid-1960s analysis of social thought offers a useful means to appreciate its importance to our story.[24] A noted conservative thinker, Nisbet was something of an anomaly in a discipline that, since at least the early 1960s, had overwhelmingly stressed its radical credentials. But, as one of his obituarists has remarked, Nisbet's work was 'so resolutely unfashionable that he regularly came back into fashion'.[25] In the face of a growing influence of Marxist thought in mainstream sociology, Nisbet pointed out that, despite Marxism's influential employment of the term alienation, its actual usage in sociology displayed a content that was distinctly un-Marxist.[26] This content was related to a conflict between 'tradition' and 'modernity'.

Nisbet set out a group of paired concepts that he believed epitomized the conflict. 'Traditional' concepts, such as community, authority and status, sat in outright opposition to modern concepts, such as society, power and progress. Nisbet emphasized that the archetype of community 'both historically and symbolically' was the family, and that the nomenclature of the family was prominent in every expression of it.[27] In fact, although he didn't draw it out fully, the family clearly played a fundamental role for his thesis in general. 'Traditional' social organization was structured by authority relations that were deeply embedded in social institutions from the family through neighbourhood, parish and guild, and integrated throughout the social body. The form of this authority was primarily personal, whether it be that of the patriarchal head of the family or the sovereign. It was characterized by 'personal intimacy', 'emotional depth', 'moral commitment', 'loyalty' and 'duty'. Nisbet described how on this view the rationalist image of progressive society, along with associated philosophies of contractual relations and utilitarianism, separated the individual from deeper ties of community that provided social cohesion and secure personhood. He added that when people are separated from long established social institutions 'there arises, along with the specter of the lost individual, the specter of lost authority'.[28] It was this kind of view that provided the content for the term 'alienation'.

Nisbet's rendition of his concepts as opposites is overdrawn. It might be more accurate to view them as coordinates that social and political theorists attempted to place in new combinations in order to reconcile 'the traditional' and 'the modern'.[29] With this understanding his characterization of alienation goes to the heart of the COS's concerns. These centred, as I've said, on 'the idle'. It was around this issue that a regeneration of moral treatment took place. For the COS, as we will see, the Family wasn't only an organizing principle for the reassertion of rationality in the minds of the mad, it was the organizing principle for the emergence of rationality in each individual of the wider community. One implication of this was that ideas and practices that viewed the individual as necessarily separate, self-interested and rational failed to appreciate the role of moral authority in individual make-up. If two of the principal gains of progressive history were rationality and the self-motivated individual, then those 'dregs' that appeared to be settling at the bottom of society surely lacked the moral authority through which mature reason became instilled. They had not emerged to full reason. Their rationality was somehow stalled in its development.

This attempt to reassert moral authority in the community, on the basis that the self-sustaining rational individual wasn't simply foreordained, was clearly political. With this in mind, it seems a little strange that when Foucault described how the Family was employed at the Retreat as an organizing principle of Reason for its structuring of madness, he didn't explicitly examine the relation of this with the long existing analogy between authority relations in the political order and those of the family. The shadow is clearly there at the Retreat. We've already noted Samuel Tuke's quotation from Montesquieu. So, when Foucault made the important point that the role of 'disalienation' performed by the Family was one of an internalized structuring of madness, we tend to overlook that the Family also introduced a political order into this domain. Madness, the family and political theory were all linked by the concept of authority. Towards the end of the nineteenth century the COS re-imported this family paradigm to wider society, retaining and developing both its political and psychological imperatives.

* * *

Up the road, Wills confessed to Kydd and asked if there was any reason why he shouldn't have a go at the Fellowship? Kydd didn't see why not, but he didn't think Wills would have much chance either.

'Have a try if you like', he said. He picked up the letter.

'Here – you'd better have this. You seem to think it's yours anyway. You might lend it to me sometime when you've no further use of it'.

The response to Wills' application contained a questionnaire that amounted to a complete personal and family history. 'Name the last three books you have read'. Wills thought of putting Kant's *Critique of Pure Reason*, Freud's *Interpretation of Dreams* and, 'for entertainment', Burton's *Anatomy of Melancholy*. He and his fellow students had good laugh about that, but, in the end, he put the 'very unimpressive truth'. Years later Wills could only remember that one of them was a P.G. Wodehouse. A lucky choice as it turned out. He later found out that Walter Pettit, the Assistant Dean of the University was a Wodehouse fan. 'Here's a Limey with a sense of humour. A man like that deserves to be encouraged'.[30]

* * *

The household, and the family, had long been employed as symbolic representations for the political order. Gordon Schochet has described how from the end of the seventeenth century political debate in England became distinguished by an attempt to replace the familial metaphor with a contractual one. The rational and conventional state was replacing traditional authority.[31] But he also noted that traditional, 'non-rational', authority was preserved in the household.[32] This is important because the quest upon which the COS embarked in the later nineteenth century found its axis between the liberal shift to contractual metaphor and the enduring familial metaphor of traditional authority. The best way to see this is to show how it was reflected in the thinking of the married couple who were among the most important theorists in the COS, Helen and Bernard Bosanquet.[33]

Once a pupil of T. H. Green, Bernard Bosanquet was, in his time, one of the most popular British Idealist philosophers. British idealism drew on German idealist philosophy in the work of Kant and, especially, Hegel, as well as on classical Greek political thought. Probably its most notable achievement was an influence on social policy through its emphasis on the inter-relationship between the individual and society, and a notion of the 'common good'. Bernard Bosanquet's brand of Idealism provided a theoretical foundation for the COS. Like British Idealists in general, Bosanquet opposed his political philosophy to utilitarianism and natural rights contract theorists. But much more so than Green, he adopted Hegel's philosophy as the major framework of his approach.

Political 'contractarians'—notably, the likes of Hobbes, Locke and Kant—had derived a basis for legitimate authority from the idea of mutual consent. This involved the assumption that individuals in 'the state of nature' had rational autonomy. It followed that this was the essential prerequisite for the production of a social compact. On this view individuals were assumed to voluntarily consent to erecting an overarching authority by means of which they gave away some of their liberty in order to preserve the rest. Utilitarian theories developed a *laissez faire* bourgeois individualism out of such theories. This was surely helped by the fact that contractarianism could accommodate materialist or idealist philosophies, atheist, quasi-atheist or religious. The general assumption, however, was of a rational earthly order that could be apprehended by 'Man's' rational capacities in an overall scheme ordained by God. Here, reason suffused everything and was available in static form for separate individuals who came together only in their separate interests.

What Hegel did was kick this rational God into motion—an aggressive act, a pretty aggressive philosophy.[34] Hegel's thought, as the philosopher John Herman Randall noted, took the form of an attempt to 'embrace all past philosophies, and to include them in a drive toward intellectual imperialism'.[35] And as God reeled into his own teleological flight he inevitably dragged the human individual with him. God was reason, and reason was on its own self-prescribed journey of self-realization. The unfolding of reason was a historical process, and the minds of individuals were intrinsic elements. The rational individual didn't pre-exist social and historical context—it emerged as part of them. The individual was necessarily social and couldn't exist unrelated to the whole.

And so, in concise form, the Hegelian-driven response to contractarianism was this: contract theories were wrong because they saw individuals as atomistic self-interested units. They failed to account for the progressive nature of reason, and so they paid no regard to the formation and movement of the mind. To contractarians the moral order of human relations was an add-on. But the emergence of political order couldn't be separated from the emergence of a moral and psychological order. The individual only came to recognize itself through human relations and the emergence of a moral order. It followed from this that neither the state nor any other sovereign authority could be properly understood as simply an authority whose rights and powers were founded on the consent of individuals. As the individual arose within a moral society it couldn't be said to base its relationship with sovereign authority in terms only of contract. Instead, it was through the moral and relational order that what is recognized as an individual emerged.

Bernard Bosanquet summed up much of his social and political thought in his *Philosophical Theory of the State*.[36] He cast the fundamental issue there in terms of what he called 'the paradox of self-government'. This was, in fact, composed of two related paradoxes:

> The paradox of Ethical Obligation starts from what is accepted as a 'self', and asks how it can exercise authority or social coercion over itself; how,

in short, a metaphor drawn from the relations of some persons to others can find application within what we take to be the limits of an individual mind.

The paradox of Political Obligation starts from what is accepted as authority or social coercion, and asks in what way the term 'self', derived from the 'individual' mind, can be applicable at once to the agent and patient in such coercion, exercised prima facie by some persons over others.[37]

Bosanquet used this as the root from which to grow his Hegelian-inspired political philosophy. It was clearly aimed against contractarian theories. In terms of our story of moral therapy extended to the community, however, it marks something more fundamental. Aimed at the social and political community as it was, it marks a kind of calibration of reason. For the likes of Samuel Tuke in the asylum, the lines of combat were clearly drawn: reason on one side, madness on the other. But, for Bosanquet, 'the paradox of self-government' in the wider community, suggested that reason itself must be measured.

We've seen that, via Hegel, Bosanquet had claimed to show that individuals were necessarily social rather than self-contained atoms, and that reason was neither a static phenomenon in 'nature' nor in individual minds. Mind had a history that predated the individual, and moral authority was co-extensive with the emergence of the individual. It followed, for Bosanquet, that, in order to resolve the paradox, 'We must show, in short, how man, the actual man of flesh and blood, demands to be governed; and how a government which puts real force upon him, is essential...to his becoming what he has it in him to be'.[38]

Again, Bosanquet substantially followed Hegel in taking one of the most radical contract theories and finding in it a fruitful idea that might resolve the 'paradox of self-government' by unifying it with the Hegelian 'Geist'— the Great Mind, or Spirit. In this way a contractarian theory that claimed to hail liberty, egality and fraternity was hitched to a hierarchical construction of the Family and history.

'Man is born free, and everywhere he is in chains'; Bosanquet took Rousseau's celebrated opening to *The Social Contract* and immediately stripped it of any radically individualist pretensions.

We expect such an opening to be followed by a denunciation of the fetters of society, and a panegyric on the pre-social life. And there can hardly be a doubt that these sentences, along with a few similar phrases which stick in the memory, are the ground of the popular idea of Rousseau, shared by too many scholars. But how does Rousseau go on? Here are the succeeding sentences. 'How did this change take place? I do not know. What can render it legitimate? I think I can tell'.[39]

Bosanquet used this to show that Rousseau's aim appeared to be not so much to cast the chains asunder as to render them legitimate. The question Rousseau raised was a version of 'the paradox of self-government'. It could be relayed as 'How could people remain free while living together in society?'. Or, more specifically: 'How could the will of the autonomous individual be reconciled with the political will of the community without the former forfeiting some or all of its own authority?'.

Rousseau began in contractarian fashion by stating that 'Since no man has any natural authority over his fellows, and since force alone bestows no right, all legitimate authority among men must be based on covenants.' No one could alienate their will to another to do with as they pleased.

> To speak of a man giving himself in return for nothing is to speak of what is absurd, unthinkable; such an action would be illegitimate, void, if only because no one who did it could be in his right mind. To say the same of a whole people is to conjure up a nation of lunatics; and right cannot rest on madness.[40]

His famous, or infamous, answer was that there was a certain kind of authority to which individuals could submit that was, in effect, no kind of authority at all. If each surrendered himself to the authority of everyone else, then how could anyone be coerced? An egalitarian sounding refrain then, of each for all and all for one.

But Rousseau made a distinction between the 'will of all' and the 'general will'. The general will was the expression of each individual committed to the body politic and thus each committed to everyone, as well as themselves. The 'will of all', however, just represented an aggregate of each individual's private interests. Rousseau seems to have thought that through direct discussion the general will would prevail. But, even so, he had to resort to the notion of a 'true' and 'false' self that accorded to the 'general will' and the 'will of all', respectively, in order to pull this off.

But Bosanquet found here the point of origin, and justification, of his own Hegelian notion of the general will. Unlike Rousseau, he saw the individual and society as part of the teleological whole. The individual couldn't realize itself outside of social institutions and the state; in fact, the historical evolution of these institutions represented an accretion of reason that was itself an expression of the 'general will' and, therefore, the individual's 'true self'. He concluded that 'The General Will seems to be, in the last resort, the ineradicable impulse of an intelligent being to a good extending beyond itself, in as far as that good takes the form of a common good'.[41]

Viewed in this light, self-government need not be a paradox. The paradox depended on a clear opposition between each individual, and between individuals and society, but now it could be seen that only through the general will could a real 'life worth living' become manifest.[42] Even though a person feels their individuality to be distinctive and to hold a definite position among or against others, a deeper look will reveal that only through assertion of the self into something beyond, into the greater self, can individuality truly express its real self.

> The centre of gravity of existence is thrown outside him. Even his personality, his unique and personal being, the innermost shrine of what he is and likes to be, is not admitted to lie where a careless scrutiny, backed by theoretical prejudice, is apt to locate it. It is not in the nooks and recesses of the sensitive self, when the man is most withdrawn from things and persons and wrapped up in the intimacies of his feeling, that he enjoys and asserts his individual self to the full. This idea is a caricature of the genuine experience of individuality.[43]

Such was Bosanquet's description of 'true' self-government. But what, in fact, is happening here? Once thrown, authority is up for grabs. And, having presumed to throw it from the unsuspecting individual, Bosanquet has proceeded to fetch it and pick it up for himself. Their selfhood is in his hands now. There is always an authority at the other end of a selfhood that has apparently been thrown elsewhere.

Bosanquet contrasted this real self, or real will, with 'the casual self', or the 'indolent', 'selfish' will, which he associated with 'rebellion', 'incompetence' and 'ignorance'.[44] One example of it, he argued, was of 'an impulse of sensual passion'.[45] This was a partial and inadequate self-will. It wasn't necessarily unnatural or without meaning, but its impulses required something more.

> If we compare them with the objects and affections of a happy and devoted family life, we see the difference between a less adequate and a more adequate will. The impulse, in passing into family affection, has become both less and more. It is both disciplined and expanded.... In the family at its best the will has an object which is real and stable, and which corresponds to a great part of its own possibilities and capacities. In willing this object, it is, relatively speaking, willing itself.[46]

This wasn't a random example. There was no room here for any Rousseauan ideas about liberty, egality and fraternity. The Hegelian unification of history and 'the great mind' pointed to a very different image of the individual and its relationship to the social order. It also pointed to an existing locus that could be used to explain both—the family. Hegel had looked here, and so did the Bosanquets.

* * *

Wills was finding the accommodation at Greenwich House difficult to deal with. The place was a settlement house in the style of Toynbee Hall. It had been opened back in 1902 by Mary Kingsbury Simkhovitch. She was well thought of in social welfare circles. Wills reckoned she was an efficient matriarch, 'forthright, intelligent, and well-informed' whose 'little husband' was a 'kind of appendage'.[47] Apparently, Simkhovitch founded the House in Greenwich Village with the aim of breaking away from the 'lady bountiful' image of charitable work by integrating with the neighbourhood.[48] When Wills was there most of the other twelve residents were women. Each night they were expected to attend dinner. At his first dinner Wills found himself in the company of half-a-dozen 'superbly groomed young duchesses' wearing 'wonderful evening gowns, elaborate coiffures' and with 'a slightly bored air of sophistication and savoir-faire'. After a week the strain was telling. He wrote in his journal:

> Came back to Jones St. Completely miserable. I hate every soul in this beastly institution. They're all horribly well to do. Super educated & refined & clever. They come & live in excellent apartments (the girls do at least) wearing expensive clothes & eating expensive food. In the very midst of the most poverty stricken area, & generously condescend to do good to the poor. And, greatest crime of all, they are perfectly well intentioned, giving me therefore no just cause to hate them for.... And probably – almost certainly – when I have left the place I shall sentimentalise over it.... Thoroughly dissatisfied with myself – wish I had never started getting educated & mixing with these beastly well-to-do 'social workers'...[49]

*   *   *

The Bosanquets, in effect, took what they considered to be the authority relations of the personal family and made them the formative structure of 'the self', of the community and therefore of 'self-government'. We will see that this understanding is reiterated, with further modifications, in the twentieth-century mental hygiene movement. But I need first to elaborate the Bosanquets' version.

'The idle' may have represented a form of alienation to the Bosanquets, but that didn't mean the Bosanquets had lost faith in progress and its perceived corollary of the self-sustaining individual. Their aim was to return 'the idle' to the path of progress. In this respect, the principal issue with the Bosanaquets and the COS, as Steadman Jones has suggested, is that they were trying to reconcile political economy based on the rational self-sustaining individual, with the desire to retain a moral community with affinities to the 'squire'.[50] This was why they founded their social philosophy and philanthropy on an axis between market liberalism, with its contractual metaphor and exaltation of the individual, and an older, but enduring, familial metaphor of traditional authority. Their means of doing this was to retain the fundamental organizing principle of the Family in its political, but also its psychological, aspects, while its direct structuring of the wider community was attenuated. Thus, rather than its 'traditional' components of authority and hierarchy, founded on intimate, affective relationships, extending to every interaction throughout the structure of society, the Family took the form of a forcing-house for the rational self-sustaining individual of the market economy.

*   *   *

Part of the deal was that Wills had to do a few days a week 'Field Care Visiting' with the New York COS. Pettit reckoned it amounted to dealing with enquiries by women about syphilis in their husbands, and asked him if he thought he could stick it. Miss Ivins, the Field Work Director, told him it would consist of taking pregnant women to clinics and that sort of thing. Wills knew he'd hate it.[51]

The field work was complemented by a class at the New York School of Social Work taken by Miss Ivins. It only took a few weeks of it to wind Wills up.

> The class in Social Case Work gets me down. To hear these youngsters discussing the 'life situations' and 'behaviour patterns' of people old enough to be their parents! The case under discussion this afternoon was a boy who wouldn't go to school and whose father was a banker. They were discussing a proposed interview with the mother. One girl said, 'I should want to know exactly how Mrs F regards her husband. What she thinks of him as a father; what she thinks of him as a husband; what she thinks of him as a man,' – 'Yes', I interrupted rather pepperily – 'and what she thinks of him as a banker.' Shocked laughter![52]

Apparently, Ivins thought it was funny and told Pettit it was 'a perfect commentary on their "history getting"', leaving Pettit 'roaring with laughter'.[53]

The fieldwork was even worse. Barely a couple of weeks later Wills had to call on couple who had asked for help. They were living in two rooms. The husband had been out of work, but now had a temporary job. The wife was nearly nine months pregnant. They owed two weeks' rent and were threatened with eviction. Wills was supposed to do a lot of investigating, 'and generally pry into all their secrets'. How come they'd recently left Canada in such a hurry? And how come they'd chosen to do it when the wife was pregnant? But he bottled it. They were such nice people he couldn't bring himself to do it. He spent an hour trying to cheer them up instead. Surely that was time better spent than forcing them to reveal faults in their character? Wills thought so. 'Gosh but it <u>was</u> depressing!'. Now all he had to do was admit to the COS that he hadn't collected any information *and* persuade them to help the couple.[54]

* * *

If a sentence can sum up the way in which the symbol of the family served the Bosanquets and the COS, it might be this one:

> The authority which all adults like to exercise finds a beneficent outlet in guiding the action of immature wills; and children who weary when left to the caprices of their undisciplined natures, find strength and contentment in a rule which is autocratic without having the impersonal rigidity of external law.[55]

This was written by Helen Bosanquet in her book *The Family*. Substitute 'the mad' for the word 'children' and it might well have been written by Samuel Tuke. Just about everything is encapsulated in this one sentence.

First of all, it refers to something existing: a real social institution. Every reader would be able to relate to it. But though it is a social institution, it is different in degree to others. The family isn't part of the contractual relations of society. It's more important than that—it forms the material and ethical basis of individuals' existence.[56] It's a moral order that acts as the fabricator of responsible individuality. Helen Bosanquet put it like this:

> In so far as the authority of the parent is based upon a greater maturity of reasonable will, it must always exist until such time as the will of the child is itself rationalised and matured... There is no tyranny involved in this when the purpose and aim of the parents includes the welfare of the Family, for then they are but guiding the will of the child to attain an end which it is as yet incapable of conceiving and attaining for itself.[57]

The traditional family is a patriarchy. But, in the preceding quote, Helen Bosanquet makes no distinction between male and female authority—only that between adult and child. Though she mentioned neither, there is a kind of combination here of a main tenet of the original moral therapy with a main tenet of Rousseau's Social Contract. At the Retreat, moral therapy de-legitimized physical force and control in favour of mental authority. And, as we've just seen, in the political sphere, Rousseau emphasized that force alone couldn't confer any right of legitimate authority. Helen Bosanquet used this kind of idea to challenge the patriarchal foundations that imbued both, and that lay at the heart of the archetype of the Family. She wrote, 'The only true and firm basis for authority must be one which finds a response in the natures of those over whom the authority is exercised; and the power of the purse, like that of brute force, elicits no response, only subjection'.[58] She therefore supported women's suffrage, and women's individual right to a wider education and fuller role in community life. In her hands the distinction that the organizing principle of the Family pointed to was between 'mature' and 'immature' wills, and there was no reason to disdain adult women as the latter.

Helen Bosanquet attempted to resolve the issue of sovereign authority in the family in a typically British Idealist style. She asserted a notion of affectionate loyalty in which two wills would merge finding individual self-assertion through common purpose. Twin authorities—a kind of condominium if you like. But, her feminism was only partial; she conceded that, ultimately, the man held overall authority. So equal, but not quite.

Authority and loyalty are related. For Helen Bosanquet, the modern family had progressed beyond submission to tyranny, to the kind of submission that could be expressed as loyalty.[59] Loyalty applied to all the family. Children shouldn't be 'entirely subservient' to parental command or family custom. Their interests and welfare were important to future progress. A child's response to this combination of affectional and rational interest went beyond mere gratitude. It had a feeling of spontaneity about it, of shared interests and sense of responsibility. And, crucially for the Bosanquets' anti-contractarian political philosophy, loyalty existed as a part of the moral order within which the rational, self-responsible individual emerged. Loyalty was active long before the individual became fully founded.

The Bosanquets emphasized the Family's role in the inculcation of the mental discipline of work. They associated it with the development of reason, individuality and moral purpose.[60] Helen Bosanquet claimed that among those who must earn a living to sustain themselves 'it is the institution of the Family which is the principal motive to work'. She made the moral imperative clear in her following comment that 'Nothing but the combined rights and responsibilities of family life will ever rouse the average man to his full degree of efficiency, and induce him to continue working after he has earned sufficient to meet his own personal needs'.[61] The Bosanquets themselves were, however, under no such moral imperative. They lived on an inherited income.

* * *

Miss Mertz of the Riverside COS must have been getting hacked off. Every Thursday morning she held a case conference for the four students on placement: two women, a priest called Kinsella and Wills. Kinsella was intending to do some form of social work and so was studying for the New York School of Social Work diploma. Wills liked him. 'He is a typical priest in that he loves the good things of life, and to go out to lunch with him means to have a good lunch'. Kinsella was 'well grounded in his faith', too, and this meant that he wouldn't budge on certain points that weren't quite in agreement with the COS view.

> He, in these conferences, batters at the C.O.S. from the point of view of Orthodox Catholic Theology (for the C.O.S. is not only non-religious, it seems to Kinsella and me to be definitely irreligious at times. They are, for example, always telling their clients the most awful, deliberate lies, justifying them with all sorts of specious arguments about the clients' welfare). I, while broadly disagreeing with K., also criticise the C.O.S. just as vigorously (but perhaps with a greater spice of humour) from the point of view of the Socialist Philosophy. We have great fun, and learn nothing, except to like each other more, and the C.O.S. less. It is a kind of unholy alliance, which Miss Mertz must find rather distressing at times. When things get too hot for her, she'll say to one of the poor females 'what do you think, Miss doings?' which gives her a nice little rest, because both the poor females swallow the C.O.S. whole.[62]

* * *

According to the Bosanquets each successfully raised individual took a mentally ingested Family authority with them into the wider community. On this view, the wider adult community didn't need the nurturing affectionate authority of each personal family. But, even so, the Family's general form, whereby rationality presided at the apex of authority, providing hierarchical structure and distribution of status, clearly imbued the Bosanquets' description of the social order. Bernard Bosanquet used a distinction between organization and association to describe how the structure of each individual mind was related to the structure of the social order. He used the analogies of a crowd and an army. The so-called mind of a crowd, he argued, was really no mind at all. It was merely the casual association of separate units on an 'extended and intensified scale'.[63] Each person had only a superficial connection with another, sharing nothing beyond immediate feelings and senses. The communication within the crowd was the contagion of excitement and emotion, it couldn't lead to concerted action, and reasoning and criticism were out of the question. The crowd was a mere mass and not an organization. An army was also made up of an extended group of individuals. But the relationship between them was organized hierarchically. As in a crowd, influences must pass between all the men. But the immediate association between each individual wasn't the driving force of the army. Instead, it was the hierarchy of command and rank, from general, through officers, to men. The army was organized hierarchically according to its purpose.[64]

A hierarchy of mind expressed in a hierarchy of the social order then. This was a description of societal development in terms of a hierarchy of 'place' and 'function': each human character finding self-realization within an idealist notion of society as an organic whole differentiated into duties and functions that were supposed to express the common good of the community.[65] Each individual mind was, likewise, an organization rather than an association, and thus a grouping of hierarchies within hierarchies organized as an organic whole and marshalled by logical capacity.[66] But this wasn't simply a description of the ongoing march of reason, expressed in the order of each mind and of society. It was a means of measurement; a means of valuing what *should* be and weighing it against what *shouldn't* be.

The description of 'the crowd' wasn't really just an analogy for the Bosanquets. This 'mere mass', with its superficial association dominated by the temporally immediate and environmentally proximate, its 'contagion' by emotion, and its low level of intelligence and responsibility, was one and the same phenomenon as those people that the Bosanquets and the COS set out to reform; the people I have, for convenience, termed 'the idle'. 'The mass' and 'the idle' were as children to the Bosanquets. They lacked a proper sense of past and future, were unable to organize, plan and defer, and, instead, lived in 'the passing moment', seeking satisfaction without responsibility.

We've noted that the term 'individualism' encompassed the idealist doctrine of the rights of man with its egalitarian implications, but that from the mid-nineteenth century it held derogatory connotations associated with egoism and idleness. The Bosanquets drew together their antipathy towards 'the idle' and 'the mass', not only with utilitarian contractarian theories, but also socialism and communism.[67] This was driven by the Bosanquets' emphasis on the founding role of the Family for the structure of the mind and of society. But it was also informed by concerns about the centralization of power in the modern rational and contractual state, and its claim to legitimacy through mass popular power. Although Bosanquet believed the state to be the regulator of the social whole, he considered its essential capacity to be the enforcement of external actions. Actions performed under compulsion couldn't be considered true aspects of the will.[68] Adopted out of submissiveness or selfishness they lacked moral value.

For the Bosanquets, 'Economic Socialism' and communism were a case in point. Any conceptualization of the state that sought public ownership and collective distribution of property entailed a 'fundamental aggression on family unity and parental responsibility' and would only act to 'favour the existence of human beings without human qualities'.[69] It would be fatal to character and the development of the moral individual. Such distribution of resources to the poor failed to make a distinction between poverty and pauperism; thus, it weakened the character and familial resources of the poor at the same time as it removed the morally disciplining power of 'less eligibility' in the workhouse under the Poor Law. Ultimately, the attempt to replace the class status and function necessary to the community of social interests with complete equality in society amounted to the promotion of infantilization in individuals.[70] 'Economic Socialists' and communists were thus 'advocates of the child-ideal'.[71] Their collectivism was paradoxically associated with egoism, or 'Moral Individualism'. It attempted to use compulsion on individuals for the social good, but only ended up promoting egoism through a denial of the fundamental role of the family order.

* * *

Wills was taking a keen interest in the socialist scene. Within days of arriving in New York he had found 'the ideal coffee house'. It was called 'Proletics' and run by the Communist Party. Wills found it the only cafeteria where people weren't in a hurry. It was 'full of long haired men with dirty fingernails, & short haired women with naked faces'. It must have struck a contrast with Greenwich House. Wills enthused about the atmosphere. 'It has paintings on the wall of Lenin & mythical workers, & everywhere animated discussions are going on. Someone told him he heard a patron of Proletics say to another – 'Do you believe in God? – Go on – take either side!'.[72]

* * *

Everything was history to Hegel—likewise for the Bosanquets. If the Family was the principal vehicle of individual self-realization through time, it, too, was of a historical nature. Helen Bosanquet took a keen interest in this nature. She borrowed Frédéric Le Play's historical typology of families, and followed him in performing a manoeuvre that Arland Thornton has described recently as 'reading history sideways'.[73] That is, she took geographical differences in family organization to be historical and progressive changes. Le Play has sometimes been called 'the Karl Marx of the bourgeoisie' for his concentration on material changes in family organization and occupation, and his associated promotion of one particular family type that appears close to the bourgeois ideal. The label is a little unfair in narrowing down the scope of Le Play's huge sociological effort, but it certainly fits Helen Bosanquet's use of him.

She relayed Le Play's three types of family: the Patriarchal, the Stem and the Unstable. The first was dispatched as historically obsolete. But the other two had contemporary relevance. Unlike the multi-generation, static, Patriarchal family, the Stem family was smaller, simpler, economically more mobile and with a more balanced authority between its heads and members.[74] It represented the 'modern' family type that the Bosanquets, and the COS, eulogized:

A proletariat residuum is impossible where all the young people who go out into the world are trained to habits of labour and obedience, as well as being strong and capable; the natural asylum of the home for the mentally and physically feeble is a far surer precaution against the marriage and propagation of the unfit than any recognised system of public control; while the firmly rooted belief that family life involves a home and property, however humble, prohibits the thriftless marriages which lead to pauperism.[75]

The third family was the Unstable family. Helen Bosanquet noted that Le Play found this type among the poorest regions. It was described as exercising little or no authority. Its members cared little for home, they offered nothing to the community, and the children 'drift into the world undisciplined and untrained'.[76] Of this family she remarked:

They are indeed at the root of most of our social difficulties. They are like baskets with holes in them; they let the old people drop out at one end, and the children at the other, to be picked up by the State, or take their chance of passing charity. And not infrequently the basket falls to pieces all together, and the whole family has to be sorted out into workhouses, asylums and prisons.[77]

It was these families that were characteristic of 'the idle', and which the Bosanquets and the COS set out to re-moralize.

But if the state was considered a blunt instrument when it came to the moral life of the individual, what should social reformers do? Clearly averse to power centralized in the state, and any egalitarian idea that mass participation legitimized its pre-eminent status, the Bosanquets and the COS promoted philanthropy as the necessary means to deal with 'the idle'. Voluntary social work performed by the 'educated classes' out of an ethical sense of duty would re-instil, through personal relations, the moral sense of 'the idle'.

The COS developed what amounted to domestic moral therapy for those whose 'character' it believed could be raised. Originally called 'friendly visiting', by the 1890s the term 'casework' was used interchangeably.[78] This work with individuals and their families was intended as a means to re-engage the character and sense of duty of those paupers not irrevocably part of the 'residuum'. In keeping with their idealist notions of each individual's reciprocal relationship with the social and moral order, the Bosanquets and the COS promoted a notion of 'reciprocity' in charitable work. But, despite the undeniable emphasis on real social engagement with the underprivileged that this term implied, it was, nevertheless, founded on a hierarchical understanding of mind, history and society that was anything but reciprocal in any sense of equitable mutual exchange.[79]

Jane Lewis has shown how Helen Bosanquet used G.H. Stout's influential psychology to explain the way 'man' was distinguished from lower animals by his progressive wants.[80] Lower animals were prevented from progressive development by the determining role of their instincts, but, in humans, 'progressive wants' enabled the rational pursuit of interests to develop. However, according to Helen Bosanquet, the problem was that some people were too satisfied with the basics of eating, drinking and sleeping. In terms of idealist thinking, these people had, in effect, deviated from progressive history. Their failure to develop progressive wants led them to be ruled by their 'habits', much as instincts ruled animals. The Bosanquets and the COS reckoned, however, that social work could correct bad habits and raise individuals to a better standard of character. History itself had a progressive character and, through personal relations, 'the idle' might have their characters brought back into accordance with it. Thus would be resurrected a sense of duty and 'emulation of social superiors'.[81]

This approach, in fact, reveals the ambiguous stance the COS held to other elements of modern state power commonly seen as problematic by theorists and social reformers. One was the rationalization of power—the processes of ordering, measuring and systematizing. The Bosanquets' social philosophy emphasized personal relations of authority constituted in the family, and so they opposed any idea that the personal volition necessary for moral behaviour could be accessed through the blunt instruments of administrative procedures or statistics. For Bernard Bosanquet people needed to be understood as individuals, 'not as abstractions, but as living selves with a history and ideas and a character of their own'.[82] This was why a personal relationship was required in social work—one of reciprocity, but also of authority.

But the Bosanquets and the COS nevertheless contributed to the rationalization of political power in two distinct ways. First, part of the basis of their philosophy was a strict application of the Poor Law through 'less eligibility' in the workhouse. People who were destitute should be dealt with under this administrative, regimented and communal regime. The reason for this apparent contradiction was that these people were characterized as having relinquished the very moral and rational volition, with its foresight and self-control that was associated with the family order. They had 'accepted the status of a child'.[83] They were the 'unhelpable'.[84]

This definition of people as either 'helpable' or 'unhelpable' entailed a further rationalization of power. Beatrice Webb, of the Fabian Society, claimed that COS social work was 'sentimental' and inefficient. But the COS argued that it stood for 'scientific' charity. The first element of this was thorough and objective investigation of individuals and their families as a means to discriminate between those people who were the recalcitrant 'residuum' and those capable of a restoration to good citizenship.[85] In any case, given that the Bosanquets and the COS saw the family as the sphere within which the rational mind of objective order and self-discipline was inculcated it must have seemed a corollary that social work should embody this rational order. As one of the leaders of the COS, C.S. Loch put it: 'Organisation implies order and method, sacrifice for a common end, self-restraint'.[86] Charity itself had to be disciplined. It must be objective and organized. This entailed an examination of a person's past and present: of their character, of their community and, in particular, of their family—both its present relations and its history. So the COS, though they may not have been very successful in their particular aims, played an important role in encouraging the extension of rationalizing power, linking the domestic sphere with population and economy.

To this extent they contributed to another common concern about modern political power. This was its totalization: the ambition of state power to penetrate every aspect of the life of the individual and the population. But the COS's contribution to totalization was ambiguous in terms of the state. The Bosanquets saw the state as regulator of the 'social will', but they denied it had any ability to moralize individuals. To the extent that they contributed to a totalizing form of power, the Bosanquets and the COS limited its centralized nature by the moralizing role they mandated for voluntary organizations. But though the totalizing tendency of power wasn't to be wholly owned by the state it was, nevertheless, inseparable from it. This ambiguity is neatly alluded to in a comment made by Lord Shaftsbury, the Chairman of the Commissioners in Lunacy. In 1874 Shaftsbury had written to the Society criticizing its campaigning over criminal lunatics. He accused them of 'erecting yourselves into a grand association for the control of everybody and everything'.[87] But within four years he'd evidently changed his mind. Shaftsbury joined a COS delegation that submitted proposals for a nationwide system of segregated schools and asylums for people termed 'mentally deficient' to the Local Government Board. This delegation was, in fact, part of a movement by the Bosanquets and the COS towards closer ties with state power.

After all, if Bernard Bosanquet's dialectic of mind was remorselessly marching onward, what was it marching against? Through a Hegelian lens, it was marching against itself. The great mind was forever adjusting itself on the antagonisms it produced. In terms of social philanthropy, the result was 'the residuum'. These were the dead wood of the dialectic; so much discarded debris left in its wake. The COS increasingly claimed that this group was heavily made up of people they labelled 'feebleminded'. They associated them with people already categorized as 'idiots' and 'imbeciles', but claimed them to be of a 'higher grade' and far more numerous in the community. Along with a derivative organization called the National Association for the Care of the Feeble-Minded, the COS agitated for legislation to provide permanent segregated care and control. This would be nationally organized and partially funded through public revenue.

Detailed histories have been published of this campaigning and the associated events that ultimately led up to the passing of the 1913 Mental Deficiency Act.[88] In a nutshell, its aim was ascertainment, supervision and detention. Under the Act all county and county boroughs in England and Wales were to provide institutional provision, arrange community supervision and ascertain the local population of people deemed mentally defective.[89]

But, in terms of our tale of moral therapy extended to the community, what did this wholesale categorization of people as feebleminded amount to? An admission that COS casework and the theories upon which it was based was a failure? Not at all. It was confirmation that casework had been hampered by the occurrence of so many people who were incapable of self-governing their own minds adequately. Self-government couldn't be resurrected because it hadn't ever been manifest in the first place. 'Mental deficiency' was equated with stalled phylogeny. The 'mentally deficient' were primitive throwbacks. The Family authority wouldn't work on them and wasn't worth trying because they wouldn't develop to rational adulthood. In fact, these people amounted, in effect, to the anti-people of the anti-family. They had no 'real self', only a 'false' self. Here were the ultimate 'egoists': rampant individualists without moral constraint. This was a permanent 'childhood', a never-has-been-and-never-will-be-adulthood. It would always require the external authority and control of 'mature' minds so that its inherent waywardness could be kept in check.

Helen Bosanquet's fine words defending women against accusations of immaturity compared with men utterly deserted her when it came to the people—men, women and children—she considered feebleminded. These were almost literally a race apart. They completely lacked self-control, foresight and responsibility owing to a 'low order of intellect' and 'degradation of the natural affections'.[90]

> It would be hard to attribute this intellectual failing entirely to absence of anything to express; sometimes, I am convinced, there must be actual suffering from the inability to give articulate utterance to the mental chaos within. Nevertheless, we are forced to recognize that, on the whole, these people are as undeveloped – or as degraded – on the side of their affections as of their intellect. The most striking proof of this is the looseness of the family tie, and the absence of all feeling of mutual responsibility between parents and children and brothers and sisters.[91]

Deficient intellect equalled brute emotionality equalled barely recognizable feelings.

It was around these views that the Bosanquets were able to draw together their idealist description of the onward march of 'mind' with Social Darwinist descriptions of 'the survival of the fittest'. Bernard Bosanquet took issue with such theories because he reckoned they reduced 'man' and society to materialist biological evolution, and ignored the pre-eminence of the evolution of mind. To him, 'the survival of the fittest' in human society was the survival and development of the most reasonable. It was a matter of character and ideas. But, from this position, Bosanquet could, in fact, align himself with much of the thrust of Social Darwinist ideas regarding 'the residuum' or 'the unfit'. He maintained that 'the struggle to realise the conditions of true family life in its moral and material senses' was the true 'fight for survival'. Those people clearly incapable of sustaining family life should be segregated compulsorily to prevent the proliferation of their degrading moral influence on wider society.[92]

The legislative result was the 1913 Mental Deficiency Act: a highly coercive and interventionist measure for public 'welfare'.[93]

* * *

So maybe this is the direction that Foucault's line about the 'dialectic of the family' takes. It has extended across the community. By means of history it has encompassed people beyond 'the mad'. These people, too, are considered to lack authority over their own mental states. For some, a form of domestic moral therapy is created. But others, it's claimed, fall permanently short of sufficient reason to benefit from the Family's moralizing authority. As we'll see, a movement for mental hygiene will be founded in the twentieth century that will take up this trajectory, and the dialectic of the Family will continue its remorseless grind.

Diabolic dialectic of reason, on it marches, like a demented Prussian general.

* * *

3

The Mental Hygiene Movement's Emotional Contradictions

'I see the King sent for her. Ain't she a wonderful woman! My word she is!! And I ought to know!' It was 1937 and 'H... C...' had telephoned the offices of the Central Association for Mental Welfare (CAMW) to congratulate Evelyn Fox on her receipt of a CBE. He had met her 24 years earlier at his Special School and she had befriended him ever since. The CAMW president, Lord Justice Scott, quoted 'H... C...'s' words in his own tribute to Evelyn Fox, adding, 'Could any words more happily convey the delight, the admiration, and the pride of our whole organisation than those simple sentences in which H... C... let his heart flow?'.[94]

Evelyn Fox had been the organizing secretary of the CAMW since its inception in 1913. In that year, the organization set up an office at Tothill, Westminster.[95] It was a rented office, which, on opening, contained a borrowed typewriter and a donation of £20.[96] Despite its small beginnings, this organization was the earliest of a cluster of organizations that formed the institutional nucleus of the interwar movement for mental hygiene. It set itself up as a central training and co-ordinating body for the voluntary organizations that were envisaged as central to the operation of the Mental Deficiency Act.

Despite H...C...'s salutations, perhaps the location for the CAMW's office was appropriate. In the Middle Ages Tothill had been a site where necromancers were punished. By the seventeenth-century Lazaretto pest-houses had been built for victims of the plague: 'Many a torch or lanthorn-lighted group of mysterious-looking figures have borne the litter of the stricken to this then solitary spot, not so much with hope of recovery, as from fear of spreading the dire infection by retaining them within the frighted and unhealthy town'.[97] Under the 1913 Mental Deficiency Act local voluntary organizations could appoint themselves experts in surveying the local population and assessing their mental competence. The hub of the CAMW's activity lay in enabling ascertainment and certification, or measures of community control, for those deemed of insufficient intellectual capacity.[98] Its activists held the flame for an approach to people that cast the ascertainment of intellectual deficit as the royal road to the prevention of 'social inefficiency'. They were volunteer arbiters of the division of intellect. People whom they believed to be mentally deficient and in need of institutionalization were to be notified to the authorities for certification.

Under Fox's forceful leadership, the CAMW rapidly became one of the leading voluntary organizations working for 'mental hygiene' in the community. But the pre-eminent focus on so-called mental deficiency was no sooner enshrined in legislation than it began to lose its singular status, even among its promulgators. Mental deficiency was no longer the sovereign of 'the idle', it seems. The president of the CAMW, Leslie Scott, relayed the admission in this fashion:

> Whereas in the old days we fondly hoped that the solution of the problem of the unfit, the degenerate and the social misfit, would be found in the proper control of defectives, we have now discovered that the mental defective forms but a small percentage of that great army of failures of civilization which is found in every country in the world. The subnormal, the unstable, the unbalanced, the temperamentally defective, the victim of certain forms of physical illness, of bad inheritance and environment ... all these make a call on us as human beings, not only by reason of their own misery and of the sorrow they cause to their family and friends, but from their inability to take their place as citizens.... They fail to recognise, or they are incapable of recognising, the accepted standards of the community in which they live and of which they form a part.[99]

Here, in essence, is expressed the ambit and mission of the interwar mental hygiene movement. Scott's division between 'the old days' and the present, referred more or less to the periods before and after the Great War. The new view of 'social failure' that he referred to never doubted the fundamental rule of intellect, with its necessary demarcation. But what also seemed needed was a means of assessing and weighing the apparent multiplicity of causes of 'social inefficiency' that surrounded this demarcation. As we will see, in this process, once again the Family played the role of organizing principle and methodological device.

* * *

Wills was late for his 'community organization' class. It was already in progress. As he hurried in to find a seat, Pettit, the lecturer, took a playful swipe: 'These foreigners, they come over here, and can't even get to their classes on time.'

Wills looked down at his Ingersoll, 'These cheap American watches', he replied.[100]

* * *

Along with the CAMW three other organizations can be said to have constituted the hub of the mental hygiene movement in Britain. These were the National Council for Mental Hygiene (NCMH), the Child Guidance Council (CGC) and the Tavistock Clinic. The CAMW's concentration on 'social inefficiency' and a host of social problems believed related had led to it attending to a variety of mental issues beyond deficiency of intellect.[101] It was through this that it combined with the other three organizations. These were all founded during the 1920s, and, although they considered mental deficiency an important issue, their concern with 'social inefficiency' focussed particularly on psychological disorders among individuals in the community.

The Tavistock Clinic was one of the most influential of the new clinics attempting to treat these 'minor disorders', or 'functional nerve disorders'. Hugh Crichton-Miller founded it in 1920 to offer psychotherapeutic treatments and provide out-patient treatment facilities for people who couldn't afford private fees.[102] Leading figures at the Tavistock were directly involved in promoting the formation of the NCMH.[103] This was set up in 1922, mainly by respected members of the Medico-Psychological Association. The NCMH gave itself the role of promoting and co-ordinating the already-existing organizations working for the study and treatment of mental illness, mental deficiency and psychological problems in industry. It urged prevention and early treatment in the interests of the health of the community.[104] Along with the Tavistock Clinic, the CAMW and the NCMH believed mental adjustment during childhood to be crucial to the prevention of later mental problems in adulthood.[105] One result of this emphasis was the creation of the CGC in 1927. The CGC's first offices shared the same building used by the CAMW and Evelyn Fox became its first Honorary Secretary.[106]

Two other organizations that were founded later were also associated with the mental hygiene movement. One was the Home and School Council, established in 1929. The other was the Institute for the Scientific Treatment of Delinquency, created in 1932. Their importance to our story will become clear in the next chapter.

All of these organizations promoted social work as an important ancillary profession necessary for good mental hygiene. In particular, they supported the creation of a profession called 'psychiatric social work'. By 1929 an Association of Psychiatric Social Workers had been founded. This provided a specific training and promoted the interests of the profession. Along with the CGC it received 'seed' money from the Commonwealth Fund of America and adopted the general mental hygiene treatment model promoted in the USA.[107]

* * *

Wills' feelings about the Charity Organisation Society (COS) were getting worse by the week. The whole country seemed to him to be dominated by a belief in the free market and the power to hire and fire. And this even in a desperate capitalist depression:

> Let me place on record, for the relief of my feelings, my belief that this is a monstrous country.... The only amelioration of these days is the so called employment agency, to which the opulent employer pays damn all, and the starving workman pays $10 if he wants a job!... And they'll all starve unless some private individual or organisation chooses to dispense charity; and they can't get that charity unless they tell some inquisitive, well dressed young female from the C.O.S. absolutely all that is to be known about themselves – age, marriage data, maiden name of woman, school and class reached, ditto for all children, ditto for all brothers and sisters, previous address & employers & so on & so forth ad infinitum. It is unGodly. It stinks in my delicate English nostrils. It is foul & abominable & loathsome & rotten & unrighteous & beastly & altogether unsavoury. And, damn it, it's bad form. I was sent this afternoon to get 'particulars' from one of God's people whose husband is out of work, & who is the mother of 5 little brothers & sisters of Jesus Christ (excuse the language). I came back without and was sent again for (among other things) the maiden names of the wives of the married brothers of the man & woman. I spent two minutes in the house, my courage failed me, & I escaped on the pretext that the husband was out. Which means, as I have simply got to get the information, that I've got to call again in the morning.[108]

* * *

In the USA, the term 'mental hygiene' had become associated with an institutionally organized movement and associated practice during the first decade of the twentieth century. Clifford Beers' role in the formation of this movement has been documented several times.[109] Mental hygienists in both the USA and Britain made due reference to his pioneering role.[110] Beers' quest to found a movement for reform began with the publication of *A Mind that Found Itself*. This was an autobiography about his mental breakdown, his years in mental hospitals and his eventual recovery. But it was also a sharp critique of the abusive treatment and poor conditions he had experienced. Determined and articulate, even before the book was published, Beers had gained the support of influential people. Among them was the psychologist William James, who praised Beers' manuscript and wrote a foreword to it when it was published. Beers also impressed Adolf Meyer—one of the leading psychiatrists of the time. In 1909, he and Meyer became the leading figures in the founding of a new campaigning body, the National Committee for Mental Hygiene.

But with these powerful new allies there came a significant change of emphasis. Beers' original intent was to improve the care of people experiencing mental troubles, but the doctors he joined forces with were more interested in directing attention to wider society. They wanted to focus on prevention and early treatment of mental troubles. This would promote a wider remit for psychiatry that would intimately connect psychiatry and its aims with notions of good citizenship. Beers was persuaded. As Roy Porter has put it, 'The mind that found itself was a mind that realized that reason really must work on the side of medicine, psychiatry and the authorities'.[111]

* * *

They must have poked up at him like so many badly kept gravestones. Sticking up at the end of the bath like that they were a magnificent spectacle of long-term neglect. Several toenails were ingrowing. Wills had decided to avoid working on an essay and manicure them instead. Already he was impressed with himself.

> ... I'm making v good progress. My left great toe is quite handsome, and I never tire of admiring it! If I am not careful, I shall acquire a kind of Narcissistic foot fetishism. Indeed, I have often suspected myself of narcissistic tendencies – I love to sit about in the nude, if there is no fear of being surprised in that costume. But happily – or unhappily – I am seldom free of such a fear![112]

* * *

One important theoretical strand that informed the mental hygiene movement came from the 'New Psychology' that had been promoted in the USA since the mid-1880s. This was associated with the work of G. Stanley Hall, William James and John Dewey. Both James and Dewey criticized elements of Bernard Bosanquet's work, but, even so, in terms of our story there were important shared strands.

First of all, just like Bernard Bosanquet, the New Psychology castigated an older psychology that had considered the mind in abstract isolation. Advertising (and capitalizing) 'the New Psychology' in 1884, Dewey announced that the 'philosophy of clearness and abstraction' had been a general failure, 'save for its destructive accomplishment'. The mind was not atomistic and it couldn't be understood separately from its environment.[113] This view became dominant. By 1908 in Britain, William McDougall was recognising that: 'On every hand we hear it said that the static, descriptive, purely analytic psychology must give place to a dynamic, functional, and voluntaristic view of mind'.[114]

Similarly to Bosanquet, this understanding entailed criticisms of utilitarianism, and its notion of pre-formed rational individual minds. The New Psychology emphasized human development through history; a teleology of growth, adjustment and individualization. In the hands of some, such as G. Stanley Hall, this had a strong hereditarian basis, and was used to describe childhood as a recapitulation of racial development. Here, ontology repeated phylogeny. For our story, what was clearly lost with the 'New Psychology' was any notion of this moral and material evolution being tied to a Hegelian dialectic.[115]

But, Hegelian dialectic or not, the key shift that underpinned moral therapy's transposition from asylum to wider community remained. The original moral therapists had confronted a failure of rational authority over mental states in people considered mad. Bernard Bosanquet had attempted to provide a theory that would do the same for certain people at large in the community. These people weren't considered mad; they came under several epithets, of which I've chosen 'the idle'. Bosanquet had done this by denying the notion of a rational, self-sufficient, atomistic self. This self wasn't somehow pre-social, nor was the moral order somehow an add-on, chosen through the self's own egoistic interest. Instead, the individual was a historical accomplishment that emerged within a moral and relational order. It didn't pre-date it. The New Psychology retained all this. Thus, divested of a Hegelian dialectic, the great barrel-organ of history nevertheless rattled on. And, in fact, as we shall see, along with it trundled the 'dialectic of the Family'.

Instead of Bernard Bosanquet's Hegelian linkage of the individual, society and history, the New Psychologists favoured the biological notion of 'adjustment'. Drawing on this term's use in association with those of 'organism' and 'function', they described mental life in terms of a unitary, organic process, developing and adapting in relation to its environment.[116] This approach could unite psychology with biology and, in the process, an evolutionary understanding of humans and their minds, with an ethical understanding of human moralization. All this dovetailed neatly with existing concerns about 'social inefficiency', and it powerfully influenced responses to the twin concerns we've seen highlighted in COS activity: the 'residuum', which they had come to term the mentally deficient, and those among the socially inefficient who ought to be able to be reformed through a form of domestic moral therapy.

* * *

Soon after his arrival at the New York School of Social Work, Wills was told by Pettit that a group of English women had been specially trained in psychiatric social work there the previous year. Wills must have been left coming up for air. Why hadn't she told him?

Soon after Wills had received the news that he had been accepted on the Willard Straight Fellowship he had written dozens of letters trying to get funding for his travel. Eventually, Evelyn Fox had replied, wanting details and asking if he could visit her in London. She'd remained standing throughout his interview with her. A biographical sketch describes her as 'short and round faced, with rough curly hair' accompanied by a 'strident' voice and a 'downright manner...tempered by the merriment and devilment in her eyes'.[117] Did she have mischief in mind here? He found her 'cold and discouraging'. She didn't ask for any more information than he'd already given in his letter. After their meeting Wills was left wondering why on earth she'd asked him to come. If he'd expected to receive encouragement or advice, he'd received none. Worse, it had cost him a week's wages to travel to London and see her.[118]

Fox had been involved in arranging the training that Pettit had mentioned. Wills realized she must have been. Why didn't she mention it to him? He never found out.

* * *

For the mental hygiene movement, just as much as for the Bosanquets and the COS, the Family provided an organizing principle for the moral, psychological and social order. The nuclear family remained envisioned as the epitome of civilization and the primary institution in the production of citizenship. But under the mental hygiene movement this authority became simultaneously exalted and opened up to greater analysis.

What the mental hygiene movement found in this examination of the family was the importance of emotionality. This derived originally from the New Psychology's discussion of instincts underlying human behaviour. In Britain, McDougall's influential *Introduction to Social Psychology* built on the New Psychology and described humans as inheriting primitive instincts which, along with their accompanying emotions, constituted the basic impulses of human behaviour.[119] Mental hygienists adopted this general view. But the interwar impact of psychoanalysis mediated the movement's concentration on emotionality in terms of instincts.

Other than the CAMW, the institutions at the core of the interwar British mental hygiene movement were influenced by psychoanalysis from their foundation. Although the historian Mathew Thomson has recently cast doubt on the supposedly pre-eminent effect of the Great War on the popularity of psychoanalysis in Britain, it clearly stimulated interest in the apparent need to understand unconscious motivations and the therapeutic value of 'talking therapies'. In fact, professional interest in ideas of the unconscious and the psycho-neuroses became widespread. By 1924, for instance, A.F. Tredgold, the prominent psychiatric expert on mental deficiency and member of the CAMW, who was no advocate of psychoanalysis, lamented that, 'junior and inexperienced doctors' had seized upon it with 'avidity', as had educationalists and 'a large section of the general public and certain sections of the public press'.[120]

Despite Tredgold's criticism, the mental hygiene movement combined elements of psychoanalysis with elements of the New Psychology. What this fusion emphasized was the centrality of emotionality to all behaviour and personality development.[121] Emotional irrationalities were present in everyone to varying degrees. The essential claim was that rational thinking had finally grasped the fact that emotional experience underlay all growth and adjustment. Humans were dynamic organisms. Emotional experience was a necessary component of this.

It has become commonplace among historians and sociologists to refer to a great extension of psychiatric knowledge and influence during the twentieth century. This process saw the 'psy professions' (as some have called them) elevate their sights above the asylum walls and direct their gaze to the wider community in an ever widening 'psychologization' of society.[122] Undoubtedly, the mental hygiene movement was important in this process. But a historiography often fully preoccupied with establishing the extent and power of this growing 'psychologization' greatly ignores the way in which outlines and categorizations of history were deployed as an authority to support this process. History was deployed as part of the means to define the essentials of personhood, its mental health and its deviations.

Similarly to the Bosanquets and other nineteenth-century theorists, mental hygienists cast the past as the twin development of rationally marshalled individual minds and 'civilized' society.[123] They also echoed the Bosanquets in seeing the family as the vehicle for this process of human individualization. In his *Introduction to Social Psychology*, McDougall had praised Helen Bosanquet's promotion of a psychological understanding of individuals, and her associated use of Le Play regarding the development of 'the stable family' and its basis for a stable modern community.[124] But mental hygienists described this 'progress' in terms of emotionality, as well as rationality. Along with the evolution of the family had evolved the interiorized individual. For Emmanuel Miller, the director of the East London Child Guidance Clinic and a leading figure in child guidance, the institution of marriage and the family seemed to be an 'ancient document on which the history of man's emotional and social development has been written over as in a palimpsest'.[125]

Mental hygienists performed a similar 'reading history sideways' manoeuvre to the Bosanquets. Existing tribal peoples were deemed to represent earlier 'pre-civilized' forms of family and authority. According to Miller, 'primitive societies' brought emotionality under rigid group control early on in a child's life. Consequently, they thwarted individuality and freedom. But Miller described 'civilized' society as having superseded the 'group mind' with the individual mind. Under so-called civilized society the individual intellect now marshalled instincts and emotions. This was achieved through the temporary, and less rigid, authority of the modern family.

This depiction, in fact, had important consequences for mental hygienists' use of the Family as an organizing principle. One of them was that if individual intellect now controlled and rationalized instincts and emotions, it never fully supplanted them. They remained principal factors in personhood. The 'primitive' resided inside everyone. A form of recapitulation theory commonly informed mental hygienists' explanation of this 'primitive' residue. The psychoanalyst Ernest Jones wrote that the child effectively condensed 'a hundred thousand years of mental evolution' as it endeavoured to adapt itself to 'civilized standards'.[126] Like him, mental hygienists contended that, for adults, a personal appreciation of the details of this recapitulation was obscured as if by a mist through the obligations and interests of later years.[127]

Family authority was amended accordingly. The 'primitive' and 'irrational' couldn't simply be superseded by rational authority. Recalcitrant emotions couldn't just be suppressed by an act of 'will'. They needed an expert understanding that would enable them to be crafted and accommodated in the interests of each person's development and that of society.

There was a central claim encapsulated in this. If rational thinking had finally accepted that emotional experience underlay all growth and adjustment, it also claimed that life was progress or it was stagnation and regression.[128] For mental hygienists, mental deficiency and certain forms of insanity represented forms of 'stagnation' and 'regression' that probably couldn't be avoided. Other mental conditions, though, need only be temporary deviations from progress and adjustment; but, whether temporary or not, all these people represented failures of adjustment and social efficiency.

* * *

Wills had lots of reading to do for his course. More than one of his tutors had recommended Miriam Van Waters' *Youth in Conflict*.[129] He read it in a few days. Her line from Walt Whitman's *The Grand Sea* impressed him: 'We keep only that which we set free'.[130] By the time he'd finished her book he'd decided it was 'excellent'; 'in spite of the woman's <u>most</u> irritating habit of vomiting the definite article at every possible place, and many places where it was not possible'.[131] Even so, he went to the length of writing out a passage in his journal. It was a quotation from Tolstoy.

> ... men think there are circumstances where men may deal with human beings without love; one may deal with things without love; one may cut down trees; make bricks; hammer iron, without love. But you cannot

deal with men without it. Just as you cannot deal with bees without being careful. If you deal carelessly with bees, you will injure them and will yourself be injured. And so with men.

Yes, Wills was most impressed with *Youth in Conflict*. But great books strike to the heart, and in this case Wills couldn't avoid being pricked.

She then goes on to discuss the ideal person for work among delinquents, and raised a question in my mind that has often troubled me before. I wonder how far I am interested in this thing because it presumes power over other people, which I confess is very sweet to me. And will this desire – this necessity almost – to dominate others ruin my attempts to put into operation my ideas?

$* \quad * \quad *$

Johannes Pols has described how the interwar mental hygiene movement in the USA attempted to promote its agenda as a matter of public health.[132] In Britain, mental hygienists attempted to build on the 1913 Mental Deficiency Act as the basis for a similar aim.[133] They continued to agitate for greater institutional provision and social control for so-called mental defectives in the cause of public health. But, their contention that the emotional components of growth and adjustment were as important as the intellectual provided a far wider remit for a system of prevention and early treatment of mental disorders in the community.[134] In fact, the mental hygiene movement's general claim to be able to teach 'man' how to 'live at peace with himself and society' suggested grandiose schemes of social reform.[135] Maurice Craig, the president of the NCMH, put this in characteristically functionalist terms: 'A nation is composed of units, and the harmonious working together of these units leads to a greater stability and happiness, and both of these are the special care of the National Council for Mental Hygiene'.[136] This vision of function and fit for the social order is reminiscent of Bernard Bosanquet's moral philosophy of duty and place. But, unlike his approach, mental hygienists often used their description of the historical process of individualization as a means of differentiation in the hope of distributing people to roles in society according to their mental ability and temperament. So what mental hygienists commonly proposed wasn't only to differentiate individuals for appropriate mental treatment, but to differentiate and distribute people to 'appropriate' areas and levels of the social order. The rapid development of mental testing between the wars provided mental hygienists with an apparently scientific means to assist with this task. One of their first uses had been as an apparently 'scientific' measurement of mind that could inform the diagnosis of mental deficiency in those cases they considered 'feebleminded'.[137] For mental hygienists, mental tests helped measure a person's capability for adjustment. They were able to employ them as a means to rank the minds of people in terms of a 'healthy' functional fit with society. In fact, long before the socialist and sociologist Michael Young coined the term 'meritocracy' in the 1950s with satirical intent, the mental hygiene movement was attempting to fabricate one.[138]

The influential psychologist Cyril Burt relayed the guts of this proposed meritocracy unapologetically in a BBC radio broadcast:

> The state, in fact, must erect a double ladder – a ladder whereby the intelligent can climb up to their proper place, while the less intelligent, from whatever sphere, drop down to their own true level. In this way, while the nation helps the individual, the individual will help the nation.[139]

An impression of how measurements of intelligence informed the classification and distribution of minds in terms of a mental hygiene of society can be gleaned from Table 3.1 (taken from a work by Cyril Burt), supplied by the psychiatrist R.D. Gillespie in his discussion of 'mental hygiene as a national problem'.[140] He introduced it as showing 'the influence of intellectual level on economic efficiency'.

It's unclear here whether 'economic efficiency' was intended to refer to personal economic remuneration or to some contribution to the economic efficiency of society, or both. But, even so, this hierarchy of employment categories represents an implied hierarchy of 'worth'. And this is compounded by the explicit fusion (at the 'lowest' levels) of employment

Table 3.1   'The influence of intellectual level on economic efficiency'

Vocational category	Proportion of total pop. (%)	IQ (%)
1. Higher professional	0.1	165
2. Lower professional	1 per 1000 3 per 1000	140
3. Higher business positions and highly skilled workmen	12	125
4. Skilled workmen; most commercial positions	27	110
5. Semi-skilled workmen; poorer commercial positions	36	95
6. Unskilled labour etc.	19	80
7. Casual labour (feebleminded)	3	70
8. Defective adults in institutions (imbeciles and idiots)	0.2	50

classifications with medical categories of incomplete or arrested development of mind. This table presumably represented not only present reality
as Burt and Gillespie saw it, but also a frame of reference that informed their
proposals for vocational training to avoid the problem of the 'misfit' in society. This functionalist vision wasn't without its contradictions for the meri-
tocrats though. As a speaker at a CAMW conference in 1926 put it, if mental
hygienists were to eliminate the problem of 'sub-normals', 'the problem of
domestic service, for example, would become even more acute than it is'.[141]

Table 3.1 is inevitably informed by mental hygienists' description of
evolutionary stages of human progress and their recapitulation in each individual's development. It links 'the primitive' with 'the civilized' through
its hierarchical scale. Below a certain level lies the destiny of institutionalization. So emotionality may have been the privileged site of therapeutic
attention for the mental hygiene movement, but intellectual capacity was
the common denominator.

* * *

Wills was getting some 'dope' on a boy for the child guidance service he'd
been placed at. The boy's mother lived in Brooklyn. He found her address
on the third floor of an apartment block. His knock at the door brought out
a 'rough looking man' in his shirt sleeves:

> "Well?"
> "Is Mrs Shostakovitch in please?"

The man closed the door without a word. As Wills later recorded, his training
so far hadn't covered a situation like this. 'I was not sure what was the correct
procedure for a good PSW [Psychiatric Social Worker]. Knock again? Go away
and come again another day? Wait a bit and see what happened?' He opted
for the latter. Two or three minutes later the door reopened.

> "You still here? Say, what do you want mister?"
> "I'm looking for Mrs Shostakovitch."
> "Oh you're looking for Mrs Shostakovitch?" (With elaborate mock cour-
>   tesy.)
> "Well, why don't you come right in?"
> "Thank you."

'And I just went right in, to find three other uncomely men sitting round
a table with cards and glasses, smoking cigarettes. They all stared at me
woodenly. It was like a scene from a movie'.

"This guy says he wants Mrs Shostakovitch."

"You don't say? What you want Mrs Shostakovitch for Mister?"

"Well – er – I wanted to talk to her about her boy who's away at school at Dobbs Ferry."

"Her boy! You a dick?"

"Oh no, no."

"Well who are you then? What do you want?"

"I'm a social worker. I want to talk to Mrs Shostakovitch. Do you know when she'll be in?"

"Oh, a social worker. He says he's a social worker Al."

Wills was 'beginning to feel distinctly uncomfortable by now and wondered what was the most dignified way of retreating…'

"Limey too aincher?"

"Yes."

"Well I'll tell you sump'n Mr Limey Social Worker. Mrs Shostakovitch don't live here see?"

The man got up, skirted the table, stuck his nose in Wills' face and repeated, "Mrs Shostakovitch don't live here, got that? And if anyone tells you she does, you can tell them from me she don't. From me see?".[142]

It may not be possible to scarper with dignity, but Wills did his best.

*   *   *

Bosanquet had attributed the ordering and self-government of the rational individual mind to the ordering and authority of the Family. But he also used the analogies of an army and a crowd to contrast this structuration of the mind to the mental chaos that was its opposite. The crowd was mere association. It had no hierarchical organization; it wasn't a 'social mind' so much as a superficial connection of separate units, joined only loosely and inadequately by immediate sense impressions. Thus, its 'intelligence' was low and characterized by 'passing ideas and emotions'.[143] William McDougall also used the analogy of an army to describe a healthily functioning mind in contrast to an association without hierarchy and authority.[144] Some prominent mental hygienists likened this mental organization and management to the captaincy of a well-ordered ship and crew.[145]

Standing on the bridge, he controls all the different sections of the crew, the engineers – his instinctive driving forces, the deck hands – his executive abilities, the stewards – his capacity for social adjustment, and the navigating officers – his intellectual and purposive side. A mutiny in any one of these departments, or a lack of co-ordination between them, means maladjustment, and, if acute, insanity.[146]

An army or a navy then. They depict hierarchies of authority stemming from one central hierarchy. Authorities within authorities, like Russian dolls. The original authority, however, is clearly that of the Family. In historical terms the Family and the rational self-supporting individual have emerged together. In biographical terms, each individual recapitulates this history, and is moulded within the authority of its family. Rationality and the moral authority of conscience originate through this external authority and become established within the individual mind. Burt's distinction between the families of 'ordinary' children and delinquents is instructive:

> The ordinary child in an ordinary home is a member of a small and self-contained society, cared for by the united efforts of both father and mother, and possessing at least one other relative of his own age and out-look to play with him, to grow up with him, to keep with him, and so to some extent to regulate his ways, or at least to report on any serious fault. The delinquent child, too often, is devoid of all such benefits. He leads an existence warped, onesided, incomplete; and lacks the most natural check against lawless behaviour.[147]

There were obvious connections to be made between the Family, the structuring of the individual mind, and the structuring of the social and political order. J.A. Hadfield, a leading light at the Tavistock Clinic during the 1930s, put this clearly:[148] 'The problem of the individual, like that of the state, is how to co-ordinate freedom with authority... In other words, authority exists to secure liberty, which is precisely the rule we discover psychologically in the individual'.[149] In both cases—that of the individual mind and that of the people under government—the problem was essentially the same: the twin demands of 'native tendencies' for freedom, and the equally important requirement of an authority to restrain and craft liberty. Hadfield explained that, just as the psychoneuroses were a consequence of an inappropriate fit between 'native impulse' seeking expression and the authority that must shape and guide it, so, too, were political disorders. He described them as 'manifestations of mass neurosis'; a conflict between 'spontaneous impulses' of the people and an authority that had failed to restrict and direct them adequately.[150]

In like fashion to Bernard Bosanquet, mental hygienists dragged the Family authority so that it overlay the social and political arena. And so radical notions of liberty, egality and fraternity were subsumed by the Family authority, with its hierarchy of status, informed by a presumed hierarchy of rationality. For Hugh Crichton Miller 'normal influences' worked on the race from top down, while 'mob hysteria' and its regressive impulse operated from below upwards:

> ... where the higher intellects fail in providing the vision and passing it downwards, where they are not able to suggest the well-balanced solution, springing from creative ideas and harmonized with the lessons of history, there regression will take place. Mob hysteria, which wastes all the lessons of history, because it cannot learn, and wastes much more that is valuable to society in its attempt to realise the one goal that it sees, will rule.[151]

The same basic premise was expressed, in his usual elegant prose, by Cyril Burt:

> Man belongs by nature with the sheep, the deer and the chimpanzee – animals that roam in flocks or forage in herds – rather than with the solitary animals, like the lion or tiger, that set out on their depredations alone. He is born with a gregarious instinct; he is, as Aristotle defined him, essentially a social creature. The term does not imply that he lives in communities for motives resolutely rational – for better self-defence or for economic gain; nor that his natural inclination is to think first of the good of his tribe. It means simply this: that, by virtue of his hereditary constitution, he is compelled, quite blindly to begin with, to seek out and to remain with others of his group. For the deep thrill of crowd-excitement, the emotional surge that we all experience when massed together in a throng, we have no very apposite word; but to the acute uneasiness that every one feels, so long as his sociability is left ungratified, we give the name of loneliness ... The unreasoning character of the impulse explains a curious paradox: that the social instinct may be the origin of many anti-social actions ... Everywhere in the human world, the ethical code of a crowd lies far beneath that of its component individuals: its morals are not the sum of the morals of each unit, nor yet their average, but their lowest common denominator, the outcome of the motives shared by all; and these motives, in turn, will be the crude and universal instincts.[152]

So, here again, as with the Bosanquets, the individualist and the masses are two elements of the same phenomenon. As J. R. Rees, the deputy director of the Tavistock, wrote, 'The psychologically adjusted adult must have an individuality, and yet he must not be an "individualist"'.[153] The 'individualist' and 'the mass' were outside history, outside the family, outside moral regulation, outside authority. Childhood, 'the primitive', madness and the 'lower classes' were linked here. They were expressions of the immediate over the temporal, of the emotional over the rational. They lacked order and direction, discipline and deferment of gratification in the interests of a higher good. They represented stagnation or regression instead of progress. 'Lower' and 'higher': lower instincts, higher purposes; lower 'races', higher 'civiliza- tion'; lower minds, higher minds. This wasn't a recipe for radical democracy. To seek egality, liberty outside the formative structure of the Family, or fra- ternity without the Family's relations of status, was to seek regression to 'primitive' childhood. Political approaches proposing liberty, equality and fraternity were, therefore, relegated under the moral order of the Family.

In keeping with this stance, the mental hygiene movement was appre- hensive about the perceived levelling effects of centralized state power legitimated by mass mandate, and concerned that the state was too distant and impersonal to adequately moralize people. Accompanied by processes of rationalization, totalized throughout society, such power held the poten- tial to undermine the Family's role in the moralization of the individual and the stable order of society. Mental hygienists seem, in effect, to have touted themselves as the professional authorities within society who could mediate these aspects of power in the interests of healthy moralized citizens able to find an appropriate place in the social order. But the mental deficiency sys- tem had been closely attached to state power since its establishment in 1913, and, increasingly, the voluntary organizations that comprised the move- ment operated in conjunction with local authorities and local education departments.[154] Meanwhile, mental hygienists' emphasis on the importance of personal familial relationships and engaging with 'the whole person' actu- ally increased the penetration of certain elements of rationalizing power into the family and around the individual.

Inevitably, the formative role that the Family played for mental hygien- ists encouraged refashioned ideas about political democracy. Hugh Crichton Miller and Cyril Burt (who were quoted earlier) both clearly expressed this. In his 1933 radio broadcast Burt attacked the same ideas of a social con- tract in political philosophy, as had Bernard Bosanquet: 'The state is not to be understood as a kind of limited liability company, whose members have signed an agreement for mutual aid and protection; it has arisen, not from artificial convention, but by natural growth'.[155] There was no rational indi- vidual that existed separately from the moral and political order, contracting into it through its own volition. In fact, Burt went on to question the voting system. 'Men have lately begun to wonder whether the principle of "one man one vote" is working quite as well as they hoped', he said. 'Is it fair to count the number of heads without stopping to consider their contents?' he asked.[156] And he answered this with a classic description of a 'meritocracy':

> It was, of course, a risky argument to base the equality of political rights upon a supposed equality of intelligence and brains. The conclusion may be sound enough in practice, but the premiss is a theorist's delusion. Indeed, one of the best reasons for assuming that all men are equal is that, by granting them equal opportunity, we shall more readily discover

who are the best. And the best will be discovered, not in any one class as judged by birth or social status; they are to be found sprinkled about in every layer of society, like sultanas in a bread and butter pudding.[157]

Hence, Burt's idea of a healthy individual adjustment being facilitated by a kind of 'double ladder' whereby each settled at the level of society to which their intelligence was suited. Hugh Crichton Miller went further. In 1933, he claimed that '... democracy, if it is to survive, must give greater opportunity to the more competent voters, and less opportunity to the less competent voters'. Crichton Miller considered that education should be aimed at producing better citizens for democracy, and this relied on enabling 'independent thought'. But, like Burt, Crichton Miller hailed democracy only to subvert it. Higher education, in his view, ought to produce the most independent thinkers and therefore it followed that these people would be 'better voters' and 'entitled to a much larger responsibility in a democratic system'. Admitting that this was easier said than done he nevertheless claimed that modern psychology now offered a means to attempt it. As if to highlight the poverty of his own argument, however, he concluded with the bizarre statement that:

> Certainly we may agree that unless a community is educated for self-government, it is much more likely to profit from a dictatorship than from the semblance of democracy which permits the demagogue to achieve autocracy. And this is the democracy which is based on the outworn shibboleth of 'one man, one vote'.[158]

For Crichton Miller then, a definition of democracy that ended up bearing little resemblance to the common understanding of the word. For Burt, and in all likelihood for the majority of mental hygienists, a less extreme redefinition of democracy. But one that, nevertheless, viewed the egalitarian underpinnings, expressed in the equivalence of voting rights, with suspicion; under such a system, how could the 'more rational', the 'more reasonable' minds prevail? And without them, how could progress?[159]

But yet mental hygiene also expressed a softening of approach towards many people considered 'socially inefficient'. The leading psychiatric social worker, Sybil Clement Brown, displayed this in her attempt to reconcile an acceptance of much increased measures of social welfare provided by the state with the continued mental hygienist focus on individual failings. She maintained that social workers had always been engaged with two different aspects of this—on the one hand general standards of societal and individual welfare, and, on the other, individual casualties. She used the analogy of traffic regulation, which at that time was itself only in its infancy. The former aspect was equivalent to what we would now call the highways department.

It involved organizing traffic flow with the least possible friction: where to put traffic lights and one-way streets, where to put a policeman on points duty. This assumed a general standard of efficiency among individual drivers associated with their capacity for manipulating their own vehicle. The second aspect was akin to dealing with the road casualties of this system. These 'required special salvage' action to deal with the immediate 'obstruction caused' and 'analysis of the causes of the accident in terms of failure of the driver or the machine'.[160]

> We have often been guilty of arguing with the victim of the accident still suffering from shock, about the value of our new automatic signals and his stupidity in not paying attention to them. We have overlooked possible refinements in the art of restoring him to health, in our annoyance that he has not been a credit to our own superior powers of organisation. We have sometimes blamed him for obstruction without considering the relationship between other urgent calls upon him, and how these have cut across the rules that we have made; or we have gone even further and argued that there is malice in his failure to comply with our demands.[161]

She was quick to note no greater fallacy than assuming that social problems could be solved by attending solely to individual adjustment. 'All case work, however individualised', she observed, 'will depend on the soundness and efficiency of social institutions. Many individual problems may be altogether prevented by providing a more adequate standard of living'. But that said, Clement Brown's essential point was to emphasize her view that social planning and provision had proceeded without enough attention to understanding the processes that lay behind adjustment and maladjustment. And this required delving into the instinctual and emotional life of individuals.[162] In this it was the family that held the formative and most pervasive influence.[163]

And, once again, this became translated into the mental hygienist meritocratic vision. Optimizing each person's ability to drive their own machine was the way to provide proper 'equality of opportunity'.[164] And, as other psychiatric social workers noted, this entailed that a person must find adjustment to themselves and their role in society through an acceptance of their particular limitations. Unsurprisingly, gauges of intelligence and tests of personality provided an important backbone informing this assumption.

So some might drive Rolls Royces, some might drive Austen Sevens, some might ride bicycles and some might ride nothing at all. And all would be fair and equal in the meritocratic society. It's still the prevailing ideology of our times.

* * *

Wills had decided to submit some old essays written at Woodbrooke College to his tutor for appraisal. One of them was on the question, 'What are the hereditary factors in juvenile delinquency?' Part of its argument was:

It is as absurd to talk of a criminal type as it is to talk of, for example, an income-tax-paying type. The liability to be a criminal depends upon the present state of the law, just as does the liability to pay income-tax. The criminal of today may, indeed, be the saint of tomorrow, as history has often shewn.[165]

* * *

For mental hygienists then, the social order was one of 'function and fit', of reason and the family hierarchy structuring the individual mind and the order of society—the parent and the child, the ship's captain and his crew, the marshal and his subordinates. Many other people held different perspectives of course. Perhaps the more telling analogy for this modern bourgeois age was the hotel restaurant and its staff. George Orwell thought so. In *Down and Out in Paris and London* he described this social order as it existed in a large Paris hotel where he had worked.[166] Orwell described a hierarchy of around a hundred workers servicing perhaps two hundred customers. Here was a system of station and rank in which each 'class' of worker knew their role and in which their limited authority declined the lower their level: proprietor, manager (responsible for discipline), the *maître d'hôtel*, the *chef du personnel*, the cooks, the waiters, the laundresses and sewing-women, the apprentice waiters, and, finally, the *plongeurs* and the chambermaids. Orwell was a *plongeur*, 'a slave's slave': hard physical work and punctuality, 10 or 15 hours a day for pitiful pay.[167] He later asked himself the social significance of a *plongeur's* life. Why did this drudgery continue? People tended to take for granted that all work was done for a sound purpose, he noted. Their assumption was that, just like many other unpleasant jobs, the *plongeur's* job was necessary. 'Some people must feed in restaurants, and so other people must swab dishes for eighty hours a week. It is the work of civilisation, therefore unquestionable.' But was it really? Orwell admitted that the hotel organization achieved a certain efficiency built on the twin principles of ruthless discipline enforced by each station over those below and a genuine pride of each worker in their position. Even the *plongeur*, he noted, managed a kind of pride through the virtue of being equal to any amount of labour. But what was this efficiency? Cleanliness, for example, was sacrificed to 'punctuality and smartness'. The workers cleaned what they were ordered to and neglected the rest. The customers received clean looking and nicely arranged plates, but saw nothing of the clatter and filth behind the scenes. The patron swindled the customers with inflated prices for food made from inferior raw ingredients, and each level of staff swindled the hotel or their lower level colleagues. For Orwell, the pointlessness of this structure and its tasks was obvious and the reason for its perpetuation was equally obvious— middle and upper class fear of 'the mob'. Better pointless work and discipline than that the poor masses should gain more liberty, as more liberty for them would threaten the liberty of the classes above them.[168]

This was a stark and very different description of 'function and fit' for the social order to the pacifying visions of the mental hygiene movement.

*   *   *

Still working for the child guidance clinic, Wills decided he'd made a break-through. That afternoon he'd carried out a home visit and, for the first time, felt that he'd really been able to apply the techniques that the Social Work school had been trying to teach him.

> It was with a Mrs Brown, for whose son at the village I am compiling a social history. As I sat there and talked to her, I really was able from time to time to get outside of myself, and criticise myself objectively. I made what is known in case work circles as a 'good contact', got a lot of 'dope', and there's more to come. Mrs B. said she didn't attach much importance to this heredity business, and gave me an astonishingly good account of the boy's developmental history. Mr Brown took me on one side later and said he wanted to see me in private some time to tell me about his wife's relatives, who were just like the son![169]

*   *   *

Several authors, such as Jacques Donzelot and Nikolas Rose, have applied the theory of power/knowledge that Foucault developed after *Madness and Civilization* as a means to show how the family has been opened up to increasing scrutiny by 'psy' professionals through the twentieth century. They describe ever multiplying conduits of power ripping open what had once been claimed as hallowed area of privacy. Dissected it lies open to the famous 'gaze'. This dissection is endless of course, and the gaze is equally endlessly finding new surfaces for its knowledge. It's an intellectual gaze. It's a remorseless and unremitting power.

Yet, throughout our tale of moral therapy extended, we have seen that the family wasn't only a target of power. It played a fundamental role as organizing principle and methodology. Ironically, the root of this tale has been Foucault's earlier theorization of power and reason in *Madness and Civilization*. But we have emphasized the element of authority in this power structure and traced out a trajectory for the 'dialectic of the family' that Foucault then claimed extended from moral treatment into an unknown future.

We've adopted Foucault's original argument that, under the rubric of the family, moral treatment made madness childhood and reason an adult destination against which madness must experience itself as falling forever short. But by following how the Bosanquets and the COS brought a form of moral treatment to the wider community, we've traced how history came to be deployed as the animating force through which the rational individual evolved. Thus provided with a wider arc, a form of moral treatment might be asserted that could encompass both internal mental states and the social order. This entailed a notion of phylogenic and ontogenic development from 'the primitive' to 'the civilized'. So 'the mad', 'the primitive' and childhood were linked. Under the mental hygiene movement this historical progression remained. Individual minds were to be classified and ordered according to a particular understanding of developmental time, and mental hygienists held out a vision of a social order built on this. But, for mental hygienists, while the rational individual emerged through 'the primitive', it never quite shook it off. As childhood appeared to recapitulate phylogenic development, mental hygienists paid detailed attention to its progression through 'the primitive' to the rational 'civilized' individual. This progression needed to be understood, accommodated and guided; its content was emotional and it had a meaning that needed to be deciphered. 'The primitive' and its emotional content thus linked 'the mad', rational adults and children; it was their common core. A common core suggested a common language. Mental hygienists attempted to engage with this language and, in the process, their appreciation of authority relations altered.

In 1932 the Association of Psychiatric Social Workers produced a booklet entitled *Psychiatric Social Work and the Family*. This attempted to describe the primary role played by the family in individual and social development. In keeping with the imperatives of the mental hygiene movement that we have so far unfolded, the individual personality that it described was not individual in the sense of a separate isolated unit. It derived from family relations and continued to be imbued with them throughout its life. The booklet argued that the importance of the family lay in the need to balance two general emotional drives: family relations needed to gain a 'just balance' between a requirement for security through 'love relationships', and a need to assert power, independence and separateness. This was an invocation of human duality common among mental hygienists. What seemed required was a healthy and progressive balance between parental authority and this duality of emotional expression. The process was one of emergent development, of something advancing through progressive contradiction. In effect, it was a dialectic.

Mental hygienists deployed this view in their advice to families on child development. They also applied it to adults, as well as children, in their approach to care and therapy. After all, as *Psychiatric Social Work and the*

Family pointed out, '[an] adult is only a child of a larger growth'.[170] And, if the child resided in the adult, so too did 'madness' in the child. Sybil Clement Brown remarked that 'The resemblance between the symptoms of mental disorder and the normal behaviour of the young child are so marked that it is surprising that they did not arouse any scientific curiosity until the latter part of the last century',[171] A statement by Emmanuel Miller gives substance to this:

> The life of a child is largely magical and animistic, and partakes of the psychotic mechanisms which we are only ready to see in fully-fledged adult cases. The pan-psychosis of childhood is not a malignant state – it is a process of development from mental autonomy to objective relationships...[172]

Thus, the child and 'the mad' were linked via the evolution of 'the primitive', and these were linked with the apparently rational adult. As Clement Brown put it:

> Those concerned with the adult mental patient, though they have no specific knowledge must be impressed with one outstanding characteristic. They are living, as we say, in 'a world of their own'. This does not mean that their mental processes are entirely different from our own. Indeed, one of the main discoveries of the last fifty years has been that most of us are to some extent, or sometimes, 'insane'.[173]

Psychiatric Social Work and the Family echoed this, emphasizing that social workers needed to access what was important about their client's situation through 'knowledge of emotional dependence and independence and emotional needs in ourselves, and in all developing human beings'.[174]

In support of his argument that moral therapy at the Retreat deployed a familial authority through which madness was made to experience itself as forever childhood, Foucault made use of an incident recounted by Samuel Tuke. He described a madman who threatened to throw a stone towards the superintendent in whose company he had been walking. This incident was relayed in our first chapter, but it's useful to repeat Tuke's description of it here:

> The superintendent was one day walking in a field adjacent to the house, in company with a patient, who was apt to be vindictive on very slight occasions. An exciting circumstance occurred. The maniac retired a few paces, and seized a large stone, which he immediately held up, as in the act of throwing at his companion. The superintendent, in no degree ruffled, fixed his eye upon the patient, and in a resolute tone of voice, at

the same time advancing, commanded him to lay down the stone. As he approached, the hand of the lunatic gradually sunk from its threatening position, and permitted the stone to drop to the ground. He then submitted to be quietly led to his apartment.[175]

Here, in contrast, is an analogous incident described by Muriel Payne in 1929. She was intimately involved with the mental hygiene movement, working closely with two of its organizations.[176] (We'll meet her again in the next chapter.) It describes an incident experienced by Payne when she worked as Matron to the Leytonstone Poor Law Homes for children.

The next day Matron was walking round the grounds, and a stone came whizzing past her head. She turned, and there was Charlie, standing with his upturned nose towards the sky, and hands deep in his pockets. Matron called to him, but he walked away in the other direction, so she took no further notice and continued her stroll. Presently another stone came whizzing past. The difficulty was to know exactly what this behaviour meant. The only thing the Matron realized was that this was evidently a very mentally sick child, who was terribly unhappy deep down in his mind. But why should he try to hurt or kill the person who was trying to be kind to him? Had she unintentionally hurt the boy when he came to tea, or was he simply venting his feelings on her as a symbol of authority up against which he had been all his life? Or was it a method of attracting attention? Or was it simply the devil in him, as the Housemother said? For more than a week, every time Matron came into the vicinity of Charlie, he hurled something at her. The method of ignoring the action had no effect in stopping him, yet when Charlie was sent for to be expostulated with, quite kindly, he at first would not come, and then, when he was compelled to do so, refused to say a word.[177]

This is a very different understanding of reason and madness, of adulthood and childhood. In *Madness and Civilization* Foucault used the stone-throwing incident to describe how an authority 'is born'; reason asserts an authority that colonizes the experience of madness itself and makes of it perpetual childhood. But in Payne's incident things are a little different. This 'mad' behaviour was meaningful; it had a language. Shorn of Payne's intentionally ironic questioning whether the boy just wanted to gain her attention, or whether it was the devil in him, her meditation here clearly sees the dynamics of authority as a key aspect of concern. If mad behaviour was meaningful, its language and meaning was closely related to the experience of authority. On the basis of this understanding, the relation through which authority was asserted became the principal area of analysis and concern.

Helen Bosanquet had deployed the terminology of 'loyalty' and 'emulation' as key components of the Family authority. But now these became potentially suspect. In part, this was because they could suggest an inculcation of rigid codes of honour and behaviour that were inappropriate to the development of the child. A terminology of affection, trust and 'emotional security' began to take precedence over them.[178] This entailed two important shifts: a call for greater toleration of certain areas of expression and behaviour by both children, and adults considered to have mental disorders; and an interrogation of relationships of authority—in particular, parental relationships and the therapeutic relationship itself.

Was authority over dominating? Or did it take no account of the child, ignore it and leave it mentally isolated? Did it act to incorporate the child into the adult's identity? Did it discriminate between children? Was its message and transmission appropriate to the emotional language of the child's instincts and development? Did its temper develop together with the developing mind of the child? These were the kinds of questions that began to take precedence.

The educational psychologist Lucy Fildes said that 'The great disaster in the life of a child was the lack of a secure basis for emotional growth'.[179] Within this context the terminology of 'affection', 'love' and 'emotional security' began to find favour. As we've just seen, *Psychiatric Social Work and the Family* had cast the question of authority in terms of an ongoing working out of the conflict between ego and love, with the family as the authority through which it did so:

> Ego and love cannot live side by side in us at peace, except by the continuous working out of the conflict between them through which we become socialised and civilised beings.
>
> It is the function of the family to provide the milieu for the working out of this conflict . . .

According to *Psychiatric Social Work and the Family*, the failure of the family in this progressive process was owing either to an authority that had been 'too exact and omnipotent in setting norms of behaviour or development, or that love had been too protective or demanding' to allow freedom for a person to be themselves.

So what we have here is a reconsideration of the relationship between authority, freedom and love (or emotional security). This was the issue that was now emerging through the concentration on emotionality as a fundamental aspect of personhood. From the time of its emergence the mental hygiene movement had constructed a project in which the formative relation between individuals and the social order was the Family. Radical notions of liberty, egality and fraternity had been subsumed to this Family authority, with its hierarchy of status, informed by a presumed hierarchy of rationality. But could this ongoing 'dialectic of the family' be pregnant with its opposite?

4
Dialectic Rightside Up?

When autumn comes the trees lining the field dissolve into a flurry of paper leaves. Burgundy to butterscotch, they console the eye for the oncoming winter. It's a small field, about 26 acres. At the top it is met by the road from the village of Great Bardfield. If you walk into the land from here the undulating ground takes you across what was once the clover field, past the trees along its edge and down to the river Pant that runs along the bottom.

When Wills and his wife Ruth arrived winter had taken over and held the land in its sovereign power. It was early 1936. In the back of their dilapidated second-hand car sat a recently acquired lurcher. It had come from Dr Denis Carroll, a psychiatrist at the Institute for the Scientific Treatment of Delinquency (ISTD). He'd been unable to look after it. They would soon find that it had a habit of finding any and every foul substance to roll in.[180]

The Wills were about to begin work on an experimental community approach to therapy in collaboration with the newly founded 'Q Camps Committee' lead by a psychoanalyst called Marjorie Franklin and supported by the ISTD. In time, 20 and more people would be living in this field. There's not much left now except a building that's used as a chicken shed and some traces on the wind.

* * *

In July 1936, nearly three months after it officially opened, Hawkspur comprised five bell tents and a temporary hut. Nine 'members' were accommodated along with six staff.[181] By 1937 they had managed to build an office with two storeys that stood at the top of the field, along with a quadrangle consisting of a cookhouse and washhouses further down the field, and at the bottom of it a bunkhouse used for meetings and activities.[182]

Here was a community not unlike the apparent inspiration for Rousseau's *Social Contract*. It's known that Rousseau was influenced by rural communities and the Swiss cantons. One, near Lake Neuchatel, he later described as a small sovereign community: self-sufficient and unburdened by taxes or tithes. Each wooden house had been constructed by the community and each person skilled in a variety of trades. The people made their own entertainment. Theirs was a small face-to-face community of more or less equal citizens who met at intervals to legislate.[183]

Hawkspur Camp held many similarities—except the fact that many of the citizens of this community were there because they were considered 'delinquent and difficult young men aged 17 to 25'.[184]

* * *

As we've seen, the Bosanquets and the Charity Organisation Society (COS) had considered the community the necessary arena within which the individual citizen engaged with moral authority and was situated within its hierarchy. Bernard Bosanquet had promoted a form of moral therapy for the community that used Rousseau as a fundamental theoretical prop and, in the process, de-radicalized him. 'Liberty, egality, fraternity': each was taken hold of and colonized by the organizing principle of the Family. Any sense of egalitarian fellowship implied by a metaphor of fraternity was returned, refashioned, to the hierarchical authority of family parental relations. Liberty was, in consequence, restricted by this moral order within which it was nurtured, and which cast the structure of its authority across the wider social order.

At Hawkspur we see this Family begin, in a sense, to pull itself inside out. Through the authority of mental hygiene, liberty, egality and fraternity began to make a comeback. Maybe Hegel's dialectic of Reason was still spinning, and in unforeseen ways. We've seen, in fact, that there was a sense in which the dialectic was retained with mental hygiene; the perceived contradiction between the instinctual emotionality of childhood growth and parental authority was dialectical and progressive. What's more, it seems that it was this very perception that informed an antithesis of liberty, egality and fraternity.

* * *

In March Franklin wrote to Wills, 'Applications are coming in more now – A long letter from a mother in Yorkshire about her troublesome son of 20. Sounds rather a good type for the camp. He will probably be able to repair the car if he does not steal it! Probably able to pay full fees or nearly'.[185]

* * *

If the mental hygiene dialectic of emotional growth and parental authority was part of a greater dialectical story then, in our tale, 1929 seems, retrospectively, an important moment of contradiction. Nineteen twenty-nine, the year Wills travelled to New York, the year professional psychiatric social work was established in England, and also the year that the Home and School Council (HSC) was founded. For a short decade before the war the HSC seemed, to its supporters, destined to become an important institution of the mental hygiene movement. The hope was never realized. But, even so, the HSC offers an aperture through which to see contradictions in development.

The HSC saw its aims as threefold: to develop Parent–Teacher Associations, to provide them with parent education and literature, and to carry out child study and research.[186] All the central institutional embodiments of the mental hygiene movement were closely involved with the organization in its early years. Leading figures at the Tavistock, National Council for Mental Hygiene, Child Guidance Council and Central Association for Mental Welfare (CAMW) sat on its executive and on its various subcommittees, as well as engaging in educational work on its behalf. In 1933 the magazine *Parent and Child* became the HSC's official organ. The first edition's statement of purpose was classic mental hygiene:

> Modern medicine and the principles of child guidance are between them laying down a series of simple rules as to how best to feed and clothe children; how to teach them good habits; how to prevent them from experiencing the feelings of insecurity, jealousy, inadequacy and frustrated ability; how to deal with them without recourse to threats and punishments; how to manage adolescence and other difficult periods in development. Thus parents are coming to rely upon the recent discoveries of science and to base their treatment of their children upon commonsense backed by knowledge.[187]

But the ideas informing this statement weren't just those of the mental hygiene movement. The HSC was also associated with the 'progressive education' movement through the New Education Fellowship (NEF). This international organization supported and encouraged the HSC's work. One of its leaders, Beatrice Ensor, was an active member of the HSC's committee, and the HSC, in fact, shared its offices with the NEF's English Section.[188] Kevin Brehony has noted recently that a significant aspect of the NEF was that it connected 'lay enthusiasts' pursuing reforms in education with influential professionals in psychology and education.[189] And it was here, at the interface between 'progressive education' and 'the new psychology', that the mental hygiene movement began to reveal a larger dialectic. Here, there began to emerge an apparent antithesis from the very structure on which the mental hygiene authority was erected.

* * *

They were begging and borrowing equipment from everywhere. Franklin had managed to obtain some crockery from her uncle. She wrote to Wills that he thought the scheme an excellent one, but that things should be called by their proper names—'a camp for young men who have lost out through their own fault'. Franklin annotated an exclamation mark.

*   *   *

Beatrice Ensor had been the driving force of the NEF since its formation in 1921. Formerly a teacher and a school inspector, Ensor was also a Theosophist. It was through the Theosophy Society's network that she organized first a group called the Theosophical Fraternity in Education and later the NEF. Ensor was especially influenced by the ideas of Edmond Holmes, the former Chief Inspector for Elementary Education, and the work of Homer Lane at The Little Commonwealth in Dorset. Holmes had been influenced by British Idealist philosophy and the idea of human self-realization as a vehicle for the unfolding of the Great Mind or Eternal Spirit.[190] His view of God as immanent within nature and each developing human soul echoed Theosophy and this led to Holmes writing a number of articles for its journals *The Quest* and *The Aryan Path*.

In 1911, soon after he retired, Holmes had written *What Is and What Might Be*. His depiction of 'what might be' in education was based on a small village school in Essex. He called it Utopia. Holmes saw here an educational enactment of his more general philosophy of life. At Utopia the children were given wide range for self-expression, including freedom for airing their opinions, asking questions and debating issues. This Holmes called The Path of Self-Realisation; he opposed it to his 'what is' in education, which he called The Path of Mechanical Obedience.

Holmes joined Ensor's group, New Ideals in Education. The shared belief of its members was that education should attend to the individuality of each child and this individuality emerged best in an atmosphere that emphasized freedom.

Holmes' description of a more libertarian and egalitarian approach to education was echoed by Homer Lane. His work at the Little Commonwealth community in Dorset brought the concentration on freedom and individuality in education together with attempts to reform delinquents. Lane gave several talks to gatherings of New Ideals in Education. Informing both Holmes' and Lanes' approach was a desire to mitigate external authority in the interests of allowing children and young people greater room for their own self-directed activity.

*   *   *

In April 1936, the first month after the camp opened, Wills wrote some stinging words in his Camp Chief Report. When he had accepted the invitation to join the venture he reminded the Q Committee they had thought £1000 the minimum capital necessary to start it. He had said that he would be prepared to begin with as little as £500. In fact, they were now starting with £400. Apart from all the difficulties this caused, Wills pointed out that it didn't look good stocking the camp with gifts, especially if they hoped, at some point, to get Home Office recognition. 'It looks like a jumble sale'. And, to add insult to injury, he hadn't received a pay cheque at the end of his very first month's employment.[191]

* * *

The Family meets liberty, egality and fraternity. Through the aperture of the Home and School Council we can spy this contradiction.

One day this summer I was riding through Letchworth when the bus stopped and two dreadful-looking old men got on to it. They were both about sixty, both very short, pink and chubby, and both hatless. One of them was obscenely bald, the other had long grey hair bobbed in the Lloyd George style. They were dressed in pistachio-coloured shirts and khaki shorts into which their huge bottoms were crammed so tightly that you could study each dimple. Their appearance created a mild stir of horror on the bus.[192]

George Orwell's displeasure was recorded in his 1937 book *The Road to Wigan Pier*. These men were, he believed, most likely some none too attractive contemporary members of the Independent Labour Party, which was holding a summer school in Letchworth. In his view they were none too attractive ideologically, as well as physically. Recently converted to socialism, Orwell lost no time in skewering the bullshit in many middle class socialists' beliefs, as well as ridiculing their often held 'natural life' and mystical enthusiasms.[193] Letchworth was, in any case, well known as a mecca for such people. The Theosophists chose it as a centre for their educational efforts and the site of their co-educational progressive school, St Christopher. It wasn't just socialism that was drawing towards itself 'with magnetic force, every fruit-juice drinker, nudist, sandal-wearer, sex-maniac, Quaker, "Nature Cure" quack, pacifist, and feminist in England', so too were all kinds of 'progressive' communities in education and social living.[194]

Norman Glaister was one of them. Another was Muriel Payne. Both were involved with the mental hygiene movement. Both were also members of the HSC and were involved with schemes that fused the mental hygiene concentration on the Family with the progressive educational emphasis on the unfolding of each unique individual within an atmosphere of 'freedom'. By looking at these schemes we can see how the organizing principle of the Family became partially inverted through its meeting with ideas of liberty, equality and fraternity.

Muriel Payne originally attended HSC council meetings as the Tavistock Clinic's representative. She was briefly its honorary secretary, served on all three of its subcommittees and later became its 'organising secretary'.[195] Payne had originally trained as a nurse. During World War I she worked at a small dietic hospital for infants. Here she formed the view that the infants' health and recovery depended far less on the medical regimen and diet than on the emotional condition of the staff.[196] While working at the Leytonstone Poor Law Homes for children she developed close links with psychotherapists at the Tavistock and subsequently took on the job of running the Tavistock hostel. Payne also seems to have had links with theosophy. After World War II she worked with Jiddu Krishnamurti, helping him set up the Rishi Valley School in Madanapalle.[197]

We met Payne in the previous chapter where I used an incident of stone-throwing, recounted from her work at the Leytonstone Homes, to show how the mental hygiene movement had increasingly come to understand madness as having a language of emotionality that was meaningful and related to its experience of authority. She was Matron to the Homes for three years from 1924 and attempted to revolutionize their organization. This is one of the experimental schemes that we'll look at.

Norman Glaister was a member of the HSC's Executive during the 1930s.[198] He had trained in surgery, but after his wife died during the 1918 flu epidemic had set out on a quest to find ways of helping improve the human condition.[199] Wilfred Trotter's influential book *Instincts of the Herd in War and Peace* provided his first inspiration. In essence, this book maintained that there was an instinct of 'gregariousness' in humans that interacted with individual experience producing 'mentally stable' and 'mentally unstable' types of people. But Trotter believed that, in order for society to keep evolving, it needed both types. The first group generally accepted the commands of 'the herd' and resisted change; they were 'resistives'. The second group weren't necessarily mentally disordered in the commonly understood sense, they were 'sensitives'. This group had been growing in society for some time and they needed to be incorporated into society so that their positive strengths of originality could complement those of the 'resistives'.[200]

Inspired by this theory, Glaister saw a way to put it into practice through involvement with a pacifist camping movement called The Order of Woodcraft Chivalry. Its Quaker founder, Ernest Westlake, intended it as a non-militarist and libertarian version of the boy scouts. The group emphasized education through outdoor life and direct contact with nature. This was informed by a recapitulationist theory that saw children as developing through all the stages of mankind's progress.[201] Glaister became one of its leaders. He also became involved in Westlake's 1929 foundation of the Forest School. Through these groups Glaister became the driving force in the foundation of an associated group called Grith Fyrd (Peace Army). This offered long-term unemployed young men self-governing camp life and voluntary education.

It was originally through Grith Fyrd that the impetus for Hawkspur Camp emerged. The organization had been referred men whose difficulties appeared to be too psychologically severe for them to cope with and, in response, the psychoanalyst Margery Franklin had taken the lead role in forming what became the Q Camps Committee. She had joined Grith Fyrd through Glaister and along with Cuthbert Rutter, the head of the Forest School, the three of them eventually made contact with David Wills in 1935.[202] When Wills met them he thought them 'woolly'. He wasn't too sure about the Order of Woodcraft Chivalry either. He and his wife used to call it 'The Order of Witchcraft Deviltry'.[203] But they saw enough in the Q Committee to recognize a common cause. So Hawkspur Camp opened as a self-governing community with him as Camp Chief a year later. This is the other scheme we'll look at. The key figures were Wills and Franklin.

* * *

The true address was Hill Hall. But Wills didn't like it, probably partly because Franklin thought it highly appropriate and amusing that, in the local Essex accent, it was 'ill all'.[204] He had his way. The Q Camps Committee decided, by vote, to call the place Hawkspur after the name of the common just up the road.

* * *

Both the Leytonstone Homes experiment and Hawkspur described the community that they set up as a 'family'. 'The members seem to have settled in very well, and the family is a very happy one', wrote Wills in his April 1936 Camp Chief Report.[205] His position at the camp was conceived as that of a father figure. At Leytonstone, Payne described her attempt to fulfil the role of mother to the institution and how her organizational changes were intended to produce an environment akin to 'one big family'.[206]

The young men who came to live at Hawkspur were selected because they were already considered to be maladjusted. At Leytonstone, Payne emphasized that the children and young adults were 'normal', but had simply lacked a family able to look after them. She wrote that 'They were not mental, or bad, or odd; they were just those children who, owing to misfortune, had been born into a stratum of society which has few advantages and plenty of kicks'.[207] But, even so, her entire account of the project was oriented around the fact that these young people had clearly come to display signs of maladjustment.

We've seen how the mental hygiene movement focussed on emotionality as a crucial component of the individual and its development, and how, in consequence, it came to reconsider the family authority that was assumed to inculcate mental self-government. Under the influence of a psychology of instincts mental hygiene described what amounted to a dialectic of development through the interaction of instinctual emotions and family authority. Mental hygienists emphasized the need, not only for the right authority commands, but also the importance of the right relations through which authority was communicated. For mental hygienists, these relations were emotional and intellectual, but (so long as intellect wasn't considered congenitally 'deficient') the emotional content was seen as pre-eminent. This prioritization of emotional relations became all the stronger the closer these relations were to infancy; the newborn child was emotionally, not intellectually, endowed.

The concentration on supposed emotional 'needs' associated with the theory of instincts, encouraged a prioritization of what was commonly called 'emotional security'. A consequent accent on greater tolerance for self-expression placed under suspicion older concepts like loyalty and honour. Couldn't a child emulate forms of authority inappropriate to its emotional development? Couldn't these older concepts of emotional ties in fact assert a dogmatic restraint or rigidity that neglected emotional sensitivities and thwarted the emerging individual? The ideal was that authority must not be colonizing, or dominating, rigid or isolating. Instead, it must speak the language of emotionality and embrace its world in order to communicate.

Both Hawkspur and Leytonstone Homes held this organizing principle of the Family and the problematization of authority relations at the core of their approach. But what transpired was a fusion of mental hygiene aimed at reinstating the 'ideal' of the family by unravelling familial relations gone wrong, with elements of progressive education that sought to release the 'divine' within each individual, and thus the 'natural' impulses for development and an organic growth within community. Under a mental hygiene eye, this emphasis on liberty and release was generally seen as too exaggerated to be appropriate to the 'normal family', but directed at children or young adults whose development already showed signs of maladjustment, it took on a new utility. These people could be deemed to need far more toleration and freedom than normal as their relationship to authority had already become distorted and antagonistic. Thus, the two main elements that came together at these experiments were, first, the mental hygiene translation of parental authority and emotional development into a need for 'emotional security', and, second, a progressive education inspired translation of psychoanalytic 'free association' on the couch to a more literal free association in lived life. So, if these experiments retained the Family as an organizing principle and methodology, it was a partially inverted one.

* * *

Wills wrote regular reports on the progress of all the camp members. In April 1937, for example, he reported that Robbie was 'regarded by some of the discontented members as an instance of someone having been made worse by coming to the camp'. Wills added that 'It is said that he has taken to using bad language, though the Camp Chief has never heard it, and that he is not as industrious as he used to be'. A couple of months later Wills wrote:

> Robbie is one of those who declined to covenant to work regularly so many hours a day and he has therefore done no regular work this month. He does however help regularly with the laundry on Mondays. In addition he refuses to take his turn in the kitchen and this is apparently due to the fact that he regards the [Camp Council] committee members as taking up a dictatorial attitude.

But, two weeks on, Wills reckoned that he'd 'settled in well'. Then Robbie went to stay for a week's holiday with a couple who'd taken an interest in him during his previous spell as a patient in a hostel attached to a mental hospital. They'd found his visit difficult because, according to Wills, they'd, 'expected him to return to the routine to which he was accustomed as a patient in the hospital, and he regarded himself as being on holiday'. The couple subsequently told Wills that:

> During the two years that the boy was at the Mental Hospital they had four times brought the father over with a view to effecting some kind of reconciliation, but on these occasions the boy 'acts like a mental defective'. He goes extremely pale, hangs his head and absolutely refuses to make any remarks at all or to answer any questions put to him by his father.

One of the camp staff reported to Wills that when back at the camp Robbie had woken one morning complaining of a very bad night troubled by dreaming of his father and mother. In the evening Robbie asked a fellow member to go for a walk with him, but he'd declined. Robbie immediately flew into a rage and tore the flannels and jacket he was wearing to pieces. Then he collapsed into 'hysterical laughter'. For three months he was very unsettled and volatile. He did no work and dropped the shorthand and typing he'd been learning. Wills tried to let him know that he and his wife were very fond of him, as were the rest of the staff. In November 1937 Wills reported that Robbie had sought him out one day to tell him of his

> ...admiration and affection for the Camp Chief and his despair because the Camp Chief had so much to put up with and 'liked so many people'. He was allowed to believe that the Camp Chief had in fact a special affection for him (which is true) but it was pointed out to him that if this were

ever suspected by anyone else it would enormously increase his difficulties which he had just been deploring. He was apparently able to accept this. The next time the Camp Chief had a day off, however, he was again hysterical and threatened suicide. In conversation with the Camp Chief next day he admitted that this was because he thought that what had been said to him the day before was insincere and was merely to placate him. He was therefore told that if ever he felt like that again during the Camp Chief's absence he should come down to the Camp Chief's house for an hour or two, it being understood that this was to be done only as a last recourse. He agreed and although the Camp Chief has had several days off and a weekend since he has not had to fall back on this remedy.

* * *

We've seen already that moral therapy's extension to the community combined psychological and political imperatives. Hawkspur and the Leytonstone Homes inevitably continued this, but they enacted them together. Self-government of the mind was combined with self-government of the community in one therapeutic endeavour. And, just as authority relations in the personal family had become questioned in terms of their appropriateness for emotional communication and security, so also was the social order of the communities at which these experiments took place. Their combination of mental hygienist ideas with ideas in progressive education focussed attention on authority that was hierarchical, rigid and dominating. We've seen the mental hygiene movement's vision of the social order. It was a modernist interpretation of function and fit; social role and social status were related to definitions of intellectual capacity and personality structure. Metaphors for the mind were recapitulated for the social order—a ship's captain and crew, a marshal and his army—order, stability, efficiency. But if, like Orwell, your role was as a *plongeur* or any of the other 'lower' functions, your view might well be far less relaxed about power and authority. Within the communities that we're examining this kind of order was challenged.

Payne provides a vivid example. When she arrived at the Leytonstone Homes the Assistant Matron, eager to impress, took her on a tour of the institution. House mothers, maids, cooks, scrubbers, porters, all were displayed before her eye, 'flown at, and left like pulp'.[208] Payne recounted circumstances rather like Orwell had described. Within the rigidly demarcated institutional hierarchy, individuals often held pride in their ability to carry out tasks to the letter. These specific duties had a certain efficiency within their narrow remit, but Payne described an impoverished appreciation of life beyond the discipline and narrow moral order of the Homes, accompanied by hidden fear, dishonesty and manipulation. All this she gave a name to; she called it 'institutionalism'.[209]

* * *

If the psychiatric social worker Sybil Clement Brown thought society's welfare system was like a highways department using the equivalent of traffic
lights and priority signs to regulate individual and social welfare, she'd
have been interested in Joseph. He habitually stole or more usually 'borrowed', cars and motorbikes. He didn't bother with a licence, he just kept on
driving—in between a couple of short spells in prison that is. Clement Brown
reckoned that the welfare system assumed individuals to have a certain ability to manage their own machine, and when they didn't 'road salvage' was
needed to clear up the immediate obstruction and find a way to make the
driver more efficient. Joseph eventually came to Hawkspur.

Here he revealed himself to be engaging and likeable, and also a chronic
fantasist and liar. It also became apparent that he didn't just 'borrow' cars
and motorbikes, he also 'borrowed' money and personal items from people he'd become friendly with. Soon he was 'borrowing' clothes from other
members and owing them money. He also 'borrowed' a number of items
from people he'd befriended outside the camp. Wills decided that if they
were to keep Joseph he needed intensive psychotherapy. Hawkspur's Selection and Treatment Committee arranged for him to stay at a hostel in
London in order to have a few months' worth of treatment sessions. Joseph
said it was a dump and refused to go. Wills said he could either go or have
the bus fare for the several hundred mile trip back home. Joseph replied that
it was a 'twist' to leave him to 'tramp back to his home' and that even prisons treated you better, which was a comment that, for Wills, unintentionally
got straight to the point.

Joseph had been brought up in a Naval Training School. These were
renowned for their harsh discipline, and Wills saw this institutional upbringing and his subsequent periods in prison as the main reason why Joseph
was such a 'tough nut to crack'. In his later book, *The Hawkspur Experiment*, Wills wrote that 'a bit of discipline' was seen as the 'common cure' for
delinquency. But, for Wills, discipline was only a means to an institutional
end. Its aim was to create organizational efficiency:

> There must be rules for carrying out everyday jobs, and people must keep
> them. That is discipline. Where I quarrel with the disciplinarian – and my
> quarrel is a bitter one – is that I resent intensely his assumption that by
> the meticulous and stern enforcement of these rules, some good accrues
> to the person on whom they are imposed. The only person who gets
> anything out of it is the person for whose convenience it is made, the
> administrative official.[210]

For Wills it was pointless to try to 'force *our* discipline, our good, upon him'.
The aim ought to be to find a means through which a delinquent could
'formulate a system of his own which is not unacceptable to society' and
which he valued because he'd found it for himself.[211] In Wills' view, the
freedom at Hawkspur was simply too much for Joseph. There were too many

things left for him to decide, too many choices. At one point Wills wrote to his 'sponsor' that:

> He has been for years trying to escape from himself (it seems to me) and the freedom of the camp life has brought him up against himself rather crudely. He got into an awful mess here – borrowed & cadged money and clothes left and right. A week ago I paid all his debts for him and he made a fresh start, but within 48 hours was in debt again... I do not altogether understand what his conflicts are, but unless he can get them resolved I am very much afraid we shall find him 'escaping' again, and finding an ultimate haven in the one place of peace and security in this world, where there are no conflicts because one's behaviour is someone else's responsibility – namely prison. I may be unduly pessimistic – perhaps he will pull up his socks and keep out of trouble when he gets a job, but the odds are very heavily against it.

Not long later Joseph went down for a long stretch.

* * *

Payne's description of 'institutionalism' took a gendered form. Deploying the mental hygiene prioritization of the family, she cast her attempt to revolutionize the homes in matriarchal terms. Her depiction of institutionalism was of maternal love repressed in the interests of the institutional organization, its guardians and the departmental authorities.[212] Payne emphasized the positive role of 'maternal instinct' and cast it against the bare and colourless rooms, mirrored in their dull hygiene by the spotless uniforms of the house mothers. She described regimented, impersonal group living in drab unmodernized buildings, with over-concentration on hygiene, and tightly demarcated tasks and roles. This was a social and moral order that crushed and 'twisted' the minds of the staff, as well as the children. The matrons were 'stunted, repressed, and soul-starved'.[213]

As spokespeople for the COS the Bosanquets had upheld Poor Law institutions to be a necessary bulwark against the degeneration of moral character. But here, in this Poor Law institution for children, the strict moral order and its accompanying obsession with 'hygiene' was pitted against both maternal instinct and 'the light that modern knowledge has thrown on the child mind'.[214]

* * *

Franklin had the ability to look marvellously dishevelled. Unruly hair, clothes that looked like they might have come from a second-hand shop, and a varying assortment of bags, purses and the like that would now and then shed their contents across the floor. The prim and neatly turned out Wills couldn't seem to see the irony. Years later he recorded a friend's description of her as an 'old carpet-bag'. 'Not kind, but deplorably apt', he commented.[215] Wills, the critic of surface appearances, the anti-militarist and enemy of rank and uniforms, couldn't appreciate that here was a woman where, to a large extent, what you saw was what you got, couldn't see that some ideal of preened womanhood might, in reality, have been for his benefit and not hers, couldn't accept her as she was.

But, for all that, Wills admitted that, while he saw her as the bane of his life, he could never quite separate himself from her. In the first place, despite her agnostic views, she seemed to hold an identity of belief in all the matters that were most important to him. In the second, he felt sorry for her and, in the end, fond of her.

Franklin was born into a family of bankers. But though they were rich Wills reckoned that they didn't think her stable and sensible enough to be trusted with capital.[216] Presumably, they must have thought her stable and sensible enough to be trusted with people's bodies and minds because she trained in medicine instead. As a young junior medical officer at Portsmouth Borough Mental Hospital, Franklin had become interested in how mental illness was affected by a patient's environment. She noted improvements that derived from positive, sympathetic and encouraging care. But these improvements often seemed short-lived. Franklin thought that appropriate psychoanalytical help might have maintained them.[217] Her interest in the environment was undoubtedly stimulated by the fact that her mother was Honorary Secretary of the Parents National Education Union and Franklin had been trained in its principles. She retained a life-long interest in education, especially the progressive movement. Her psychiatric training included work under Adolf Meyer in New York and she subsequently trained in psychoanalysis under Sandor Ferenczi in Budapest. An early member of the British Psycho-analytical Society, she also joined the Howard League when it formed in 1921 and was one of the earliest members of the Institute for the Scientific Treatment of Delinquency.[218] Some of its medical members sat on Hawkspur's selection and Treatment Committee, while Franklin acted as its secretary and convenor.

It was through this role that Franklin drove Wills potty. She would often send several letters a day to him with thoughts, suggestions and questions that he should answer. Sometimes she sent telegrams too. In the middle of a largely sodden field, with a camp to build by hand, few staff, hardly any modern tools and a gradual accumulation of delinquent, unmotivated, unhappy, disturbed and sometimes very disturbed men to support, Wills struggled to deal with her demands. She would occasionally visit the camp at weekends and Wills reckoned that she often used to sit at the meal table staring penetratingly at a particular camp member. He claimed it gave the members the willies and they used to say she was a witch.

* * *

Payne's description of the social order of the Homes was of a centralized power, cascading down through rigid layers of authority, totalized through the life of the institution and justified in the name of rationalization and efficiency. We've seen that mental hygienists expressed concerns about these aspects of social power. But where they commonly viewed popular power, in terms of mass participation, as an associated concern, at Leytonstone and Hawkspur this was apparently inverted. Popular political participation became, theoretically, the antithesis to centralization. Meanwhile, it held the potential to become the measure through which concerns about rationalization and totalization could be weighed.

Payne described children at Leytonstone Homes whose minds had been 'twisted'. They were simultaneously distrustful of authority and yet shaped by the institution's inappropriate version of it. These children had a complete lack of confidence in anything staff said, displayed fear if they thought they had done anything wrong, terror of speaking the truth and no appreciation of justice.[219] Some were angry, insolent and hard, but also clearly hurt. Others were unnaturally withdrawn. All were mentally isolated from the kind of relationships that could cater to them emotionally. Wills described the young men admitted to Hawkspur in much the same terms. They were 'emotionally deranged or disordered'.[220] They had either suffered from inappropriate and misapplied authority in their families or in Homes, Borstals, prisons and other institutions. Because of this experience all of them behaved antisocially and with little concern for others. At the same time they displayed signs of deep dissatisfaction with themselves, often accompanied by self-destructive behaviour.

How could 'twisted minds' be unravelled? The mental hygiene movement had inherited a model of the mind which described the development of each individual, from infancy to adulthood, as paralleling the evolution of civilization. This was cast as the reciprocal rise of intellect and the individual. But, wielding a revised version of psychoanalysis, mental hygienists had added stages of emotional development to this model. Children and adults seen as showing signs of 'maladjustment' were believed to have failed to satisfactorily pass through these stages. These people were partially mentally inhabiting earlier stages. They needed somehow to be brought into a new relation with them and so move beyond them. Psychotherapeutic practice associated with mental hygiene in effect attempted to free up these blockages so that development could take place. Under the influence of progressive education, Hawkspur and Leytonstone used a kind of lived free association. But the trick to be pulled off was to find a way to do this while also providing an environment of emotional security. On this view, one without the other wouldn't work.

But how could relatively untrammelled liberty be emotionally secure and therefore lead to mental re-adjustment? One central element in the attempt was a drive towards egalitarian relations. Authority that was hierarchical, dominating or rigid was seen as failing to understand the language of emotionality. If this articulation of authority aimed to produce healthy adjusted citizens it was misguided. What were required, instead, were relationships that were open to emotional expression, allowed trust to become established and in which a just authority could be openly negotiated.

Rigid, twisting, emotionally illiterate authority would be replaced by egalitarian relationships. At Leytonstone Homes measures were taken to diminish the institutional order of rank: 'this attitude of superiors to subordinates, which riddles so many Institutions of every description, and thus kills all happiness'.[221] Uniforms were dispensed with and titles of rank abandoned. But, partly because these were homes for children, the egalitarian approach was not pursued as radically as at Hawkspur. Here, all titles and measures of enforcing 'an artificial respect' were also dismissed.[222] Indeed, at Hawkspur, the entire day-to-day life of the camp took on an egalitarian dimension. Staff and members lived together and shared the work between themselves without any privileges of status.

Wills later wrote that, 'Most of our members needed to be given *security*. We aimed at providing it on a deep level through the medium of affection. This tends to be a slow process, and members were meanwhile constantly seeking it on a more superficial level through the medium of authority'.[223] Wills elsewhere variously described the aimed-for relationship of emotional security as one of unstinting 'approval', of 'affection' or simply as 'love'.[224] Likewise, Franklin described it as 'love and concern'. This relationship was fundamental; it informed all of the other measures associated with Hawkspur, and gave them their worth.[225] Many years later, Wills put it this way:

I hold the view – then hesitatingly and tentatively but now fervently – that it is hardly possible for a man to become really friendly with another if there is a gross disparity in their positions, either socially or in the hierarchy of their group, whatever that may be. The odd exception there may perhaps be, but more often than not these exceptions will be found to be due to nothing more than the erection of mutually acceptable facades which have little to do with real feelings.[226]

Here, equality of status was an essential ingredient in the construction of meaningful fellowship. And both were necessary for the resurrection of mental health.

*   *   *

In early November 1936, Mr Hobden, of the Russell Cotes Home and School of Recovery in Dorset, congratulated himself on having found a most suitable young man to take up the post of his 'Second Assistant'. He wrote to the young man saying that he saw no need to take him on provisionally for a trial period as he'd proposed. The present occupant of the post intended to leave and it was simply a case of waiting until the position was vacant. Apparently, it had been suggested to the young man that if he wanted to pursue social service he should try to gain training at the Quaker Woodbrooke College. Hobden disabused him of the value of this suggestion:

> Now if you really want to be spoilt for Social Service go up to Woodbrooke and you will most likely come away unfitted for such work. I have seen many from Woodbrooke and but few improved by the course. Now my friend, ask yourself this serious question – what do the Teachers know of S/S [social service]? They have never done the work, they have read books, and what help are these. You cannot teach such work in lectures. Temperament is not a science and can never be reduced to one. Get all the practical experience you can, but do please keep away from academic atmosphere.
>
> P.S. All successful S.workers are qualified only by natural aptitude for the work. A passion, a desire that cannot be quenched . . . [227]

But Alasdair, the young man in question, wasn't what he seemed. He wasn't really a member of staff at the new Q Camp experiment at Hawkspur, as he'd claimed to Hobden. Wills had to write and explain. A tricky thing to do, especially while remaining true to his professed beliefs in unstinting approval and affection. He wrote that Alasdair's residence at the camp was owing to having previously been in trouble with the police 'by letting his imagination run away with him'. And he added that 'I don't know if he had allowed you to gather that he is a member of staff. This is, of course, not the case, though he is a highly respected member of the camp, and President of Camp Council'.[228]

*   *   *

If Hawkspur's small-scale self-governing camp is reminiscent of Rousseau's inspiration for his Social Contract, its attempt to promote greater liberty and fraternity within egalitarian relations perhaps has more direct affinities with the then contemporary ideas of R. H. Tawney. In recent years the sociologist Nickolas Ellison has described Tawney's ideas as representative of a strand of egalitarian thought associated with the Labour Party that he calls 'qualitative' socialism. Ellison follows the political scientist Bernard Crick in seeing egalitarianism as the cohering idea that holds together and gives meaning to socialist values of liberty, equality and fraternity. He places 'qualitative' socialism as one of three strands of egalitarian vision that have vied within the British Labour Party: 'Technocratic' socialism saw equality in terms of economic power, 'Keynesian socialism' saw redistribution within a mixed economy as the most promising means for greater equality, and 'qualitative' socialism envisioned a re-moralized society through egalitarian fellowship and fraternity.[229]

Tawney had studied at Oxford when T. H. Green's idealism was still dominant, and the influence of British idealism is evident in his political thought. His Christian socialist theorizing emphasized the moral element in social and economic concerns. He believed that people's lives should perform a function for the common good, and saw socialism as a stage in humankind's development to maturity.[230] This necessity of community and fellowship was undermined, however, by capitalist relations that produced class differences and privileges. Tawney attacked the way capitalism used people as means to ends and its subordination of some people to others. For him, a belief in God must entail a firm conviction in equality. He insisted that all people should be thought of as essentially equal; they constituted ends in themselves. Wills held the same view as Tawney here, and for the same reason. It was a view that went further than some 'meritocratic' idea of 'equality of opportunity', which, as Tawney pointed out, was just the opportunity to become unequal.

There are other similarities too. Tawney saw liberty as 'equality in action'. He didn't accept the idea that liberty and equality were antithetical. Instead, he emphasized that egalitarian social and economic conditions promoted greater freedom for the populace as a whole.[231] Freedom wasn't the consequence of an absence of regulation; it was the other side of an equality in which people came together in fellowship to agree and construct their environmental, social and economic arrangements. Freedom and egality were inevitably relative, but that didn't mean that the terms therefore had no substance.

The experiments at Hawkspur and Leytonstone Homes displayed very different understandings of the connection between liberty and mental health to those then being promoted by the mental hygiene movement. During these same years the movement was pressing for a comprehensive system of mental hygiene for the community. As the previous chapter noted, this was cast as a public health issue. A Royal Commission on Lunacy and Mental Disorder set up in 1924 (ironically, prompted partly by concerns over wrongful detention in asylums) provided mental hygienists with the opportunity to influence legislation.[232] Its report reiterated the mental hygiene philosophy that services should be oriented around prevention or early detection and treatment of mental disorders.[233] The subsequent 1930 Mental Treatment Act introduced the important innovation of 'voluntary patient' status (albeit in limited form) in relation to this aim. As the late Clive Unsworth pointed out in his 1987 analysis of the politics of mental health legislation, this voluntary status wasn't intended as a recognition of the freedom of patients to decide whether to enter or leave hospital in their own interests.[234] Instead, it served to reduce the penal character of entry to mental hospital under certification, while simultaneously promoting a notion of social responsibility to submit to public health measures necessary in the interests of the community as a whole. The hope was to encourage people suffering mental stress to submit to early treatment. Mental hygienists' comments on this legislation confirm Unsworth's view.[235]

But the therapeutic approaches developed at Hawkspur and Leytonstone entailed a more genuine interpretation of the term 'voluntary' and its association with freedom of choice. In the process they transformed freedom and choice from the hoped for result of successful therapy to important elements of therapy itself.

At Hawkspur, Franklin described how it appeared that, for many of the members, certain stages of development had been left out. The nature of the camp allowed for 'a lessening of assumed adulthood, and re-living through a younger developmental stage', thus enabling a more genuine maturity.[236] At Leytonstone, Payne reported that the introduction of a much freer and more egalitarian atmosphere, centred on approval and affection, and with little imposed regulation, had similar initial results among the children in the various houses. At first the youngsters were bewildered and suspicious, and then they increasingly reacted by running wild. In Payne's view their behaviour expressed phases from infancy and childhood that had been stifled by inappropriate authority. They were becoming more 'infantile', but they were gradually working back through stifled developmental stages. This included a slow shift from hostile or unnaturally reserved reactions to kindness and sympathy, to the expression of cravings for attention and emotional outbursts.[237]

In an attempt to ameliorate some of the wild behaviour that greater liberty accompanied by emotional support had released, Payne introduced a form of self-government. Hawkspur attempted this from the start. We've seen that the terminology of 'self-government' lay at the heart not just of the original moral therapy, but also of the Bosanquets' moral philosophy, and of the twentieth-century mental hygiene movement. In the original moral therapy, a person's loss of authority over their mental states represented 'alienation' and a failure of mental self-government. Later in the nineteenth century, Bernard Bosanquet had brought together this definition of self-government with that of a country being governed by its own people. He outlined what he called 'the paradox of self-government': How do you have authority over yourself? And if this is, in fact, possible, how do you then embrace an external authority without yielding your own? Bosanquet reckoned he'd resolved the paradox by invoking Rousseau's notion of the general will and interpreting it through Hegel. The rational, self-sustaining individual mind was a social and historical emergent. The vehicle of this emergence and its organizing principle was the Family. An incremental internalization of parental authority engendered mature rationality and therefore mental self-government. Under this rendition, the Family cast its shadow across the social order as a hierarchy of rational ability providing the structure for distribution of status and function. This hierarchy of heads represented an order that was fundamentally inegalitarian and antidemocratic.

The mental hygiene movement had emerged between the wars with an associated vision, despite its introduction of emotionality as an essential element in the development of minds and society. But within the communities developed at Hawkspur and Leytonstone the partially inverted organizing principle of the Family placed a priority on popular power through moves towards more egalitarian and democratic organization.

Payne didn't describe her system of self-government in detail, but it relied on the older children acting as leaders, and on the Housemother or Father being able to inspire and guide without showing they were doing so. In fact, it was never fully introduced across the Leytonstone Homes. This was partly owing to lack of staff and partly because Payne felt that many of the children were too young to implement the idea fully. After two years' effort she eventually managed to create a Children's Court of Justice and a Children's Council for the whole institution, but their remit was limited.[238]

At Hawkspur the system of self-government was called 'shared responsibility' and was much more fully elaborated. Franklin had coined the phrase to denote the fact that the members didn't govern all aspects of the role and running of the camp. Wills put it like this:

> The sphere of the Camp Council's authority is a limited one – it is limited to the domestic affairs of the camp, and is concerned primarily with people's relations to one another, and the day-to-day conduct of family life. Further, the Council consists not only of members. As we all live together, we all have a voice in the affairs of government.[239]

This was a system that evolved as it went along. In fact, that was part of its point; just like freedom, it was always a work in process. This wasn't in the sense of a cumulative learning by staff and members how best to organize themselves. It wasn't the central aim that self-government should work smoothly and without any problems. The point was that through an engagement with the practice of self-government members would come to take some responsibility both for their own behaviour and the social organization of the community. It was when the system wasn't working particularly smoothly that members tended to experience this moral effect at its most penetrating level.

Wills wrote that one of the main characteristics of Hawkspur was that few, if any, of its members actually wanted the responsibility of ordering their own affairs. Most would have preferred authority to have lain elsewhere, and taken or evaded orders as they could. But the system of self-government denied them this option.

For the staff, there was a number of useful results from this. Members were confronted with the fact that their actions, or very often inaction, had real effects on people. As Wills described later, 'The discussions...which took place with monotonous frequency on such subjects as, "Why hasn't so and so emptied the shithouse bucket when it's his bloody turn?" brought into clear perspective, and in a very practical way, the social consequences of failing to accept one's responsibilities'.[240] At the same time members were able to see that their own views on other people and the running of the camp could be listened to and could effect change. The actual process of the Camp Council, which rapidly adopted a weekly sitting, could also be cathartic, allowing personal animosities and frustrations to be expressed and defused.

*   *   *

In 1937 Wills wrote to one father concerned about his son's apparently poor progress. Wills tried to explain the way in which he thought Hawkspur's self-governing camp life was likely to help his son. He explained that:

> It is probable that failings in his character have been pointed out to him in the past mainly by those in authority over him. He is now having them pointed out to him by his equals, that is by his fellow members. I am afraid that this makes life very difficult for Gerald... The sort of thing that I mean when I say that his fellow members point out his failings is that they are very quick to bring it home to him when as frequently happens he takes one attitude about a thing with me and another attitude about this same thing with them. He has had this very much brought home to him lately and it is likely to make a much greater impression on him than if the same thing were said by someone in authority by whom he would expect to be corrected.[241]

*   *   *

For Wills and the Q Committee the Camp Council was primarily a means to allow the members to work out their relation to authority; to unravel the twisted relations to forms of it that they were assumed to have. An important element of this was that the Council provided a type of authority that was largely acceptable to the members. Although emotional security was to be provided by the informal relationships of acceptance and affection it was realized that the Camp Council provided an often much-needed basic level of authority. Wills described this as offering a certain amount of external discipline that could offset the constant emotional pressure of having to discipline oneself in a society of far wider freedom regarding work and behaviour than members were used to.

Though Wills didn't describe it like this, there was a clear distinction being made here between the notions of being 'an authority' and being 'in authority'. 'Traditional' discipline—the kind of order that Leytonstone and Hawkspur were reacting against—entailed a combination of both 'in' and 'an authority' for those subjected to it. On the one hand, to be 'an authority' is to be assumed to have special knowledge or abilities that are not available to those who are subject. This authority 'knows best'. To be 'in authority', on the other hand, is to have been provided with a circumscribed remit by others. We can note an important distinction between the two forms of authority. The first, 'an authority', as the philosopher R.B. Friedman has shown, presupposes an inequality between people *before* the authority relation; it informs the construction of the authority relation. The second, 'in authority', rests on a recognition—tacit or otherwise—that we all have personal views and desires, but that, at the substantive level, no agreement can be found. Therefore, a remit for authority is agreed whereby clashes of views and desires can be controlled. The assumption informing this authority is that all who contract into it are equal. No individually expressed opinion is recognized as constituting 'an authority' that others must listen to and obey.[242]

By transferring large areas of the 'in authority' to members, and making any assertion of being 'an' authority within the lived life of the camp a matter of subjective opinion, Hawkspur strongly reinforced the intended aim of communicating a genuine respect for each member and faith in their judgement. All were, at this substantive level, of equal status before the constituted authority.

But this division of types of authority also offered another advantage for the therapy that Wills and his colleagues theorized. The constituted authority, however tardy it was at times, allowed the staff to continually offer affection and approval, even when misdemeanours were being sanctioned by the Camp Council. Added to this, with the Camp Council 'in authority' through direct democracy, the field was left open for anybody to acknowledge anybody else as 'an authority' in some way. This could happen in all sorts of, largely short-term, ways. But what was hoped for was the kind of acknowledgement of 'an authority' that was embodied in the psychoanalytic theorisation of 'the transference'. Staff sought signs of this attribution of authority—of a parental authority figure—as a means to promote therapy. In effect, the aim was to make use of this position of authority in order to erode it. By recognizing a transference in lived life, and building on it to express loving approval in an egalitarian and open environment, a new set of relations would emerge and dissolve earlier authority relations.

* * *

Wills offered a poetry group for interested members. He wrote of Julian that 'He is convinced that he is an artist, though he is a little doubtful about what his medium might be!'.

Julian was utterly miserable at Hawkspur from the start. Tall and 'willowy', he swanned around in 'an "arty" style of dress, giving the general impression of effeminateness'. Charming manners and the best of intentions were accompanied by an apparently complete inability to do anything for himself, and a 'firm belief that he was an exceptional person for whom practical things ought to be done by somebody else'.[243] He thought it very unjust that he had to mix with such rough people at Hawkspur and do things for himself that he'd been accustomed to servants doing. But Julian had fallen on hard times and this was the result. Very aesthetic, very sensitive and physically weak, he couldn't cope with the camp life. Wills reported in the first week of his stay that he couldn't carry one bucket of water, couldn't bring himself to wash in such rough surroundings and was clearly not getting enough to eat. Added to these problems he was very upset by the aggressive manner of some of the other members. Wills and the staff tried to protect him and nurture his interest, but he must have found life incredibly hard.

Still, Wills and the staff encouraged his writing and musical interests. Not long after Julian had arrived at Hawkspur, Wills wrote of his poetry that 'The prose seems to be stilted and pompous – rather Jane Austenish – and it is very doubtful whether he can yet produce anything publishable'. But he came to be quite impressed with some of Julian's work.

Simon had arrived at Hawkspur with a recommendation from the previous institution he'd lived in, that he be given 'strict discipline and constant supervision'. Still a teenager, Simon had immediately fallen in with the 'antisocial element' of Hawkspur camp and within a few months had become very aggressive, with Wills complaining that he 'bullies some of the quieter members of the camp unmercifully'. A lot of this bullying was physical, as well as verbal. A blessing though was that Wills reckoned Simon had made a 'positive transference' to him—which might explain why Simon attended the poetry group. Nearly four months after he'd arrived at the camp, Wills remarked that 'in spite of his general ignorance and uncouthness [he] is a regular attender at the poetry group, where he displays indeed a better taste than some of the more "refined" members'. The public pressure brought about through Camp Council was hard on Simon, though. At one point he was in trouble on a charge of stealing Julian's clothes. Wills report noted that 'Simon's own clothes are of the institutional type and Julian's are almost exotic'.[244]

* * *

Hawkspur's egalitarian form of self-government gave it a distinctive approach to working life at the camp. Arthur Barron, who came to Hawkspur as a student helper, summed up the philosophy. Not unlike Tuke at the Retreat perhaps, or the Bosanquets in regard to the wider community, Barron maintained that work was necessary in order to develop and retain self-respect, 'develop one's character or use one's powers'. But the similarity is only superficial. Barron added that work shouldn't be just something you did because if you didn't the community would punish you. Work was defined as 'effort directed to some end', and it didn't matter whether that end was trying to score a goal in a football match or the pursuit of an ideal. What mattered was that it was honourably tried for, that it didn't conflict with what you believed to be right, that you were able to give yourself freely to it and that it had a level of difficulty that tested your powers to the full.[245]

This description was a far cry from that of the Bosanquets, or Cyril Burt and R.D. Gillespie. Intelligence tests there may have been on entering Hawkspur, but this didn't determine a person's 'level' in the social order and the kind of work to which they would be condemned. Likewise, the odd performance test was carried out on a prospective member apparently according to the whim of the tester, but this wasn't used as an 'objective' means to discover aptitude and place people into certain kinds of work.

As Barron put it, work 'served the wider purpose of the camp'.[246] The 'Memorandum on Proposed Q Camp for Offenders against the Law and others Socially Inadequate' of July 1935 noted that 'It seems likely that one of the good influences of the camp will be the experience of doing and making things which are of social value without reference to their use for the acquisition of money'.[247] There were four areas to the work required: domestic work; the development and upkeep of the grounds, gardens and animals; building construction and maintenance; and the general and unending work associated with life in such basic conditions. This last area of work became dealt with by a squad called the G.B.A.s (General Bugger Abouts). This was created for those members who seemed unable to settle down to one area of work for any length of time and who often hindered rather than helped. Many members saw them as slackers, but the staff didn't. The G.B.A.s was an attempt to provide an outlet for their 'high-spirits'. The emphasis was on variety and they did all sorts of jobs, including felling trees, digging refuse pits and general maintenance around the camp.[248] Domestic work was shared between everyone—staff and members. It took up around a third of the working week. The other work was divided up between squads 'to each according to his ability' and not simply in terms of an equal amount per head.[249] Given that the work was shared out democratically at the Camp Council, this amounts to a notable egalitarian inversion of Burt's view that the relative weight of a political vote should be distributed according to 'ability', while work (though not its content) should be distributed according to the number of heads.

The work squads operated under a member of staff who acted as leader. This person was qualified in the general work each squad covered. But here again, rather than assume that they knew how best to do the work, and that the members required telling how to do it, each leader tried to provide space for members to discover how to accomplish a task for themselves. Work arrangements to construct and maintain the camp were part of a wider educational aim. Staff tried to provide opportunities for learning through various activities outside work. These often involved them sharing their own personal enthusiasms with interested members. Both Wills and Barron emphasized that this approach had a number of elements, but that the core was a genuine respect for each member and faith in their judgement.[250]

* * *

In October 1936, Wills wrote in his Camp Chief Report that

> It is gratifying to see an increasing sense of responsibility among the members. On their own initiative they called a members' meeting to correspond with the staff meeting which is to be held regularly in future and will bring recommendations to the Camp Council. Among a number of minor recommendations they asked that there should be an additional day off per week to correspond with the staff members' day off but that any member slacking at work should be deprived of it as they think that there must be some form of punishment for slackness.[251]

But, by October, Wills was acknowledging that it had

> ... proved more difficult than was anticipated to get essential work done soon enough. Members have jogged along all the summer without working too hard, and it was assumed that when the bad weather approached they would speed up a little for their own comfort, but this has not proved to be the case, and it has been necessary to urge them along a god deal this month in order to get the camp decently presentable. At the time of writing some are still in tents and others have moved into the day-room.[252]

In December he wrote that hardly any work had been done since the last report, except by members of staff. Camp Council rules were being openly flouted and it was a job just to keep the camp reasonably clean. Wills decided to indulge in some provocation. He told them that if there were complaints about the state of the place by any of the people who visited it would be him that got it in the neck and not them. He then announced that as the members had reneged on their voluntary undertakings about behaviour in the camp, he was going to renege on the agreement that everyone got two shillings a week pocket money whether they worked or not. There would now be a rough and ready system of deductions according to work obligations not fulfilled, with those who had many deductions receiving a reduction of not more than sixpence and those the with fewest getting an extra of not more than sixpence. Wills reported that there was 'a good deal of half-serious opposition'. Next morning the sound of hammering could be heard from the bunkhouse. Well, it was a form of work. Four members were barricading themselves in. They smashed some windows, did a bit of damage to the bunkhouse and threw a note out that said they wouldn't come out until the payment scheme was dropped. But then lunchtime arrived and they all decided lunch was more important. So Wills asked the Camp Council what they were going to do about 'such lawlessness'. There was some angry discussion, and Wills admitted in his report that 'the staff were roundly abused, often with some justification'. Rodger, the leader of the revolt, had let rip at Wills saying that he and the staff were worse than the members because 'they thought themselves so bloody perfect, but were really no better than anyone else. The CC was the worst of the lot because he talked a lot of blather, and never did anything'.[253] Rodger reckoned the members could come up with a better way of working things, which, of course, Wills was only too happy to challenge them to do.

Influenced by Rodger, the members came up with a scheme that claimed to favour anarchy as the best method. So they abolished the constitution and rules. Wills reported in January 1937 that this hadn't proved very successful, though he also noted that work had 'started with some measure of enthusiasm after the boredom of the Christmas holidays' with 'more in fact being done than had been the case for sometime'. But, even so, Wills thought he might still need to take some action to regulate the camp if the members didn't do so. He reckoned that because of 'a complete misconception of the true meaning of the word' some of the members enjoyed the feeling of freedom from responsibility for anything that 'anarchy' gave them, despite the muddle and discomfort.[254]

Anyway, other members wanted to return to the earlier way of running the camp, but with far fewer rules. By the next month they had called a Camp Council to put their proposals. Their proposals were accepted with staff supporting them. Wills was away at the time. After some more lapses from doing agreed work another Camp Council agreed to introduce a further form of sliding scale reductions. A similar revolt took place in November, but was even more short-lived.

Wills reflected on some of the consequences of these issues in his June 1937 Camp Chief Report:

> I think we should bear in mind two important differences between the camp and other self-governing communities with which we sometimes compare ourselves. Between ourselves and Grith Pioneers there is this important difference; that Grith Camps are composed of men who have each limited income of their own which they find goes further if pooled with the resources of other men with similar incomes, and their labour (so far as it is concerned with the production of food) has an immediate

relation to their well-being. Our members have no such interest in the camp, and are inclined to feel that as someone is paying for them they are entitled to all they can get with no return from themselves.

Again between ourselves and the Little Commonwealth there is an important difference. We have attempted to make the whole camp the sphere of self-government, whereas the L.C. extended it only to civic matters. Work was run on the same authoritative lines as it is outside, and the essential domestic work was in the main carried out by the staff, with assistance from citizens. It is probable that the responsibility that we have asked our members to accept is too great, and indeed we are probably asking more from them than we should be able to give ourselves in like circumstances. When we consider in addition that some of our members are grossly inefficient socially, and some quite incapable of work, it is not surprising that a better spirit has not been developed in the camp. I feel therefore that one stage of our experiment should be regarded as ended, and that self-government should be limited in its scope, though perhaps it should be as absolute as possible where it does still remain.[255]

In November that year Wills reported that the Camp Council's elected Committee had tried hard to make it clear that everyone was expected to work and that the absence of punishment didn't mean there was no obligation to do anything.

* * *

So the experiments at Hawkspur and the Leytonstone Homes expressed a partial inversion of the authority structure provided by the Family: emotional security, freedom, egalitarianism—a kind of therapeutic version of liberty, egality, fraternity. Did these experiments therefore refute the moral order generally envisioned by the mental hygiene movement?

The historian Mathew Thomson has recently attempted to elaborate a distinction between what he calls 'psychological' regulation' and what he sees as a more romantic and utopian interest in psychology as a tool for the release of human potentialities. His point is to argue against a tendency in historical work to see late nineteenth-century 'amateur enthusiasm and romantic idealism' as rapidly declining in the twentieth century because of the influence of the new science of psychology.[256] However, in doing so he suggests an opposition between 'psychological regulation' on the one hand and 'radical experiment' on the other.[257] His point that radical and amateur interest in psychology persisted into the twentieth century is surely correct. But there seems to have been as much integration as there was opposition. As we've seen, our movement for mental hygiene, for example, was involved with, and promoted, the schemes that we've looked at in this chapter. It did so by incorporating their approaches within its general theory.

We noted earlier that the general view of the mental hygiene movement was that children or young adults considered maladjusted might benefit from greater than usual tolerance and freedom as a means to unravel distorted or stalled stages of emotional development. This was the main point of theoretical engagement with elements of progressive education. Inherent in this stance was the shared assumption of a relationship between individual and social development. Thus, the mental hygiene emphasis on measurements of intellect could mediate for whom such therapeutic approaches were appropriate. This helps to explain why relatively 'conservative' figures closely associated with the establishment, such as Cyril Burt, could support highly radical approaches to the treatment of so-called maladjusted children. Burt, as we've seen, essentially advocated the ascertainment of mental capacity as the royal road to organizing a hierarchical society of function and fit. But he nevertheless supported Russell Hoare's village colony for adolescent delinquents, called Sysonby House. This colony operated for a few years during World War I. Hoare believed society to be dominated by coercion, fear and competition. To him, most adults hopelessly limited themselves and children with rules, regulations and conformity. His response was to reject all forms of punishment for children and to attempt to resist all imposed authority.[258] Hoare's claimed aim was to produce rebels against present society, and it's difficult to reconcile his iconoclastic views with Burt's public and professional image. Yet, Burt advocated Hoare's approach and 'free-discipline' colonies in general for certain delinquent children.[259] He wasn't alone among establishment figures in advocating these approaches.

The reason is that those children and young adults deemed able to benefit from 'free-discipline' and the hoped-for re-integration of earlier stages of development were those considered of a certain 'type' and of acceptable intelligence. As we saw in the previous chapter, the mental hygiene movement may have privileged emotionality as its locus of therapy, but the common denominator was measured intellectual capacity. Their theoretical position was based on the inseparability of the progress of 'mind' with that of the individual and of civilization.

* * *

Franklin had to reply to one Hawkspur member's father who had complained that he had heard worrying reports about his son's personal untidiness since being at the camp. He seems to have suggested removing his son and finding somewhere offering more training and discipline. Franklin wrote that his son showed 'social aspirations' and that 'frequent neglect of his personal appearance ... would be bound to reflect on his reception socially, and this appears to me the best, because the most normal, stimulus for greater care'. She explained her thinking like this:

The achievement of improvement through his own efforts is the only kind that I feel is of value to him, even if it does not reach as higher standard as we could hope. Unless he is to be <u>permanently</u> under external discipline, I think that an environment in which he was subjected to orders from outside would not be helpful, as it would not help him to achieve independence in the world. It is likely to throw him back into a passive condition of childishness, even if it were an obedient and conforming child.[260]

* * *

Contrast the progressive radical experiments at Leytonstone and Hawkspur with, for example, the modern 'progressive' care and training offered at the Epsom Manor Mental Deficiency Institution. The Superintendent described its organization in an article written on request for the CAMW's journal *Mental Welfare* in 1924.[261] In it he decried prison-like regimes and presented his own institution with its 'relative freedom' as the best means to improve behaviour and conduct. The Manor Institution contained over a thousand residents. Here, the sexes were allowed to mix at the weekly dance and at other organized occasions. Girl Guides, Brownies and Scout associations were formed, and freedoms and privileges allowed for them. The wards for most male (but not female) residents over school age were unlocked. These residents were allowed to come and go to work, and enter the gardens and recreation grounds at will.

And yet the picture this Superintendent painted of his institution seems anything but un-prison like. He expressed this clearly when he wrote: 'Our residents have been uninstitutionalised and are only gradually becoming so; experience goes to show that it is only a matter of time and organisation in which normal brains generally outwit the abnormal'.[262] In the 1980s the sociologist Robert Castel echoed Foucault in claiming that nineteenth-century moral therapists aspired to 'total tutelage of the patient'. It seems as if such aspirations were retained in this kingdom for the 'mentally arrested'.[263]

The 'freedoms' described were the substantive rewards for acceptance of one's definition and for behaviour considered responsible. As such, they were constantly open to being revoked. The Superintendent described the residents as having primitive instincts 'unschooled in inhibition', and therefore requiring constant surveillance and discipline. The main method of control consisted of classification and a system of rewards and privileges. Both sexes over school age were divided into four classes, each class carrying different rewards and privileges. Residents could be promoted or demoted through the classes according to their behaviour. Continual records were kept by staff containing judgements of residents' attitudes and conduct. Unpunctuality, misbehaviour or laziness received negative marks and could result in loss of privileges or demotion to a lower class. This removal of privileges was widely defined. It included confinement to bed and 'on the very rarest occasions a simplified diet'.[264] Positive marks were awarded for efficient work in the institutional workshops or wards, and good general behaviour. The guide and scout organizations represented further elements of this system of control and training through promotion and demotion. They represented the highest rungs of the institutional hierarchy. Once accepted into their ranks members were allowed extensive parole to go about unattended in the institutional grounds. The lodges could also occasionally invite outside friends to visit, so long as their names were submitted to the Medical superintendent for approval.[265] He noted that it was only from the ranks of the Guides that selected women had been 'permitted' to work as domestic service outside the institution, either daily or on licence. Similarly, it appears from his remarks that it was only from the ranks of the scouts that some boys were allowed home for the day unaccompanied.[266]

Here, then, was a system of constant surveillance, judgement and control; a relentless 'gaze' with a stick in its hand. Only a few years earlier Franz Kafka had written a short story called 'In the Penal Settlement' in which inmates were placed inside an 'inscription machine' that engraved upon their bodies the nature of their crimes.[267] It might well have been written about the people detained in such mental deficiency colonies. Just as at the penal settlement, these people were inscribed over and over again with the 'nature' of their 'condition', and forced to confront it as the condition of their being. This was a world in which inmates were under perpetual judgement. Who wouldn't be cast into a continuing uncertainty and fear of being found at fault in such a place? Here, liberty and, therefore, citizenship and rights were restricted and defined by self-appointed experts. Any desire for the kind of liberty or choice that others in society took for granted was dependent in these circumstances on a person accepting their nature as defined by those in control. By acquiescing in this way a very few might follow the means designated for making themselves 'responsible citizens' and achieve release—or, more likely, limited freedom on licence. Another possible option was escape. Quite a number appear to have tried it.[268]

In fact, over a decade after this article that claimed to decry prison-like regimes, mental hygienists proposed mental deficiency legislation should be brought into line with the 'voluntary' procedures endorsed in the 1930 Mental Treatment Act. The Feversham Committee, founded under the auspices of the Ministry of Health, proposed that a system of voluntary admission should be established that would largely do away with the need for certification. It's a proposal that has gone un-commented by researchers. But, just as the 1926 Commission didn't intend 'voluntary status' as recognition that patients should decide whether to enter or leave hospital in their own interests, neither did this proposal. Unsworth noted that 'voluntary status' signified a reduction in the penal character of entry to a mental hospital and an associated assurance that this was in the interests of the community as a whole. The Feversham Committee's 1939 recommendations strikingly bear this out regarding mental deficiency. We've already noted that the 1913 Mental Deficiency Act was an unprecedentedly coercive and interventionist strategy. But the Feversham Committee complained that it had, in fact, been based too much on the principles of the 1890 Lunacy Act and therefore 'governed by the fear of illegal detention'.[269] It reckoned that 'too great an insistence on the liberty of the subject' defeated its own aims.

Thus, people were only brought to institutions after they'd been neglected, cruelly treated, become criminals or inebriates. It contrasted this rendition of the workings of the Act with the proposals of the Radnor Commission that had preceded it. These, the Committee said, had proposed that mental disability and not poverty or crime should be the criteria for state care and control. But this is a curious argument. First, it assumes that all people potentially categorizable as mentally deficient will inevitably become abused, criminal or drunken. (Left unsaid was the continuing fear of sexual licence and sexual crimes.) It's difficult to see how the Committee could justify such claims. They appear more like prejudices that were a consequence of their theory of 'mental deficiency'. Second, the Committee's statement seems to tacitly admit that poverty could get somebody certified under the Mental Deficiency Acts. Finally, the conclusions that the Committee drew from these contentions reveal clearly that their understanding of the term 'voluntary status' amounted to a means to get people to submit more easily to the 'care' and 'treatment' that mental hygienists professed to be urgently needed as a matter of public health. Thus, the Committee emphasized that 'The principle which actuated the Commission was that the community should assume control of defectives at an early age and continue that care as long as was necessary'.[270] This is essentially a call for indefinite detention mediated by the psychiatric profession. This impression is reinforced by the Committee's added comment that certification 'should be reserved for cases where the consent of the parent or the patient cannot be obtained'.[271]

Was the dialectic of the Family not operating here? Already thrown out of the Family, these people were, it seems, to have no part in its partial inversion.

* * *

Wills had fallen out with Franklin over some letters containing confidential material that had been 'found' by a member and displayed around the camp. The letters were Franklin's and it seems that she had accused Wills of leaving the material accessible. This clearly ignited his general frustration with the deluge of correspondence he received from her, and its poor presentation. He reckoned that this was the reason confidential matter hadn't been secured properly. Wills wrote:

I have been wondering for some time whether I could without offence offer you a few elementary rules for correspondence. I think the time is now opportune.

1) Write on one side of the paper only
2) Do not write on odd scraps
3) Pin the sheets together
4) If you must number the sheets, number them accurately
5) Confine confidential matter to separate sheets
6) Avoid verbosity

... Your writing is at the best of times is not easily read, but when it flows on to both sides (but not consistently so) of all sorts of odd scraps of paper incorrectly numbered, it is difficult enough to read them in the first place and an almost unthinkable labour to read them again to answer them and then time to sort out the confidential matter for filing.

... I confess that I was extremely angry when I first read your letters. I realise, however, that they were probably written under considerable stress and my only feeling now is one of resentment at your apparent assumption that you are the only one who has any interest in preserving confidences.[272]

In reply, Franklin made the surely valid point that she had considered all the contents of her letters to Wills confidential. She was obviously hurt. She wrote, 'They will, however, in future be shorter & less frequent & in general I shall interfere very much less in the internal affairs. This no doubt will be an advantage'.

5

Developing in the Womb of the Old?

The bud disappears in the bursting-forth of the blossom, and one might say that the former is refuted by the latter; similarly, when the fruit appears, the blossom is shown up in its turn as a false manifestation of the plant, and the fruit now emerges as the truth of it instead.[273]

So said Hegel. But which is the blossom and which the fruit? At Hawkspur the need for emotional security coupled with an associated need for lived emotional expression resulted in the promotion of a form of liberty, egality and fraternity. As we've seen, the mental hygiene movement accommodated this radical approach with its project as a useful, but specific, measure applicable to certain 'maladjusted' young people and adults; it was a means through which individual maladjustments could be unravelled and stages of emotional growth re-done. But Hawkspur certainly also contained an implied radical critique of the social order. It carried a strong implication that the relations of radical democracy—maximizing liberty, equality and fraternity—were therapeutic because they constituted important elements of the conditions under which emotional health could not only be regained, but also sustained.

*   *   *

It was 1939 and Wills had received a letter from an ex-Hawkspur member. Clive had found a job and was living in his own digs. He told Wills proudly that he had to ride several miles to get to work instead of just rolling out of his bunk and starting work when he liked, as in the old days. But still, he added, hard work hadn't killed anyone yet. The real reason Clive was writing though was because he wanted to know why Wills hadn't kept his promise about writing to him. He reckoned that as Wills had made that promise he ought to keep it. Clive gave his best wishes and asked to be remembered to Ruth, the staff and members, and especially the G.B.A.s (General Bugger Abouts). Then he signed off telling Wills not to forget his promise. Wills soon replied.

I was quite under the impression that I had kept my promise about writing and that you owed me a letter. However, I may have been mistaken about it so here we are now.

I see you have changed your address and I am not quite sure whether I can read it so I hope this letter reaches you. Let me know if it doesn't!

G.B.As have ceased to be. They have been transferred to other squads where they have to work instead of B.A ing....

Wills signed off by saying that they were having their anniversary celebrations at Whitsuntide, and suggested Clive try to come for the weekend.[274] But, as the termination of the G.B.A squads suggested, the war was bringing enforced changes to the Q Camps venture.

*   *   *

Q Camps was a small outfit and Hawkspur a relatively short-lived experiment. Although associated with the mental hygiene movement, the methods, let alone any wider implications that might have been drawn from them, were seemingly on the margins. Added to this, war would soon engulf the nation and radical ideas that mixed pacifist, libertarian and collectivist lifestyles weren't likely candidates to prosper. But, by looking at some of the important wartime activities taken on by the mental hygiene movement, we can see that the kind of ideas expressed at Hawkspur had a wider effect. Partly, this was because they were open to different interpretations. They certainly oscillated and vied with far less radical ideas about authority, self-government and the social order that were closer to those with which the mental hygiene movement had emerged. In fact, if there was such a thing as a dialectic, with the onset of war it seems to take on more of the appearance of an ideological spinning-top. But still, that's no doubt in the nature of a dialectic in progress.

One way into showing these wartime dynamics is to take a look at a book published just before war broke out by the psychiatrist and psychoanalyst John Bowlby, and the economist and Labour politician, Evan Durbin. *Personal Aggressiveness and War* was an analysis of the causes of aggression and international conflict. The text was unusual in several ways. In the first place, it was an unusual departure for a Labour politician to place so much emphasis on psychology as a central element of the socialist vision. (Durbin went on to reiterate this emphasis in his most substantial work *The Politics of Democratic Socialism* published the following year.) Similarly, it was unusual for a work that made so much of mental hygiene principles to be so explicitly socialist in its vision.[275] This was an important development for a mental hygiene movement traditionally hostile to socialism, especially as Bowlby would rapidly become an important influence on it. Up until the general strike in 1926 Bowlby had been politically a supporter of the Conservative party. After it he shifted to labour, building friendships with Durbin and Hugh Gaitskell in the process.[276] From that time too he trained in psychoanalysis and in 1936 began working in child guidance at the London Child Guidance Clinic. Here, he was strongly influenced by the psychiatric social workers' casework approach and theorization of emotional relationships in the family.[277]

Personal Aggressiveness and War displays the key components of the mental hygiene version of moral therapy that we've traced. Bowlby and Durbin linked together 'the primitive', madness and childhood. In so doing, they also performed the 'reading history sideways' manoeuvre, as earlier mental hygienists had done, and before them Helen Bosanquet. Existing indigenous peoples were designated 'primitive' and interpreted as examples of earlier stages in development towards both civilization and the civilized individual. In the case of Bowlby and Durbin, the linkage of 'the primitive', madness and childhood took the form of an identity between apes, children, 'primitive peoples', and neurotic or psychotic symptoms in 'civilized man'.[278] Again, like earlier mental hygienists, they claimed that what linked them was a primitive, instinctually derived core of emotionality. This core endured in the rational, 'civilized' adult and, according to them, was best observed and understood through psychoanalytic studies. The particular aspect of this core that Bowlby and Durbin were interested in was that which found expression as aggression.

Bowlby's work is usually described as a mixture of psychoanalysis and ethology.[279] But, in the context of our story, we can see that this early book appears as a kind of compromise between the influences of progressive education and psychoanalysis. It's like an amended version of Freud's *Civilization and its Discontents*.

Instead of seeing aggression as a primary instinct, Bowlby and Durbin described it as a derivative of frustration. It was through the frustration of 'primitive' primary drives that a tendency towards aggression was understood to emerge. Bowlby and Durbin believed that this frustration was, to some extent, inevitable and related primarily to the family sphere. Here, again, they reiterated the kind of dialectical development of the self in relation to family authority that mental hygienists expressed between the wars. There was an inherent duality of the human personality. But in the same way as we've seen other elements of the mental hygiene movement had begun to do, they aimed to allow more leeway for one side of this duality. Thus, they wrote 'children are as naturally co-operative and sociable as they are selfish and anti-social'. They placed this duality in a dynamic development critically related to parental authority. And they maintained that the weight of human impulses was towards peaceful sociability and co-operation.[280] The key was to amend parental authority so that it didn't distort these impulses. In keeping with this contention that co-operative and sociable instincts predominated in human development, Bowlby and Durbin maintained that aggression and war emerged because of a minority of faultily socialized individuals. 'War, like crime', they wrote in classic mental hygiene fashion, 'is the result of the existence of anti-social minorities'.[281]

The means to ameliorate these antisocial tendencies was, for Bowlby and Durbin, twofold: cure and control. The cure would take generations, as it relied on 'a new type of emotional education' that would amount to a widely disseminated 'change in the technique of parental control'.[282] What was required was that parental authority should be mitigated in the interests of minimizing the frustration that infants and children experienced and which, as they developed, became repressed and often projected, or displaced. It was these processes that, Bowlby and Durbin argued, led to the group expression of frustration as aggression and, ultimately, war in 'civilized' adults. They argued that:

> The restraint of impulse is so frequently carried out on principle – as a desirable form of 'discipline.' Parents believe that children ought not to have what they want – the denial of impulse will make good character. We hold that the opposite of this is true.[283]

Frustration in childhood was inevitable to some extent, but the point was that it shouldn't be exacerbated by suppressing and punishing the resentment that would naturally be expressed. The essence was that children should be allowed to express their feelings of aggression even though, of course, 'acts of irremediable destruction' should be prevented.[284] Thus, just like Hawkspur, the emphasis was on a combination of freedom and emotional security. Moral development was associated with a gradual emergence of the self through increasing sociability. It wasn't something imposed through the suppression of instincts by a sovereign authority. Bowlby and Durbin were influenced by the psychoanalyst Susan Isaacs' attempt to allow a maximally free and unrestricted environment at the Maltings School.[285] (In fact, she'd written an advice column for the Home and School Council's *Parent and Child* under the pseudonym Ursula Wise.) They claimed that modern analytical psychology had shown the 'rapidity and strength with which the social and affectionate impulses of the free child develop' and that co-operation was 'the overwhelming impulse of human life'. They linked this to the need for greater freedom of expression:

> It is only within the circumstances of freedom that social habits and a spontaneous desire to co-operate can flourish and abound. 'Spare the rod

and spoil the child' – as a quiet and convenient member of the familial group. Spare the rod and make a free, independent, friendly, and generous adult human being.

Just like we saw at Hawkspur, we see here a partial inversion of the Bosanquets' rendition of family authority. This construed freedom in restricted terms as a component permanently subordinate to the family structure of knowledge and status inequality. Bowlby and Durbin wrote:

> Instead of violent and ungovernable anger, inordinate selfishness, and vanity, the child that is not afraid to express its feelings is likely to exhibit affection, independence, sociability, and courage, more rapidly and more naturally than a repressed child. Familial life with them is not a nightmare of disorder, or the false calm of strong discipline, but a moderately peaceful and very lively society of free, equal, and willing co-operation.[286]

All this is very similar to much of the core theory that informed Hawkspur. But so far we've only noted Bowlby and Durbin's 'cure' for aggression and war. This was a long-term aim. In the meantime measures of 'control' were essential for wider society. It's in the perceived need for these measures that Bowlby and Durbin's theory reiterates more closely Freud's Hobbesian understanding of the individual and society. And in this mixture of approaches to 'cure' and 'control' can be seen another reconfiguration of the coordinates of social and political power that we've been tracing—its centralization, popular base, rationalization and totalization.

The possibility of war and existing socialist responses to it clearly informed Bowlby and Durbin's thesis. Their approach certainly displays very different conclusions to those arch pacifists at Hawkspur. Bowlby and Durbin definitely didn't want to bring about a pacifist generation. But in typically mental hygienist and psychotherapeutic fashion they didn't just strongly disagree with pacifism, they diagnosed it as 'neurotic'. Echoing Freud's view of the notion of 'universal love', they claimed pacifism was 'as profoundly neurotic as the manifestation of transformed aggression itself'.[287] In their view the emotional environment that surrounded each child would only gradually shift as each generation of parents brought with them 'a less warped and aggressive personality'. In the meantime wars and conflicts only served to strengthen the fears and hatreds between nations and simultaneously encouraged repressive childhood upbringing.[288] So what was required was a strong government that could use measures of force to restrain the minority of the population who would not live peacefully with others. Likewise, a strong nation should resist the aggressive tendencies of other nations.

Thus, if the parental role in the individual family was to relinquish '"discipline" and the restraint of impulse' in wider society a quasi-parental order was to retain them. As Durbin later put it in *The Politics of Democratic Socialism*, 'It would take decades to affect the course of political relations by emotional education... What therapy cannot cure, government must restrain'.[289] Helen Bosanquet had long ago claimed that familial authority couldn't be had by force and submission. Likewise, we've seen that Rousseau had much earlier applied the same kind of logic to government in wider society. Bowlby and Durbin returned to Hobbes. For the foreseeable future in the wider society of relations between adults, strong government must prevail. This was partly to provide support for the family to create the right emotional atmosphere for personality development and partly to restrain the 'primitive' emotionality still apparently driving a significant number of individuals. One of their conclusions was that Hobbes had been very largely right. They wrote 'without government, and in a state of nature, man's life, thanks to his animal passions and rivalries, tends to be "solitary, poor, nasty, brutish, and short"'.[290]

This Hobbesian view of a tacit (and irrevocable) contract between individuals and a centralized state authority was mediated by an emphasis on intellectual and technological leadership. Bowlby and Durbin certainly forcefully advocated the further penetration of the family realm by a psychologically informed centralized and rationalized power. This was to be wielded on behalf of the people by psychological experts in association with social and economic technocrats. Therefore, despite the socialist imperative of their work, Bowlby and Durbin relegated any perceived benefits of greater egality and democracy in the wider adult community to the need for 'the primitive' elements of personality (in this case tendencies for aggression) to be ameliorated as far as possible. Thus, they argued that, in as much as any society claimed to be democratic and peaceable, this was because of the type of people who populated it. These people were the kind of people who could 'support the responsibility, freedom and toleration' that democracy required. As Bowlby and Durbin succinctly put it, 'They are peaceful and democratic because of the kind of people they are'. The same argument, they contended, was likely to apply to equality.[291]

We've seen in the previous chapter that the Labour Party has been described as having three competing, though inevitably overlapping, visions of equality: technocratic, Keynesian socialist and qualitative. Though Durbin might be most closely associated with the technocratic aim of centralized public ownership, he has also been described as an unusual synthesizer of ideas and approaches.[292] In this case, there appears to be a peculiar compromise (or perhaps subordination) with the 'qualitative' egalitarian vision associated with the likes of Tawney. The idea of an equality of dignity and worth built on 'right relationships' in a community of 'fellowship' and 'fraternity' seems here to be accommodated but limited to the family sphere in the interests of personality development.[293] It seems that, for Durbin and Bowlby, this held out the possibility of ultimately creating enough people with the kind of personality structure that could sustain the qualitative egalitarian vision across wider society.

In the meantime—and it was going to be a very long meantime indeed— expert leadership and psychological expertise would be required. For Durbin, at least, the aim of improving emotional health via family relations was combined with that of improving the general level of intelligence in the population through eugenic methods. Sharing views not uncommon among socialists and feminists at that time, he supported sterilization of 'the unfit' and positive eugenics through artificial insemination.[294] This was a view that had been expressed by significant figures in the mental hygiene movement, and notably given voice in the Central Association for Mental Welfare (CAMW) and National Council for Mental Hygiene's (NCMH) calls for legislation to introduce voluntary sterilization in the mid-1930s.

Bowlby's work would rapidly become highly influential within the mental hygiene movement, as well as on government policy. The onset of war appears to have been one of the important factors in this. But Bowlby didn't follow up on the libertarian and egalitarian elements embodied in the proposals to amend family authority emphasized by this book. He largely concentrated instead on only one aspect of familial relations. This was his insistence on the need for a continuity of relationship between a young child and its mother, or mother substitute. Any disruption to this Bowlby believed could create what he called the 'affectless character' with delinquent and criminal tendencies.[295]

Maybe it was this narrowing down of his focus that eased the mental hygiene movement's accommodation of his views.[296] In any case the movement shared his emphasis on the family as primary for moralization and citizenship. Likewise, his connection of mental disorder with social problems, such as crime, delinquency and aggression, and reduction of them to internal family relationships, was a long held mental hygienist conviction. But, that said, the movement also continued to express and develop ideas that coupled emotional security with greater liberty of expression as necessary for healthy mental adjustment. Bowlby situated this need squarely in the home, and important strands of mental hygienists' wartime work for the evacuation scheme did the same.

*   *   *

Alan had been diagnosed as having many 'schizoid' symptoms by Hawkspur's Selection and Treatment Committee. But they believed that it was too early to diagnose schizophrenia and that maybe it wouldn't develop under the camp regime. Hawkspur seemed to help at first, but after leaving the camp Alan was later certified to a mental hospital. He exchanged many letters with Wills. In one he told Wills how much he trusted him because of the many proofs of friendship he had given, one being that he was the only person who bothered to write to him. Alan wrote letters twice because he said he had no way of knowing if they were posted until Wills replied, and he feared losing touch with him. They often discussed books. Alan wrote in one letter that the only decent thing about 'that turnip' Julian was his poems 'or perms' as Julian called them.[297]

* * *

Just before the war a committee set up by the Ministry of Health, following pressure from the mental hygiene movement, recommended that there should be a unified national voluntary mental health association.[298] War prevented its full consideration, but the NCMH, Child Guidance Council (CGC) and CAMW joined forces with the Association of Psychiatric Social Workers (APSW) to form a Mental Health Emergency Committee (MHEC).[299] By 1942 the committee had been renamed the Provisional National Council for Mental Health (PNC). This included all three voluntary organizations, but not the APSW.

It was the MHEC that initially arranged to supervise the psychological and social issues of evacuation. Regional offices were set up in the 13 civil defence regions under the control of trained psychiatric workers. One of their main roles was to work closely with government ministries and local authorities. They helped organize and supervise wartime hostels, arranged billets for misplaced children and gave advice on problems of adjustment.[300] Their activities gradually extended to include Public Assistance Homes and others in existence before the war. Specially trained workers visited nurseries across the country advising on their operation, while various courses and conferences were provided on behalf of the Ministry of Health for other residential workers.[301]

With this experience the PNC attempted to set the national agenda for child care and development. It produced various pamphlets during the war directed at care workers and the public alike.[302] In these the nuclear family remained portrayed as the mainstay of 'civilization' through its adjustment of individuals for good citizenship. In the autumn of 1944, for example, the PNC produced a pamphlet on 'The Care of Children Away from their own Homes'.[303] 'Family feeling', it announced, 'is the basis of society and anything which threatens its strength attacks the structure on which civilization depends'.[304] A 1944 pamphlet promoting a 'simple "ABC" of the child's "needs"' reckoned that wartime conditions were causing family cohesion to breakdown with dangerous consequences for the development of individual mental health and social efficiency in general:

That this break up of family life has profound effect is clear enough from the thousands of examples seen. Social problems appear, such as a rise in juvenile delinquency, due in part to the absence of parental authority, and an increase of sexual relationship outside marriage, which may be ascribed to lack of good family influence, and failure of the individual to assume social responsibility.[305]

The mental hygiene movement's interwar concerns about psychologically unmediated state power were associated with this. A 1944 report on juvenile delinquency prepared for the Bradford City Council by an educational psychologist with the MHEC said that 'The assumption of responsibility for the child's education by the state has given parents the opportunity to shelve their obligations in individual character training of their children'.[306] It also recapitulated interwar fears about 'the masses' with its claim that 'high wages and easy money for the juvenile decreases his respect for property and for privilege'. This 'maladjustment' and 'immaturity' was connected to immaturity in the parents. The report claimed 'a considerable proportion of the parents of delinquent children are themselves children in mentality and social consciousness'.[307] Thus, 'the primitive', 'childhood' and 'the lower classes' remained linked just as they had been with interwar mental hygiene thinking.

But the Bradford report proved to be controversial with city councillors. This wasn't because of its patronising tone and obvious class bias, so much as its apparent attack on conservative values.[308] One thing that appears to have upset some councillors was its criticism of religious instruction in schools in relation to delinquency levels. It said that, of the 370 delinquent children studied, the proportion of those attending church schools was substantially higher than those from council schools.[309] We've seen that leading psychotherapeutic elements of the mental hygiene movement had criticized dogmatic and dominating authority in regard to emotional health and development. The Bradford report reiterated this with regard to religious schooling. It maintained that there hadn't been a properly 'unbiased investigation' of the effects of religious teaching on children's character development and behaviour. There were, it said, potentially negative effects of attempting to inculcate ideals that were 'far beyond their possibility of achievement, and completely beyond their power of emotional understanding'.[310] Echoing progressive education ideas, it maintained that morality could only be properly developed through social and relational interaction, and not by the 'passive acceptance of a code superimposed from above'.[311] The authoritarian inculcation of moral values could be counterproductive, particularly if it failed to acknowledge important differences in the various stages of emotional development.

Equally controversial, and for similar reasons, was the report's assurance that delinquency was a symptom of a lack of emotional security. 'The main thesis put forward is that a punishment for the delinquent act as if he was fundamentally the responsible person concerned is illogical', it stated.[312] This view had been expressed in child guidance and would, as we'll see, become increasingly common. Delinquency may well have been still understood by mental hygienists as a form of mental maladjustment and therefore a failure of rational authority over one's mental states, but this contention was clearly being loosened from its association with the idea that responsibility for the failure lay entirely with the individual. In keeping with the mental hygiene focus the Bradford report placed responsibility mainly on the delinquent's family and its emotional attitudes.[313] But, as we've just noted, it also castigated religious school environments for asserting a dogmatic and rigid authority that was inappropriate to emotional development. As this didn't speak the language of emotional development how could it communicate with the child? Communication should involve listening and so, along with prescribing the need for a 'detailed personal history of the child's development and his family background', the report emphasized the need for children to express their own views in a 'neutral' atmosphere without blame, so that their feelings and attitudes could be taken into account.

* * *

In a letter to an ex-Hawkspur member Wills wrote:

> You will see that we have a new address. We have moved over from the camp to here to look after evacuated children who have not been getting on well in their billets. We brought most of the Hawkspurians with us but that was not a great success and most of them have left now. We are virtually run by the County Council now but it is a means of keeping Q alive during the war and the expectation is that we will return to Hawkspur when the war is over.
>
> Write again and let us know how you get on.

But Wills didn't stay long at the new location in Bicester. He was fed up with the inappropriate premises for the children, the attitude of the County Council and the methods adopted by the staff. In a report sent to the Q Committee after he'd resigned, Wills said that the Council hadn't understood their methods. This had been exacerbated by some senior staff's disregard for principles that were considered axiomatic at Hawkspur. Mostly, he said, this had taken the form of using 'physical violence as a means of securing obedience and checking "insolence" – a word which never existed in the Hawkspur vocabulary'.[314]

Wills took up a post of warden at Barns, a hostel-school initiated by the Society of Friends in Scotland. This originally catered for evacuees who displayed difficult or disturbed behaviour that prevented them from being billeted with ordinary families.

* * *

At the end of the war, the government set up the Curtis Committee on the Care of Children Deprived of a Normal Home.[315] The PNC submitted a memorandum based on the experiences of its child welfare workers. This was later published as *Children Without Homes: How Can They Be Compensated for Loss of Family Life?* It amounted to a detailed criticism of existing childcare institutions.[316] At the heart of the critique was an insistence on the need for emotional security. This required a softening of authority commands and relations, along with a reduction of hierarchy and status, coupled with greater choice, freedom and emotional expression.

The report's author was Ruth Thomas, the PNC's senior educational psychologist.[317] She had been analysed by Anna Freud and worked at her Hampstead Nursery during the war. Her introductory comments read as an echo of Bowlby's proposals for mental hygiene action through the modification of family relationships. She wrote:

> Real progress [in child care] is only assured when it is possible for the values for which the social services stand to become in fact the values of the good family which automatically then takes the management of its own difficulties into its own hands. In the programme which we have now set ourselves, first to ensure a sound economic basis for family life... [and] adequate services in the spheres of health and education aiming at family enlightenment as well as direct therapy, lies the most profound hope for future good citizenship.[318]

Just as in keeping was her description of the criteria on which the PNC criticized existing provision. The goal of all childcare, whether in the family or outside it, was to enable the child to achieve a healthy maturity. This required an analysis of the basic needs of child development. In the view of the PNC *any* upbringing outside the biological family was inevitably handicapping to a child. Therefore, particularly in the case of institutional care, the aim should be to compensate for the loss of the 'natural' family as far as possible so that the child could eventually take its place in the community as little handicapped as possible.[319]

Family life was portrayed as the primary community through which a child gradually emerged into responsible citizenship in the wider community. Therefore, institutional care needed to fulfil four general requirements: first, a close and continuous relationship with an adult able to care in such a way as to promote trust and consequent desire to become like the adult; second, a small community life of mixed ages and sexes where family life might be adequately modelled; third, access to a wider community in which a child may participate; fourth, opportunity for freedom and diversity in occupations, possessions and leisure time.[320]

The essential issue to be kept in mind was 'how does the world of the family, foster home or Home look to the child when he views it from the angle of his needs? In short what does *he* find satisfactory'.[321] With this statement Thomas signalled the centrality of emotional security and the need to tailor authority to this. A child's *own* expression of its personal relational experience was becoming, in theory at least, central to determining whether the quality of relations was adequate.[322]

But as the report was directed primarily at existing care in institutions what transpired was a critique of existing provision. This was condemned as detrimental to child development and mental health when ordered for organizational efficiency and staff interests. Life in large groups promoted gross disturbances of development and mental health, such as bed-wetting, antisocial behaviour and emotional disturbance.[323] Most Public Assistance Homes kept children in a nursery until the age of three and then moved them to a mixed home until the age of seven, with subsequent moves at 11 or 14. This was described as 'utterly wrong from the point of view of mental health'.[324]

The PNC based its proposals for reform on some of its own wartime residential ventures. At these a 'Group System' had been employed in which children were divided into groups of four or five of mixed ages and temperaments, with separate living space for play, meals and sleep. A 'Group mother' co-ordinated each group, providing special outings, clothing and treats as part of her substitute mother role.

The PNC's call for reform also entailed a critique of the 'medical model' often employed. In contrast with the emotional relations of the family order, hospital-like regimes were considered over-preoccupied with physical hygiene, cleanliness and efficiency, and associated with an excessive requirement for order and control.[325] The report insisted that these factors distorted and deteriorated the quality of emotional relationships necessary for children's mental health and development. Like the Bradford report, the PNC declared that 'Culture is not imposed by lessons or even by a cultivated environment, but by the feelers the child puts out to draw them into himself'.[326] Consequently, the report said that children's freedom of choice and expression should be allowed for and promoted.[327] 'Children, like adults', it remarked, 'need to be able to reject or turn down what is offered them'.[328]

The PNC placed great emphasis on the child as a 'living human being' whose emotional relationships were central to their health, happiness and development.[329] Unlike task-oriented physical nursing, the need for affection, emotional security and appropriate relationships needed to be taken account of throughout the whole day.[330] This required greater scrutiny of the attitudes and activities of the staff. They were now more directly focussed on as possible causes of children's maladjustment and behavioural difficulties.[331] Staff were to be recruited who were 'capable of forming real relationships with the staff and children' in contrast to those who placed an exaggerated importance on their authority.[332]

The report, in fact, maintained that it was only children who had become passively conforming and thus isolated from emotionally secure relationships, who appeared to accept typical institutional life.[333] It added that evidence suggested current care 'may be cultivating the backward child in the unnatural setting of institutional life'.[334] Poverty of experience in children's institutions could produce the appearance of mental deficiency, it said.[335]

But the report didn't consider the institutional care of children considered to be 'genuinely' mentally deficient. The Curtis Committee didn't cover their care either. It had considered them to be outside its remit. But despite taking this stance its report, nevertheless, went to the trouble of recommending that no administrative changes should be made regarding their care.[336]

*   *   *

In a letter to friends written in 1940, Franklin wrote of a 'tentative' new venture that might be based in Buckinghamshire. It would cater for evacuated children from about the age of 11 upwards who had been 'difficult' in their billets. She suggested somewhere not too near a town, maybe a house or a camp of huts, with fields, garden and poultry. Still hopeful that Q methods could be influential despite the war, she added that 'perhaps (in spite of the disesteem in which, as you know even better than I do, I am held there) the truest example of all we used to mean by Q as a democratic oasis of toleration has moved to Scotland!'.[337]

*   *   *

The psychodynamic underpinnings of the mental hygiene movement took little interest in the care of children and adults termed mentally deficient. They were seen as constitutionally incapable of the relational experience required for full mental health. Lucy Fildes, one of the movement's prominent educational psychologists, and chairman of the PNC's Committee on Hostels, stated its opinion this way:

Separate hostels should be provided if possible for the following two groups of children: -

i)   Those constitutionally inferior, whether their inferiority is intellectual or emotional or both.
ii) The neurotic, i.e. the child of essentially normal personality make-up distorted by environmental circumstances.

This need for separation rests on the fundamental differences in make-up between the two groups of children – those who are by nature inferior, either intellectually or emotionally, need above all things adult guid-ance and support to a degree above what is considered normal. Neurotic children, on the other hand, require maximum freedom.[338]

This was a distinction also emphasized by the psychoanalyst D.W. Winnicott. In 1944 he gave a paper to the PNC's Child Guidance Services Committee on hostels for 'maladjusted' children. This prompted a mem-orandum to the Ministry of Education stating the principles on which they should be based.[339] Winnicott insisted that mentally deficient children should have separate accommodation and treatment. He gave the following reasons:

> This is not only because they need special management and education, but also because they wear out the hostel staff to no purpose, and cause a feeling of hopelessness. In such difficult work as that with problem children, there must be some hope of reward... [340]

So, it isn't surprising that the PNC's recommendations for childcare weren't applied to mentally deficient children. During the war it published articles on provision for mentally defective children in its journal *Mental Health*. C. H. W. Tangye, the headmaster of a special school for such children, con-tributed an article in 1941. In it he remarked that opinions differed over whether the expense and effort of evacuating mentally deficient children was actually worthwhile.[341] Indeed, the only negative effects of the loss of home life that Tangye referred to in his article were those on his staff evacuated along with the children.[342] Tangye, nevertheless, argued that his experience of evacuating special schools had proved to him the value of 'community life' for mentally deficient children.[343] But his idea of this was very different to that outlined for 'normal' or 'maladjusted' children.

During the preparations for evacuation it was agreed that children des-ignated mentally deficient shouldn't be billeted with private families but evacuated as a group by setting up residential schools in evacuation areas.[344] These residential schools emphasized control and discipline. One evacuated headmaster claimed gains in physical health, educational standard and gen-eral conduct at his evacuated school. He attributed much of this to the chance to apply one central authority of 'control' separate from what he saw as the detrimental effects of the children's home and family.[345] This allowed the correction of bad habits which were seen as the result of poor or indulgent home training and outlook.

Attitudes towards bedwetting illuminate the gulf between approaches. In its memorandum of evidence to the Curtis Committee, the PNC said that:

> The homeless child is particularly susceptible to [bedwetting] because of the emotional deprivations he suffers.... It is often a form of aggressive

behaviour – the child's way of registering protests when conditions make him unhappy. In almost all cases, however, it is a form of behaviour which is beyond the child's power to control consciously and can be dealt with successfully only where it is possible to get some insight into the deprivations from which he suffers and against which he is unwittingly protesting. Punishment in all cases is liable to exaggerate the difficulty and add other forms of adverse behaviour.[346]

These principles weren't applied to mentally deficient children. In July 1941 the PNC printed W. A. G. Francis' record of enuresis at his evacuated special school for mentally deficient children. Bedwetting, he explained, had quickly become a frequent occurrence. The method devised for dealing with it was based on praise and punishment. Children who wet their beds were referred to as 'offenders' and sent to the 'Camp Commandant' to be recorded, and 'encouraged' or 'censured'. Out of a total of 216 boys living at the camp about 140 had appeared on the register. Francis reckoned that this system successfully trained the large majority of children in 'good personal habits'.[347]

Children diagnosed mentally deficient thus continued to lack the 'right' to the kind of environment and emotional engagement that mental hygienists had come to believe essential for other children. There was what amounted to a conceptual trap-door in mental hygienist theorizing. Children considered to be below a certain level of intellect were deemed unable to benefit from the close family-style emotional relationships considered essential for other children. Through the trap-door they tumbled into a quite different form of 'care'.

* * *

Tommy had been writing to Wills and Ruth ever since he'd left Hawkspur in 1938. In 1941 he wrote excitedly to say he'd been called up. Among the letters exchanged was one from Wills remarking that he seemed to be seeing a lot of coastal resorts in his training. He added, 'If you see any white shoes they're mine!'. Wills explained that he'd left them at some resort or other in 1911. 'P.S.', he added, 'You're a soldier now. Why don't you hurry up and finish the war off?'.[348]

* * *

Outside the realm of 'mental deficiency' we can see that the emphasis on emotional security and associated questioning of authority relations had a wider effect. In fact, it was expressed even in work within the military. We can see this by taking a brief look at two experiments. One was a radical attempt to treat psychiatric casualties at Northfield military hospital near

Birmingham. The other a description by three psychiatrists of their work at the rehabilitation service for ex-servicemen called the Civil Resettlement Service (CRS).

These days Northfield is the most well-known experiment. This is mainly because it popularized the term 'therapeutic community', and there has been interest in retrieving some of its history among recent practitioners in this field.[349] But it's worth noting here that historians themselves haven't given therapeutic communities any important role in the history of psychiatry. In fact, two popular and broad ranging works dismiss their effect and influence.[350] For Edward Shorter they represent simply one aspect of a mid-twentieth-century eclectic approach to psychiatry: an approach born of desperation and rendered unnecessary by the later 'neurobiological revolution'.[351] For Ben Shephard, the therapeutic communities developed in the 1940s and 1950s were pampered experiments with debatable results, marginal effect and little long-term influence.[352]

But there are obvious reasons why Shorter and Shephard consider therapeutic communities unimportant. For his part, Shorter claims to provide a history of psychiatry free of 'sectarianism' and the imprint of ideology.[353] This would be a unique achievement if it were remotely possible. And, as is usually the case with those who make such claims, Shorter provides an argument that is all the more bluntly ideological for his apparent assumption that he has stripped ideology away to bear the kernel of truth about both mental disorders and the past as it relates to them. Shorter's is a reductionist account of mental problems that claims to show the ultimate success of genetics and the biology of the brain. He states from the outset that 'history' has already revealed psychoanalysis to be, just like Marxism, another 'dinosaur ideology'.[354] Given this, it isn't surprising that he relegates psychoanalysis, and therapeutic communities to the status of redundant and misguided alternatives.[355] Ben Shephard's discussion of therapeutic communities is in the context of a detailed and thorough history of twentieth-century psychiatry within the military. But Shephard's account is very much in keeping with the sort of military history that sustains a view of emotionality as weakness. In essence, his book asks which therapies were the best at stopping soldiers from breaking down or malingering. Given that sort of question it isn't surprising that Shephard doesn't judge favourably any sort of treatment that smacks of pandering to a soldier's vulnerabilities and emotions. Therapeutic communities could easily be regarded as doing that.

But therapeutic community experiments are actually very important because many of them display the partial inversion of the Family as an organizing principle of mental health and responsible citizenship that we've been tracing. Within the mental hygienist attempt to speak the language of emotionality in the interests of individual mental adjustment and societal progress, a contradictory force had emerged that appears to have pushed against the Family and its authority structure. It's within this context that the idea of the therapeutic community takes on its importance.

We can place Northfield alongside the other experiment that I've mentioned and briefly draw out some shared principles. Both were sponsored by the mental hygiene movement. The leading mental hygienist J. R. Rees, who'd taken over from Crichton Miller at the Tavistock in 1939, became head of the Directorate of Army Psychiatry in 1941.[356] His so-called 'Tavi Brigadiers' were involved with both experiments. From 1942 onwards, methods of treatment for servicemen diagnosed psychoneurotic were developed at military hospitals. Northfield was one of the most important. The psychoanalyst John Rickman considered it 'one of the biggest, costliest experiments in psychodynamics now running'.[357] We'll address ourselves to what is known as 'the second Northfield experiment' where the psychiatrist Tom Main was a leading figure. The CRS emerged from attempts to understand the psychological difficulties of officers who had escaped from prisoner of war camps.[358] By January 1946 it comprised twenty regional resettlement units intended to help men from all ranks to re-adjust to civilian life. We'll briefly discuss the approach described by the psychiatrists A.T.M. Wilson, Martin Doyle and John Kelnar.[359]

*　*　*

In 1942 John wrote to Wills at Barns in Scotland. He told him that recently he'd borrowed Wills' book *The Hawkspur Experiment* from the library. While reading it he'd become incredibly 'homesick' for the place, even though he'd been none too happy when he was there. John said that he was very happy in his present work but that he'd experienced such a strong longing to contact someone connected with Hawkspur that he'd phoned Franklin, Carroll and Glaister, but hadn't been able to get hold of any of them. He said he longed to see Wills and Ruth again sometime, somewhere and signed-off asking Wills to please write.

Wills wrote back.

I was delighted to have your letter which, as you see, I am answering by return of post.

I have been thinking of you a good deal lately, wondering whether you are still at the same address... Dr Franklin told me you had rung her up, but of course she is not much in London these days, and Carroll is in the R.A.M.C. somewhere in Scotland when last heard of... Your Hawkspurian nostalgia doesn't surprise me. The times we had in the past always seem much pleasanter than they did at the time. The memory has a habit of sticking to the pleasant things and glossing over the unpleasant. Soldiers who fought in the last war, some of them, have much the same feeling about the trenches.

He rounded off by saying that he and Ruth would always be glad to hear from him, or to try and meet in London.[360]

* * *

Mental hygienists' wartime description of childhood emotional development as requiring 'emotionally secure' relationships, appropriate to each stage of progressively widening interpersonal relations, was reiterated in the context of adult therapy at the two schemes. At the CRS repatriates experiencing difficulties in resettlement were considered to be revealing an inner need for 'security and affectionate relationships'.[361] It was claimed that separation from their homes and communities had created acute feelings of isolation, frustration and distrust of authority. This, it was argued, often took the form of an 'embittered withdrawal from social relationships'. The CRS aimed to counter this suspicion of authority, facilitate a return to a 'less regressed social attitude in the unit', and then wider re-relation with family and community.[362] Similarly, at Northfield, Tom Main described patients' progression through what he called the 'various therapeutic social fields created in the hospital'. The aim was a growth in sociability from regressed and isolated states, through small groups of interaction, to full social relationships outside hospital.[363]

Mental hygienist's recommendations for institutional child care emphasized that mental health and moralization emerged organically within lived experience throughout the whole day and couldn't be imposed by sovereign authority. This view placed primacy on the emotional content of relationships and an associated emphasis on greater freedom and choice. The wartime experiments followed these principles. But because their expression of them was in terms of therapy, rather than the upbringing of 'normal' children, their rendition was closer to the radical interpretation at Hawkspur and Leytonstone Homes. We've seen that these employed a more genuine interpretation of the term 'voluntary' than that intended in the 1930 Mental Treatment Act. At the same time they turned freedom and choice from ultimate aims of treatment to important elements of therapy itself. Similar transformations are evident in the wartime experiments.

Northfield and the CRS conceived therapy as an ongoing part of the whole day. The CRS considered freedom and choice over when and how to enter therapeutic situations an intrinsic aspect of re-socialization.[364] At Northfield, Main held to the same approach. Patients were to be 'free to move at their own choice and at their own speed within the social fields which best [suited] them'.[365]

At Northfield and the CRS, just as in residential childcare theorizing and at the pre-war experiments we've looked at, hierarchy and authority were to be reduced in the interests of more direct emotional engagement. The PNC had criticized rigid discipline and relations built around organizational hierarchy, claiming that they created isolation, passivity and mental maladjustments. It had also incriminated the prevailing medical model of care in these processes.

At Northfield Tom Main described the commonly existing mental hospital as authoritarian and hierarchical, with a rigid organization designed for discipline.[366] For him it encapsulated the 'traditional mixture of charity and discipline' and 'a practiced technique for removing [patients'] initiative as adult beings'. He argued that hospitals operated in their own technical and organizational interests of efficiency. Staff therefore considered 'good' patients to be those who'd become dependent, conforming and passive.[367] The mental hospital needed to be reorganized. It shouldn't dominate and isolate patients through its bureaucratic order and treatment regime, but seek to provide the social support and opportunities provided in 'spontaneous and emotionally structured (rather than medically dictated)' communities.[368] For Main, it wasn't just that there was an administrative need for discipline and containment; there was also an emotional one on the part of the staff. This 'need', and an associated desire for gratitude from the patient, just promoted mental ill-health. He emphasized that staff needed to confront their own emotional needs and conduct. Participation with patients in group therapy sessions was necessary, not only to encourage full community participation, but also so that staff inter-relations could be revealed and understood.

In the same fashion the CRS contrasted its open and relatively egalitarian approach with the rigid discipline of the army. Indeed, it openly characterized itself, in contrast to the 'paternal and authoritarian' military, as a 'maternal and democratically conceived community'.[369] It's a statement that offers yet another gendered idea of the basis of 'democratic' relations. Muriel Payne had given one version in her description of an authoritarian hierarchy at Leytonstone. Bowlby had implied something similar in his rendition of family relations. But the key word seems to be 'maternalist'—the mental hygiene movement can't remotely be considered feminist.

Still, with the terminology of 'democracy' and 'authoritarianism' we have two terms that will lead us into the post-war world. They're now both intimately related with therapy in mental health.

*　*　*

It was Christmas 1941. Wills had received a card from Richard. He was married with two children now. The card was signed Richard, 'Or should I say "Tom Beeley"?'.

Wills wrote back,

> Well it <u>was</u> good to hear from you. <u>Two</u> kids now ay? That sounds good... There was a cryptic note at the end of your card – what was it now? 'Or should I have signed myself Tom Beeley?' That seems to suggest

that you have been reading that obnoxious book called the 'Hawkspur Experiment', for if I remember rightly there was someone in that called Tom Beeley. You seem to think the man Beeley is something to do with you. Now what can have made you think that? Do you think what is said about Tom Beeley could accurately be said about Richard? Do tell me. I should be most interested.[370]

6
Alternative Dialectics

'Toby Ale and Stout' it reads; a singular advertisement in the heart of the throng. But they probably weren't much bothered exactly what they were drinking. The lorry load of beer barrels is pictured travelling slowly along Piccadilly Circus. On top of it are smiling women and men, some in civvies, some in uniform. In front and beyond, and disappearing into the distance, is the rest of the vast crowd that is celebrating Victory in Europe Day.[371]

The photos of the time show scene after scene of jubilant crowds streaming through the streets. These images might recall Bernard Bosanquet's distinction between a social order of organization and that of mere association, as represented by a crowd. The kind of mind and social order pointed to by the formative structure of the family could, in his eyes, be likened to an army. This was rational organization with concerted action, informed by a hierarchy of rank and function.

We've seen how interwar mental hygienists echoed this image. So it probably isn't so surprising that along with the wartime experiments sponsored by the movement, mental hygienists made comfortable use of the army social structure to develop their ideas of psychiatry as a comprehensive public health measure. Here was just the kind of social order to which the mental hygiene movement could hope to apply its methods. If our look at important wartime experiments shows a continued tendency towards placing primacy on ideas of 'emotional security' and self-expression in ways that diminished hierarchies and emphasized greater liberty within more egalitarian association, it remains the case that this was only one side of our apparent dialectic.

Mental hygienists lost little time in making use of their wartime experience to reinforce calls for a comprehensive preventive psychiatric service. One was the psychiatrist Kenneth Soddy.[372] He'd worked for the Emergency Medical Service early in the war, but had resigned along with John Bowlby over the treatment of psychiatric casualties after the retreat from Dunkirk.[373] They'd both joined the army instead. Soddy had been given the rank of Colonel in the India Command. As Deputy Director of personnel selection and chief technical officer, his aim was to 'establish modern scientific selection methods in the three fighting services and in the Indian civil service'.[374] After the war he was appointed Medical Director of the National Association for Mental Health (NAMH).[375] This was the fully incorporated version of the Provisional National Council for Mental Health (PNC).

Soon after its formation, NAMH published a booklet written by Soddy called *Some Lessons of Wartime Psychiatry*.[376] He warned in it that 'The atomic bomb bogey hangs over us, the urgent need is to improve the state of mental health of mankind before it is too late. Can there be any higher priority than this?'.[377] Along with other mental hygienists Soddy claimed that the distinctive psychiatric contribution to the army hadn't been in the traditional treatment of illness so much as through 'furthering a positive mental health policy, by selection, elimination of the unfit and by advice on leadership, on the handling of men and on morale'.[378] Like other mental hygienists he described this as innovation, but really it was the importation into the army of the ambitions and procedures they'd developed and attempted to deploy between the wars. As Soddy put it, 'The most important subjects for study were the fitness by personality and past experience of the individual for the role in which he was cast, the quality of training, the leadership, the indoctrination of the soldier and the foundation of good morale'.[379]

Soddy emphasized that the army was a 'closely knit authoritarian hierarchy' with a specific function to destroy enemies of the nation. Two psychological processes informed this. One broke down elements of personality associated with previous family life and substituted 'an idealised father figure', which was 'in this case the authoritarian society'.[380] The other used this figure to protect the soldier from the guilt associated with 'licenced homicide'. For Soddy, the psychiatrist was successful when he 'shared the life, fitted into the autocracy and adapted his methods to those of an authoritarian paternalistic society'.[381] Soddy suggested that psychiatry could apply lessons learned in the military to civilian life. He acknowledged, though, that civilian life didn't have only one special mission like the army and that it could 'survive with a far less degree of rigid organisation'.[382] He saw the civilian as both more individualistic and more emotionally mature than the soldier, with many and varying 'father substitutes'. Passions couldn't be directed against single objects as in the services. So, as Soddy put it, 'The adult civilian must deal with his emotional problems himself and cannot pass them up for solution by higher authority'.[383] Here, again, Soddy was really only repeating interwar mental hygienist views. The 1932 Association of Psychiatric Social Workers (APSW) booklet, *Mental Health and the Family* had remarked:

> The 'father image' in family life, in the community and in religion, seems to be at the present time divesting itself of authority, removing the protective guidance of a decalogue, and asking of individuals, instead of

conformity and adjustment to a social norm, the harder but more adult task of adjustment to the self, achievement of inner balance.[384]

This was the classic mental hygienist description of the individual personality as something like a Swiss Army knife. Founded in the family, it is self-adjusted and therefore adapted to all occasions. Dib dib dib, dob dob dob, always prepared, ready for anything in the 'meritocratic' society.

In a 1950 article entitled 'Mental Health', Soddy explained that the family was the centre of mental hygiene concern. Similarly to Bowlby, he presented emotionally 'mature' parents as the requisite of a 'mature' social order. But less like Bowlby, Soddy described how 'harmonious living' rested on the 'toughness... of mental fibre' that resulted from the environment of security and challenge provided by such parents.[385] Others echoed this view. For instance, in a speech to the APSW the mental hygienist D. R. MacCalman quoted approvingly an anthropological account, written 50 years earlier, of the people of Nyasaland in Africa as living in 'native simplicity'; 'Primeval Man, without clothes, civilization, learning, religion – the genuine child of nature, thoughtless, careless and contented'.[386] MacCalman reckoned that, in contrast, the emergence of civilization had been encouraged by difficult environments and that, likewise, the challenge of 'interpersonal relationships' encouraged the individual effort of adjustment to attain inner harmony.[387]

For Soddy, as with the mental hygiene movement in general, the obverse of mental health in the individual and the community was, 'mental disorder, mental deficiency and emotional maladjustment'. It's clear that, to Soddy, all these categorizations of people represented inferiority.[388] In *Some Lessons of Wartime Psychiatry* Soddy maintained that even if there was a vast expansion in the number of psychiatrists treating mental disorders in Britain, they still wouldn't be able to cope with the number of 'neurotics and misfits' as fast as heredity and environment threw them up.[389] He commented on the 'problem' of the 'unintelligent borderline defective' in this way:

> To the extreme upholders of the eugenic view it might seem logical to put all dull and unstable men in the 'front line' so that the bulk of the nation's casualties will occur among the constitutionally inferior – thus improving the natural stock in the next generation. Such a view might have something to recommend it (were an end to justify the means) had not the national existence depended on the efficiency of the armed forces. Such men are bad soldiers, and the retention of bad soldiers... endangers the security of the whole...[390]

Hardly a resounding reproach for the 'extreme' faction of the eugenicists. Less still any suggestion that the ethical basis of 'harmonious living' might be based on some notion of equal human worth. But, needless to say,

Soddy blithely claimed a few pages later that the psychiatrist in the military was 'in no sense engaged in moral judgement as to the worthiness of the individual'.[391]

This distinction between 'mature' parents enabling 'robust individuals'[392] on the one hand, and the 'immature' and 'inadequate' on the other was clearly suffused by class. In his 1950 article Soddy paraphrased the findings of a 'carefully organised survey of a highly civilized community' to depict the 'facts about mental ill-health' in the community:

> ... the lowest socio-economic groups of the community [show] the highest incidence of insanity, neurosis, children's maladjustment, mental deficiency, crime and delinquency. These groups contain the bulk of the people of dull and backward intelligence, whose difficulties in life are all the greater because inferior intelligence tends to be linked with inferior emotional stability; they also present a further menace to the community because of a disproportionately high fertility rate. These sections of the population, therefore, are of particular significance to mental hygiene.[393]

Through 'the primitive', through madness, through childhood and through class apparently emerged what the mental hygiene movement liked to call, 'the mature individual'.

* * *

Mr Arbuckle of the Scottish Education Department visited Barns towards the end of 1945. Wills had been Warden there since 1940. A record of Arbuckle's verbal report on Barns has him saying that Wills 'believed in free discipline which, in some respects, did not conform to ordinary social standards'. Arbuckle observed that 'the children need not wash their hands before meals if they did not wish to do so, no form of punishment was meted out for misdeeds and an undue familiarity between him [Wills] and them was permitted. He [Wills] was somewhat self-opinionated in that he did not consider that he had anything to learn from persons engaged in similar work'.

Arbuckle 'sensed a feeling of suspicion and mistrust' about Wills and his methods. Opinions varied on him, with some thinking that he did get fruitful results and others reckoning he was a hypocrite. For Arbuckle the truth was somewhere in between. He didn't think Wills was 'actually dishonest judging from his own peculiar standards but there was something indefinitely wrong which suggested that Mr. Wills was not suitable for dealing with children'.[394]

* * *

According to the criminologist Terence Morris the very mention of the words 'Welfare State' could, almost more than 'nationalisation', 'polarise the conversation at a dinner table throughout the six post-war years of Attlee's Labour administration'.[395] They certainly animated doctors, as the correspondence pages of the *BMJ* and the *Lancet* testify. One doctor's speech, published in the *Lancet*, denouncing governmental 'dictatorship' over the medical profession was responded to by a psychiatrist called Brian Kirman. He wrote that this doctor clearly took the view that if the principles on which the new NHS should operate were decided by the representatives of the people, then this was 'dictatorship'. Kirman went on, 'In its place he appears to advocate "government of the doctors, by the doctors, for the doctors" as the essence of democracy'.[396] But Kirman was in the minority. One correspondent to the *BMJ* warned of the potentially deleterious effects on individual personality that 'undue control by the state' and plans for social security would produce. It was the sort of view that chimed with the traditional views of mental hygienists. The psychiatrist R. H. Ahrenfeldt, who had worked with J. R. Rees during the war, replied to this correspondent that for some years 'eminent psychologists and psychiatrists' had concurred that an 'over-protective and authoritarian "paternal" State' was a serious threat to the 'emotional and social maturity of the individual and a liberal democratic form of society'.[397] Among others, Ahrenfeldt cited the psychoanalyst J. C. Flugel's 1921 *The Psycho-analytic Study of the Family*. Flugel was senior lecturer in the Department of Philosophy and Psychology at University College London. In his book Flugel had equated the effects on personality of an extension of state power during World War I with that of 'modern socialistic thought – especially its cruder aspects'.[398] According to him, both encouraged people to treat the state in the same fashion that they had learned to treat their parents in childhood. Small children accepted dependence and parental supervision of even the minute details of their lives, and this state power similarly discouraged 'emancipation from the primitive attitude of dependence on the parents' by working against individual initiative and self-reliance.

This is a classic example of concerns about power in the modern society. Centralized power has grown in the state. It's a totalized power that entails a process of rationalisation through which people's lives are ordered and inspected. It's also a blunt power that can't successfully moralize individuals. It only serves to infantilize them. The clear implication is that under socialism this would be the actuality of what might appear to be popular power. This long-standing presentation of socialism or communism as inimical to 'maturity' and mental health was also directed at the recently defeated Fascist states along with the social organization of the nations that had given rise to them.[399] But, with the landslide Labour election in Britain, and the emerging international 'cold war', concerns about the former remained prominent. For instance, Ahrenfeldt summed his letter up with an almost Charity Organisation Society (COS)-like emphasis:

> It is of course recognized that a basic standard of social and economic security is essential for the emotional maturation of the individual to be possible. But it would seem...that it is only by educating the individual to initiative, self-reliance, and independence, giving him the opportunity to improve his own and his family's condition *through his own efforts*, and insisting that he should, in so far as his capacities permit, take on the burden of his responsibility as an adult that there can be any hope of giving rise to well-adjusted and mature personalities and a lasting significant culture. In the face of contemporary events and ideological trends, it would be difficult indeed to find place for such hope outside the realm of fantasy.[400]

But, in all honesty, who were the people most likely to have to 'prove' themselves to be self-reliant and independent in this way, if not those in society with least economic and social power?

It was a common pronouncement within the mental hygiene movement, however, that post-war welfare measures meant examples of material need could most often be seen as 'symptoms' of deeper psychological difficulties.[401] Tavistock psychiatrist H. V. Dicks regarded Flugel's book as a 'Bible'. A significant figure at NAMH, he reported in 1954, for example, that marital problems weren't, in his experience, caused by economic hardship. Instead, 'unconquered childish needs' prevented the co-operation through which families could deal with hardship. He declared that 'Insecurity and environmental pressure reveal psychological weakness and are used as its alibi'.[402]

In the early 1970s the social worker Rob Holman concisely exposed the politics that such views appear to validate:

> On the whole, the economic and social machinery of society is regarded as working well, providing a tolerable standard of living for most people. Unfortunately a minority of existing families are the grit in the machine, being unable to use it themselves and causing trouble for other people. The deprivation of this minority is due mainly to their inadequate child rearing practices which fail to instil in their children the skills and will to perform like the rest of the population. If these family habits and practices can be improved, however, the minority will be enabled to achieve better education and jobs and so move out of poverty. The other sections of society wish to abolish poverty and will willingly provide the necessary resources and be prepared to incorporate the poor into their ranks.[403]

The silent sentence at the end is surely something along the lines of 'Oh yeah, of course they will, and I just saw a pig fly past the window'. But then Holman was writing in an academic journal. The implication of his argument, in any case, was that a powerful state was required to redress structural economic inequalities and redirect resources. As he wrote '... why change people if the opportunity structures of society remain unchanged?'.[404]

The image of state power and authority described by the mental hygienists above is, however, analogous to Emmanuel Miller's interwar description of 'primitive' groups.[405] It's a homogenizing and static force that truncates individuality and freedom. But it marks a regression from even this 'primitive' order in the sense that it lacks those aspects of moral and social authority that are based on intimate personal relations, built from the relational order of the family. Many appear to have feared that the family would completely lose its functions to the state.

At an international conference in 1948, D. R. MacCalman echoed other delegates' apparent views when he suggested that the functions of the family seemed to be failing. He said 'Vast numbers of individuals would appear to be growing up with insufficiently mature independence'. And he asked '[is] the family unit surrendering its functions to the wider unit of the state, which has not yet learned to exercise them adequately?'.[406] Dicks repeated the interwar mental hygienist contention that the modern family allowed the development of greater individuality and freedom than 'traditional' societies. It held the potential for 'levels of maturity and breadth of personality well above the closely regulated social limits of olden days'.[407] But, as 'shrunken successor to the traditional kinship group', it had become 'the "pint pot" into which a whole gallon of intimate human relationships has to be compressed'. Simultaneously, massive political and technological change undercut its functions and severed its social links. So, to the un-moralized social atoms that both the COS and the mental hygiene movement had focussed much of their concern and action on, was added the 'atomised urban family'.[408] Its pressure-cooker emotional relationships threatened to increase social and individual dislocation.

Dicks argued that the state's economic and employment policies were in opposition to the imperatives of mental health and development. With this contention he went on to reveal the gender prejudice that all too often underlay mental hygienist ideas. These state policies, he said,

> ... tempt the young wife and mother away, and make it easy for her to by-pass the deeper experience of motherhood by dumping the baby in a crèche. This in its turn perpetuates her uncertainty over acceptance of a feminine destiny and sharpens the conflict over sex roles in the partnership ... Where is the incentive towards the preservation of good family life?[409]

But though these kinds of views seem to have predominated, other interpretations were clearly possible under the mental hygiene emphasis on correct emotional relations for mental health defined as 'maturity'. In 1951, a speech by Alfred Torrie, the NAMH medical director between 1948 and 1951, made news across the Atlantic in the *Virginia Free Lance-Star*. 'British Called Immature Because Men Boss Women', ran a headline in its 15 May issue. It was reported that Torrie had told a meeting of marriage guidance councillors in Harrogate that emotional immaturity underlay a lot of Britain's troubles. This was mainly due to the British male treating his wife as a slave. He was quoted as having said 'We are still a patriarchal nation, a male-dominated society', and that 'Woman is just the slave, the helot, the chattel, and Britain won't grow up until women have more influence'.[410]

Such a statement was clearly unusual among mental hygienists. But, in any case, it shared the same conceptual basis that informed the movement. Post-war, many mental hygienists considered the family to be atomized and, if not totally failing, at least under extreme pressure. So how could they shore it up? How could they retain the Family as an organizing principle and methodology?

The answer was perhaps to extend the relations of emotional depth, which mental hygienists considered crucial to the emergence of the mentally healthy and responsible individual, closer to the lived world of people in the community. The Social After-Care Scheme (SACS) can be seen as one way in which this was attempted. But, in doing so, it brought with it the dialectical contradictions we've been tracing. Here, we'll glimpse again the tendency toward placing primacy on ideas of 'emotional security' and self-expression in ways that placed relations of authority into question and, consequently, emphasized greater liberty within more egalitarian association.

NAMH developed the SACS during the war (in its precursor guise as the PNC). Beginning in January 1944, it operated under the general supervision of the Board of Control, along with the involvement of the three armed services and the Ministries of Pensions, Labour and Health.[411] Originally, a group of 30 service psychiatric hospitals and Emergency Medical Service Neurosis Centres were involved. Psychiatric patients thought likely to benefit from after-care were notified to a Regional After-Care Officer operating in the relevant civil defence region. These officers were psychiatric social workers. They visited the discharged patient at home in order to help with satisfactory adjustment to civilian life. NAMH described their work as helping to link and co-ordinate the appropriate local social and medical services, and, at the same time, provide a 'friend and adviser to the patient'.[412] Even though there were few personnel available to run the scheme, the SACS had seen about 1500 people within the first six months. During 1946 the remit was extended to civilians and by the end of that year the total of cases dealt with had risen to around 10,000.[413] Though run with very small numbers of staff, the SACS was the first community service provided on a national scale for the after-care and support of people with psychiatric problems.

But NAMH was thwarted in its hopes of turning this scheme into part of a comprehensive mental hygiene measure for the community. The Ministry of Health decided to discontinue funding the SACS in the face of the 1948 National Health Service Act, which placed the power (though not the duty) to provide after-care for people suffering from mental illness under the remit of local authorities.[414] In response, NAMH pressed for local authorities to take up the scheme in their areas.

The psychiatrist T. A. Ratcliffe was closely involved with the SACS and later derivative work promoted by NAMH. He declared that 'The concept of mental health in a community would seem to demand two things – that individual members of that community should be themselves stable, secure and settled and that the community pattern itself should be a mentally healthy one'.[415] Similarly to Soddy, he claimed that the 'problem group' encompassed by the 'socially maladjusted' and the 'social misfit' represented a barometer of the mental health of the community. This group included the 'delinquent, the chronic absentee from industry, the problem family, the solitary and inadequate personality type', as well as 'chronic minor ill-health, divorce, child neglect, and other similar problems'. These, Ratcliffe announced, were the targets of the 'expert in mental health'.[416]

The archetype for mental health remained the Family. The psychiatric social worker E. M. Goldberg worked on the original wartime SACS and described it as an attempt to provide a 'mature' 'mothering' relationship that would enable the development of emotional maturity and consequently progression through stages of sociability and responsibility.[417] Ratcliffe described it in the same maternalist terms:

> ... just as the parent-child relationship should be the epitome of future relationships for the child and the path which leads him on to adult maturity and independence of personality, so the client-Psychiatric Social Worker relationship should be an experience which leads the client on until he can form his own mature adult relationships in his environment.[418]

'Maternalism'? 'Paternalism'? That's the trouble with the use of these gendered terms. In any case, we can see two related manoeuvres going on here. They're both traditional to the mental hygiene movement. First, personal behaviours and experiences, commonly considered to be expressions of mental troubles, are interpreted as varying degrees of failure to pass through successive stages of mental and social adjustment within the intimate emotional relations of authority in the personal family. All the people displaying such feelings or behaviour are considered to be developmentally 'immature' and 'infantile' in varying degrees. Second, these categories of personal experience and behaviour are associated with behaviour that is commonly understood as definitive of social problems in society, such as delinquency, divorce, 'problems' in industry and 'problem families'.

So this paternalist maternalism or maternalist paternalism (take your pick) would seem to sit easily with the status hierarchy of intellect and personality, mediated by mental hygienist expertise that was the mental hygiene movement's original aspiration. But it also expressed the contradictions that our dialectical tale has been following.

In terms of the Family as an organizing principle and method, it's clear that the SACS displays its wider and deeper application. It represents an amended rendition of family authority along the lines we've already seen at the pre-war and wartime experiments. Mental hygienists had described human development in dialectical terms as an inherent duality of human nature emerging in dynamic relation with familial authority. This duality took the form of impulses that were selfish and antisocial, coupled with those that were co-operative and sociable. Bowlby and Durbin had theorized before the war that the latter impulses were more naturally expressed, so they recommended an amended parental authority so that the expression of co-operative and sociable impulses wouldn't be distorted. SACS' theorizing expressed much the same stance, but extended it to designated adults in the community at large. It attempted to repair 'unsatisfactory relationships' in infancy and childhood that had endured in adulthood through an amended quasi-parental authority that might allow already distorted impulses to be relaxed and more sociable ones to emerge.[419] A relationship of trust was seen as crucial in creating the emotional security needed for this to take place. The role of authority was clearly under question here. The caseworker's role wasn't to be the 'authoritarian one of giving advice'.[420] NAMH stated:

The caseworker will not only refrain from imposing her will but will also avoid imposing moral values on an individual. Even when the client acts in a way which seems contrary to his own interests, the caseworker will still strive to be objective, neither condemning nor condoning but respecting the right of individuals to hold to their own standards and beliefs.[421]

The SACS was entirely voluntary.[422] NAMH commented that:

The principle of freedom of choice is very important in social work, and an individual must be allowed to decide for himself whether he wishes to consult a social worker. There must be freedom to refuse as well as accept help, and the fullest benefit from social casework will only be obtained by a client who volunteers his co-operation...the success of the NAMH after-care scheme was bound up with the application of this principle.[423]

Just as with the pre-war and wartime experiments, these principles challenged the existing organization of institutional care and disrupted the medical model, which emphasized an internal disease process. For instance, in earlier psychiatric casework and, indeed, in prevailing psychiatric clinic work, the collection and analysis of a case history constituted part of diagnosis and was a preliminary to treatment. Theorists of the SACS argued that diagnosis and treatment couldn't be separated easily; the priority of the emotional relationship subsumed both. This idea was also expressed by some psychiatric social workers working in child guidance. A booklet published by the APSW in 1946 maintained that 'In actual practice diagnosis and treatment form part of a single process...'.[424]

Associated disruptions were expressed in child guidance. Here the emphasis on medical terminology entailed a shift away from terms like 'advice' and 'guidance' to those of 'patients' and 'treatment'. But, paradoxically, the apparent need for emotional security and interrogation of the quality of authority relationships undermined the straightforward designation of the patient. For John Bowlby, workers had come to recognize 'more and more clearly that the overt problem which is brought to the Clinic in the person of the child is not the real problem; the problem which as a rule we need to solve is the tension between all the different members of the family'.[425] Increasingly, workers spoke of illness having been 'projected' onto children.[426] Dugmore Hunter, consultant psychiatrist at the Tavistock, asked in 1955 'Is the child with his presenting symptoms, really the most ill member of the family, or has he been forced into illness by a mother and father who for some reason must avoid awareness of disturbance in themselves and so provoke illness in the child [?]'.[427] These ideas made it even into the normally conservative and anodyne promotional leaflets on the mental health services published by NAMH. A leaflet written in the 1950s by the psychiatrist Jack Kahn, for instance, described the presenting problem of the child sent to the clinic for treatment as a maladjustment funnelled into him or her by the group tensions of the family.[428]

Ratcliffe described SACS style work as 'relationship therapy'. He emphasized that the 'one common factor of all socially maladjusted people' was an inability to form 'adequate stable or satisfactory human relationships with others in their environment'.[429] In fact, it became a commonplace view among mental hygienists after World War II that a core unifying feature of *all* mental disorders was failure in sustaining relationships.[430] This helps explain why the SACS and its later developments worked with people who had largely been seen as beyond psychiatric casework help. A NAMH report on the wartime and immediate post-war SACS described 30 percent of the people receiving assistance as suffering from psychoses.[431] It suggested that this showed that 'social supervision of recovering and chronic psychotic patients is a valuable social therapeutic instrument'. It added that these people were widely regarded by social workers operating the scheme as suitable for care in the community, and that 'a diagnosis of insanity' need not now be considered 'tantamount to an order requiring incarceration'.[432] The author of this report was Kenneth Soddy. In fact, his booklet *Some Lessons of Wartime Psychiatry* recommended this very SACS scheme as an existing and successful model that could be developed to provide a 'Psychiatric Social Service'.[433] So Soddy combined this advocacy with promotion of a mental hygiene programme that targeted the working classes as the main locus of mental disorder, mental deficiency and maladjustment, and described such people as intellectually or psychologically 'inferior'. It's a repetition of the way in which the movement in general had advocated radical community therapy experiments at places like Hawkspur while retaining them within a hierarchical vision of the social order built around individual mental 'ability' and personality.

Soddy's report reckoned that civilian intake for the service was distributed evenly among the social classes except for 'a relative infrequency of the lowest income group'. This suggested, said Soddy, that 'there is a threshold of cultural standard below which such services make little impact' and that 'something else should be provided for the so-called "social problem" group.[434] Yet, Ratcliffe reckoned that the post-war work included people diagnosed as 'mentally deficient', 'psychotic' and 'neurotic', along with descriptions by referring agencies that included 'an idle scrounger' and 'the least worthy case' dealt with for some time.[435] And so, under the SACS, diagnostic categories appear unreliable indicators of mental and social capabilities. Additionally, its emphasis on emotional relationships as crucial to mental health and recovery, and its blurring of distinctions of status and authority, encouraged a disavowal of the kind of traditional task-oriented care associated with nursing or treatments administered in mechanical fashion. There was suspicion about the consequences of any authority that was insensitive to the emotional contents of actions and relationships, and unreflective about the impact of force and command on fragile therapeutic rapport. After the war NAMH remarked, in reference to the SACS, that a training in nursing was not a preparation for social casework and that 'a health approach' was not always an advantage in psychiatric social casework.[436]

Soddy maintained that 'many neurotics, psychotics and psychopaths are essentially lonely people whose contact with their social environment is poor' and that psychiatric social workers in the community should 'provide a medium of stable friendship for all who need it'. But, given the overall mental hygiene scheme that Soddy and others advocated, with its condescending and pejorative terms used to describe the targets of mental hygiene, what sort of friendship was this? What seems generally implied was a fusion of friendly and parental relations. As with the other experiments we've looked at, we have relationships that aim to be more communicatively open than those usual in families, while simultaneously they are intended to be deeper than those usual in daily life.

In 1947 suspicion about the therapeutic consequences of authority relations revealed itself in an APSW debate about a Ministry of Health circular recommending psychiatric social workers be used as Duly Authorising Officers (DAOs). These officers held designated authority to initiate compulsory removal to a mental hospital. The consensus in replies to an APSW circular was that adoption of the role would be a danger to 'friendly relationships' and that the 'authoritative approach was to be deplored'. But views at a later APSW meeting were mixed. A letter from Kenneth Soddy was read to the audience. It stated that there was no need for them to get involved in such work, and added:

> In no circumstances should a Psychiatric Social Worker employed on community care in the wider mental health field, or on the later stages of After-Care of Mental Hospital patients, be employed as an Authorised Officer in the area in which she is working. To be so employed would, I think, have a very serious effect on her relationship with her patients.

Several leading psychiatric social workers disagreed with him though. One argued that it was important to distinguish between the authoritative and the authoritarian approach. She maintained that the former only meant having legal backing and wasn't necessarily harmful. Another reckoned that doing the work might make psychiatric social workers more 'clear headed' if they 'saw every procedure through, however ugly and difficult'. But another maintained that possession of the powers of the DAO would likely ruin the confidence felt by patients and families in the psychiatric social worker. The meeting failed to come to a resolution.[437]

Spilling over into areas beyond the immediate therapeutic environment, this questioning of authority was sometimes directed at the roles of professional groups, including those within the mental hygiene movement. At a NAMH conference in 1946, the redoubtable psychologist and mental hygienist, Lucy Fildes, warned that

> ... in considering schemes for the well-being of children, I would suggest that those who are in authority should consider very seriously whether the provisions they are to make are for the glorification of the authority, or for the good of the child.[438]

In 1949 Mary Capes, the Medical Director of the Southampton Child Guidance Clinic (CGC), drew attention to the common struggles for power between medical and educational professionals at local authority level. She went on to urge that those who headed CGCs should question whether they were abusing their authority within their clinics.[439]

* * *

Wills and Q Camps weren't the only people developing community therapy. In 1949 F. G. Lenhoff opened Westhope Manor. The Westhope Society for Social Education aimed to found a community that as far as possible resembled 'a family group'. It endeavoured 'to inspire confidence instead of imposing authority and to secure influence and co-operation mainly by our example'.[440] Westhope catered for 'problem children and adolescents'.

The same year that the Manor opened, however, a psychiatric social worker from the Walthamstow Child Guidance Clinic wrote to the APSW to express her concerns. She'd visited Westhope twice and wrote 'I am becoming very concerned about the number of homes for maladjusted children which are being opened and are not all they seem to be'.[441] To Lennhoff she wrote a strong letter detailing issues that worried her. She wrote, 'The abandoning of your own family to a small community of disturbed children is laudable, but I cannot see that it is necessary or desirable if the group is also of subnormal intelligence, and it is surely very bad for [your daughter] to be constantly associating with such a group'.[442]

The APSW contacted NAMH who replied that they were about to make a visit, but added that 'We have absolutely no evidence that the Home is unsatisfactory ourselves, and in our dealings with Mr Lennhoff have found him particularly understanding of the problems of difficult children and adults'.[443] A letter followed from Miss Bavin, a psychiatric social worker at NAMH. She had spoken with NAMH's Medical Director, Alfred Torrie, and visited Westhope. She wrote that 'it is felt that Mr and Mrs Lennhoff are making a real contribution towards meeting the problem of difficult children in need of special care'. She emphasized that several of the issues that the Walthamstow psychiatric social worker had raised were now being dealt with, adding that Lennhoff 'is an artist in this kind of work and as such finds it difficult to cope with regulations'.

NAMH, in fact, supported several innovative workers who made their names developing community therapy approaches. These included Richard Balbernie, a psychologist who ran a school for maladjusted children considered educationally 'backward'.[444] His work was highly regarded by NAMH.[445] Balbernie went on to develop therapeutic community style work at the Cotswold Community, an approved school for maladjusted children.[446] Here, he developed methods employed by another well-known childcare innovator, Barbara Dockar-Drysdale, who herself had been supported by NAMH.[447]

* * *

The questioning of authority relations derived from a concentration on the relational requirements of emotional development even had an effect on ideas about the institutional system for 'mental deficiency' that the mental hygiene movement had promoted. In 1950 a psychologist called H. C. Gunzburg published a critical article in NAMH's journal, *Mental Health*. Basing his argument on the need to recognize and treat emotional maladjustment, Gunzburg wrote that institutionalization could negatively affect some inmates; a view at odds with mainstream mental hygiene thinking as we've been noting.

Let's unpack Gunzburg's argument a bit though. It bears all the hallmarks of the mental hygiene ideas that we've been tracing in this chapter, but for this very reason its apparently damning criticism of the mental deficiency system is so much less than it might seem.

Gunzburg began his article by noting that a visitor to a colony for mental defectives might wonder how so many different 'types' could possibly be classified under one heading. But it's obvious from the beginning that Gunzburg wasn't about to question the principal assumptions informing the designation of mental deficiency. He wrote, 'At one extreme we find the helpless, speechless and thoughtless idiot, who is evidently in need of constant assistance'.[448] Why thoughtless? He might've been a psychologist, but Gunzburg couldn't possibly have known this. His assumption betrays his prejudice, as does his later remark that the institution's organization was determined by the needs of the 'low-grade' inmates, and thus its attempts at training were based 'near the "cabbage-stage"'.[449] Meanwhile, at the other end of the variety of 'types' observable in an institution, was a 'well-mannered and normal looking individual' who expressed himself easily and skilfully and appeared 'to score highly above many uncouth labourers outside the colony'.[450] Gunzburg's description here rests on much the same ground as Burt and Gillespie's use of IQ measurements to inform their vision of a healthy hierarchy of function and fit in society. As we saw, they merged medical categories of 'mental deficiency' with the 'lowest' levels of employment categories. And running right through this, like a steam train everybody was trying not to notice, was a hierarchy of human worth.

With Gunzburg, this hierarchy of mind transposed to a hierarchy of the social order was still intact. It was just that, using the concept of emotional maladjustment, Gunzburg was going to train a few people so that they might find a place at the bottom rung of the ladder. These people were the 'high-grades' or 'feebleminded'. Gunzburg argued that 'it has increasingly been recognized in recent years that emotional maladjustment plays a great, if not decisive part in the feebleminded person's failure in society and that intellectual inferiority should be considered only an aggravating factor in many cases'.[451] There had, indeed, been some attempt made before the war to promote work with a selected group of these patients on the basis that they exhibited 'emotional instability' in addition to mental deficiency. But the numbers believed capable of 're-stabilization' and return to the community were small, and the methods entailed a pretty basic imposition of order and discipline in the institutional environment. Gunzburg's estimate of the numbers who might benefit was equally limited. It was limited to a small number of those who were adolescent. The reason he only saw adolescents as appropriate appears to be that they were still developing emotionally.

It was here that the order of the family, with mental hygienists' amended perception of its authority relations, came in. Gunzburg wanted to introduce a separate unit that combined a Training School with small-group family-style hostel care.[452] This was natural enough from a mental hygiene point of view. Having designated some people capable of emotional development and of potentially adequate intellectual ability (albeit at the most 'minimal level') the organizing principle and methodology of the Family became essential. The surrogate parents Gunzburg envisioned for his units would provide 'the feeling of security and trust which will make attachment to their new home possible'. The family environment would create an 'atmosphere where rules and a code of behaviour are felt and absorbed and not merely learnt and known'. Through these affective ties would grow 'that pattern of acceptable responses which will facilitate adjustment to society'.[453]

Just as the wartime promotion of these kinds of principles entailed a simultaneous critique of prevailing institutional care, so Gunzburg's introduction of them to the mental deficiency system did the same. Here, he pointed out, the institution continued to be organized as an 'autonomous and fundamentally stable community for patients whose mental condition is unalterable'. Isolated and separated from the real world, its organization was adapted only to 'intellectual subnormality'.[454] Disciplinary rules and routine were based on the needs of the 'low-grade' inmates. An 'impersonal handling' of the inmates by an all-male or all-female staff left no room for initiative. But, as we've noted, this wasn't a thoroughgoing criticism of the mental deficiency system. It merely argued that the institution wasn't suitable for those young adults considered to show signs of emotional maladjustment. For the other inmates—those people designated as 'idiots' and 'imbeciles', and those designated 'feebleminded' who were either older or whom Gunzburg thought couldn't benefit—it seems that the institutional order was to remain largely unquestioned. Gunzburg accepted that custodial care appeared necessary for the majority of patients. He also accepted that large institutions of over 1000 beds were defensible from an administrative point of view, as well as providing 'material advantages for health and entertainment of the patients'.[455] It was only for the small group of emotionally maladjusted adolescents that the institution could be appropriately described, in the negative sense, as 'a special type of prison' with an 'indeterminate sentence'. And it was only for these adolescents that the terminology of 'institutionalization' likewise took on a negative form. Gunzburg wrote, 'Whilst he may to all outward appearance become stabilized or rather "institutionalized" he really becomes more and more estranged from the life to which we wish to return him one day'.[456]

*   *   *

Sometimes they transfer their hatred of their father to someone or something else of whom that father reminds them, someone in authority. Thus they may, in a religious environment, become God-haters, in a capitalist society Communists, and I presume, in a communist society anarchists.[457]

It's a typically mental hygienist statement. In fact, it was written by David Wills in his 1941 book *The Hawkspur Experiment*. But his view was that the people he interpreted as having distorted relations with their fathers weren't really interested in Communism or anarchism as political ideologies to be pursued and embodied; once you'd spent any time with them you realized they were only interested in constant revolt. Meanwhile, Wills himself appears to have increasingly considered his work allied to that side of socialism most closely associated with anarchism. His 1941 book had likened Hawkspur to the social anarchist Ethel Mannin's fictional 'Longmeadow Camp' for children, described in her 1931 book *Rose and Sylvie*, and during the war Wills corresponded with the Freedom Press anarchist bookshop. His 1945 book on the Barns hostel experiment found confirmation for its approach to education in the anarchist Herbert Reed's *Education Through Art*.[458] In the same book Wills made clear his repudiation of the State Socialism represented by the USSR. Rehearsing his fundamental disagreement with the use of punishment to maintain an imposed system of discipline he remarked that 'We see the same practice in wider and more important spheres of human conduct'. One of the worst, he noted, was the USSR where 'for crimes against the state – against discipline – thousands (if I may dare to say so of our fellow-fighter for freedom) have been executed merely on suspicion'.[459]

*   *   *

On 15 April 1950 a gangly looking man stood up and spoke to a public meeting. What he said was clearly inflammatory—at least it would no doubt have seemed so to NAMH and the government. He warned his audience that some psychiatrists (especially, he believed, in the USA) wanted to lock up anybody who had crossed the authorities. He went on to say that very often the criteria for locking people up as mentally deficient was that they were considered to be 'obviously that type belonging to a low class, the scum of society – somebody you didn't like the look of'.[460] It seems the rabble-rouser had his effect. The meeting passed a resolution condemning what it saw as the tendency to extend mental deficiency to encompass those exhibiting a lack of 'social adaptation' rather than intelligence.[461]

The speaker was Brian Kirman, author of the letter to the *Lancet* in which he had suggested that members of the medical profession try to grapple with the idea that they should be the professional servants of a democracy. He was, in fact, a member of the Communist Party of Great Britain. But Kirman could also make every claim to know what he was talking about at the meeting he had addressed. He was a psychiatrist himself and he worked as the deputy Medical Superintendent of the Fountain Mental Deficiency Institution at Tooting.

Whether NAMH and the government liked it or not, the critique with which Kirman was associated was damning and influential. The meeting at which Kirman had spoken was part of a concerted campaign against the operation of the mental deficiency system founded under the 1913 Mental Deficiency Act. It was organized by the National Council for Civil Liberties (NCCL). The NCCL has been described as at this time little more than a front operation for the Communist Party and its controllers in the USSR. MI5 certainly reported to the government that this was the case, and there were clearly staunch members of the Party at its core, including outspoken apologists for the Stalinist USSR.[462] These facts led journalist and Trotskyist Paul Foot to remark ironically of these people's devotion to 'Stalin's Russia, in which there were no civil liberties of any description'.[463] But, that said, there were clearly many Trades Union socialists and non-Stalinist communists who were involved with or supported the NCCL campaign. In fact, it was inaugurated under the Secretary-ship of Elizabeth Allen, who was a veteran of the pre-war International Peace Campaign, but had also been an active member of the Women's Liberal Federation.

So Marxists and civil libertarians came together in criticism of the workings of the mental deficiency system. It's a curious combination in a sense. Civil libertarians view civil society as the realm of the free, rights-bearing individual formally equal in law, and released from old illegitimate authorities of privilege and status based on caste. Marx saw civil society as merely created by the relations of production; an expression of capitalism's economic base. It was a form of exploitation that had superseded earlier forms:

In the modern world, every individual participates *at the same time* in slavery and in social life. But the *slavery of civil society* is, *in appearance*, the greatest *liberty*, because it appears to be the realized *independence* of the individual... of his *own* liberty... when in reality it is nothing but the expression of his absolute enslavement and of the loss of his human nature. Here, *privilege* has been replaced by *right*.[464]

Ultimately, irreconcilable bedfellows then. But, as we'll see, the actions of Kirman and others suggest something a little more genuine than simply a cynical Stalinist attempt to use civil liberties as a lever to exploit weaknesses in bourgeois democracy.

The spur for the NCCL campaign apparently came from an elderly hospital chaplain and a retired accountant. Both of these men had separately discovered cases of children they believed to have been unjustly detained in mental deficiency institutions. They reported the cases to the NCCL who, after investigation, began a national campaign.[465] In the post-war years, with the new Labour government's sweeping social legislation, health and welfare was increasingly expressed in terms of rights. The NCCL was able to employ this terminology in order to challenge psychiatric policies. Support came mainly from socialist organizations. Independent trades unions, trades councils and the Socialist Medical Association helped to organize conferences up and down the country. But the Medical Practitioners' Union also helped.[466] Ex-patients and relatives or friends of patients in institutions were encouraged to attend and make their views heard.[467]

The NCCL campaign has been depicted as a straightforward attempt to expose the number of people being compulsorily detained who weren't actually mentally deficient.[468] This was certainly a strong element of the campaign. It was argued that many people certified as 'feebleminded' were, in fact, 'normal' and should never have been certified. But socialist organizations weren't only concerned about particular instances of clearly wrongful committal. They also condemned the extension of compulsory detention powers based on medical determinations of 'social defectiveness', often using the category 'moral defective' under the existing Act. The Socialist Medical Service investigated a number of cases. It declared itself determined to fight class distinctions and denounced certification based on 'social circumstance' rather than intelligence quotient.[469] It was in this context that Kirman had made his accusations. In early 1952, he reiterated his views in the more formal language required for the *Nursing Times*:

> Most ordinary people who have thought at all about mental defectives, including magistrates called upon to sign detention orders, expect the patient...to be obviously stupid or childish. It is surprising therefore to find, as sometimes happens, people labelled as mentally 'defective' who have passed difficult examinations, speak two languages fluently or who, on tests, have proved to possess an intelligence well above average. Although most of us do not regard such people as mentally defective, quite an influential body of psychiatrists and other health workers are prepared to support the inclusion of this group of cases as 'socially defective'.[470]

Such examples of patient's abilities don't appear to have been particularly rare. In 1945 a deputy County Medical Officer reported investigations into employment in the community of people previously diagnosed mentally deficient. He noted that some earned more than the welfare worker supervising them, while others drove tractors on farms and performed other farming tasks that were 'by no means monotonous'.[471] In 1958 a social worker working at one mental deficiency institution commented (uncritically) on the number of discharged patients who returned to pay visits at the weekend in their own cars.[472] Kirman continued his article by claiming that psychiatrists and magistrates might be too ready to judge cases according to their own moral standards with little understanding or sympathy for the class and surroundings of those they were dealing with.[473] The NCCL made similar criticisms about apparent class and moral prejudices.[474]

In fact, Kirman's boss—L.T. Hilliard, the Medical Director of the Fountain Institution—was also involved in the campaign.[475] In an article for the *British Medical Journal* he wrote that

> A scrutiny of the medical certificates which originally formed the basis for the detention of these patients makes one wonder if enough care was given in some cases to the evidence on which diagnosis of mental defect was based.[476]

These views were certainly at odds with the views of many prominent mental hygienists. We've already noted the views of NAMH's first Medical Director, Kenneth Soddy. Other mental hygienists expressed associated views. The psychiatrist David Stafford-Clark, a member of one of NAMH's standing committees throughout the 1950s, produced a book called *Psychiatry Today* for the popular Pelican series. In a section covering mental deficiency he wrote that idiots were

> ... in fact considerably less intelligent than domestic animals. Their habits are simple and unformed and their emotional responses crude in the extreme ... unlike imbeciles ... they may be neither happy nor unhappy in the accepted sense of these descriptions.[477]

Of imbeciles', he wrote, 'Allowed to roam about without care or supervision they may commit murder, rape, or arson'. And of mental deficiency in general he contended 'A high proportion of the ranks of prostitutes, vagrants, and petty recidivists are found on examination to suffer from a degree of mental defect'.[478] A review by R. F. Tredgold in NAMH's periodical *Mental Health* praised the book highly as done 'superlatively well' and 'never prejudiced'.[479]

Such, apparently authoritative, views permeated more general texts on health and hygiene.[480] Curran and Guttmann, for example, in the 1945 edition of their *Psychological Medicine: A Short Introduction to Psychiatry*, blithely associated people considered feebleminded with crime, prostitution and sexual disease. They claimed that many of the people that doctors found 'maddeningly incapable of giving a straight answer or a consistent history' were feebleminded. They added though that it was worth remembering that 'minor degrees of mental deficiency are fairly common in the lower strata of the community, and that such persons are often useful members of the population not easily replaced in the performance of dull and simple tasks'.[481]

The NCCL disputed this negative depiction of people diagnosed mentally deficient, particularly those labelled feeble-minded. Its critique challenged the authority of such diagnoses and questioned the nature of the 'treatment' and 'rehabilitation' predicated upon them. As we'll see shortly, its logic extended beyond a straightforward attempt to protect the liberty of wrongly detained subjects. But, for the moment, we should note that the NCCL's campaign certainly incurred the anger of many psychiatrists and mental hygienists. In an exchange of letters between the NCCL's Secretary and the M.P. Kenneth Robinson, a member of one of NAMH's standing committees (and later Minister for Health), the latter remarked that the mental deficiency institutions had long been a 'favourite windmill' of the NCCL and that the 'wild tilting' of its Secretary ran true to form. He continued by accusing the organization of 'deep prejudice' against the mental health system'.[482] In response to the NCCL's allegations NAMH sent out written enquiries to institutions. It sought information on the extent to which people on license were recalled against the wishes of their parents or guardians, and the number of occasions children labelled educationally subnormal were transferred directly to a mental deficiency institution on leaving residential special schools. NAMH pronounced itself reassured by the responses. A special meeting of its Mental Deficiency Sub-Committee, held to consider the NCCL's allegations that had been published in a booklet, *50,000 Outside the Law*, decided that, though reform of the administration and legislation on mental deficiency was needed, the NCCL's report had been 'limited and prejudiced' and that it had 'distorted' the true picture.[483] A subsequent letter to the MP W.S. Shepherd, giving NAMH's views added:

> The [NCCL] report deals only with grievances of certain individuals alleged to have been unjustly treated either in institutions or on license, and wholly ignores the immense benefits conferred on defectives as a group by the protection and training clauses of the Mental Deficiency Act of 1913 and subsequent Acts. No mention is made of the thousands of deficients happily provided for nor the thousands more whose parents are anxiously waiting their admission to an institution.[484]

It went on to say that the NCCL's emphasis on wrongful detention as the determining factor in mental deficiency legislation was completely wrong and unhelpful: 'In fact the Mental Deficiency problem today is conditioned by the need for increased institutional accommodation where deficients may be trained; the concept of "permanent detention" is a long outworn one'.

This contention was correct in as much as the idea that the mental deficiency institution should be the hub of a system of smaller hostels and homes through which some 'high-grade' people might be 'trained' and re-enter the community had been promoted before the war. But NAMH was surely expressing some denial here over the reality of a steadily increasing number of people living in underfunded and overcrowded mental deficiency hospitals. Indeed, its remark about 'the thousands of deficients happily provided for' might well be held up against the views of several of its Mental Deficiency Sub-Committee members regarding food allowances at mental deficiency hospitals. Discussing the impoverished financial allowance for food compared with other hospitals, including mental hospitals, they had asserted that 'many defectives were undiscriminating about food and did not complain so long as they had enough in bulk'.[485]

In a brief review of *50,000 Outside the Law*, Hilliard responded to criticisms such as NAMH's:

> Some critics of the pamphlet have referred to its emotional method of presentation and have said that it only deals with the alleged evils and thus gives an unbalanced picture of the present situation. But its avowed aim is to focus attention on what the Council calls 'one of the gravest social scandals of the twentieth century,' and they urge a public enquiry into the present state of affairs.

> Mental deficiency is a subject of which most people have little experience, and it is too often treated jestingly. The pamphlet quotes authentic cases and also some letters written by defectives which will give the reader some glimpse of the human background of this difficult problem.[486]

While Stafford-Clark's book received a rave review in NAMH's journal, another book written by a psychiatrist and aimed at a popular audience received the opposite. R. F. Tredgold, son of A. F. Tredgold, and presently editor of *Mental Health*, was none too impressed with Brian Kirman's book, *This Matter of Mind*.[487] In fact, so much was he unimpressed that he devoted the summer 1952 editorial to attacking and dismissing it. A champion fencer, Tredgold has been described in a biographical sketch as a '6 foot 4 inch figure, with an imposing head that looked borrowed from the Addams family'.[488] Large head or not, he was clearly intent on skewering Kirman.

According to Tredgold, Kirman's book was an object lesson in the dangers of failing to establish a 'standard of professional ethics'. Kirman had 'sacrificed the scientific method of investigation' in his attempt to 'prove a social thesis of which he is already firmly and irrevocably convinced'. 'That this thesis is of extreme left-wing Socialism', added Tredgold, 'is immaterial'.

But, of course, it wasn't immaterial, otherwise why mention it?[489] It's clear that Tredgold strongly disagreed with Kirman's political position. But rather than engage with it he simply dismissed it as self-evidently beyond the pale. As to the apparently bogus scientific elements, all Tredgold managed in criticism was the statement 'Some of the explanations he gives in support of his own thesis or against the theories of others, are irrelevant', and the associated claim that Kirman selected only evidence that supported his case. The reader was apparently supposed to take this on Tredgold's authority without him feeling it necessary to offer a few examples.

But one of the areas that Tredgold surely ought to have been interested in, and could have briefly engaged with, was the 'science' of intelligence and mental testing. In fact, Kirman's book was itself, in part, a critique of what he believed to be its unscientific claims. His discussion amounted to a sustained and cogent attack on the role of 'intelligence' as a fundamental in the hierarchical ordering of society—a role that, along with 'emotional maturity', the mental hygiene movement had notably given it.

What after all is 'intelligence'? Kirman attacked Cyril Burt, in particular, regarding the utility and scientific basis of mental tests and took apart Burt's well-publicized claims of a deterioration in national intelligence. He pointed out that, though everyone was familiar with the term 'intelligence', it was, in fact, difficult to define, still less measure.[490] Intelligence covered a great number of qualities, including 'alertness, powers of observation, energy, speed of reaction, interest, initiative'. In turn, these were related to qualities that Kirman considered to be mainly socially determined, 'such as patience, courage, steadfastness, perseverance, consistency, optimism, kindliness (to take only a few), which determine the social value of intelligence and the use to which it is put'. Added to this, he pointed out the obvious: that it was rarely the case that people consistently made 'logical and intelligent' responses to difficult situations. He remarked, 'It is a common experience that people who at times exhibit an extremely high level of intelligence, at other times behave in a stupid, childish, or irresponsible manner'.[491] But Kirman stressed that people weren't isolated individuals. We've seen that this was stressed by the mental hygiene movement. But Kirman made better use of it in this respect. He wrote:

> Fortunately for us, we are not dependent on our own resources and all live as members of a society. It follows that for the intelligence of an individual to be useful to the community he need not make a fully intelligent response to each situation which confronts him. It is useful if he makes only a small contribution.[492]

As a component of Kirman's Marxism this suggested a more genuinely cooperative and egalitarian idea of 'function and fit' for society to the hierarchy of heads envisaged under the banner of mental hygiene.

Tredgold didn't refer to any of this. Within the Communist Party, how-ever, there had been intense debate about measurements of intelligence that had begun in 1947.[493] Much of it seems to have been in relation to main-stream schooling selection for '11 plus'. Some, like the psychologist Monte Shapiro, maintained that mental tests were important measures of progress and attainment. Indeed, Cyril Burt had engaged with this debate in articles published by the *Daily Worker*. But, in 1949, the married couple Joan and Brian Simon had separately attacked not only the tests' usefulness, but also the 'bourgeois psychology' on which they were based.[494] Burt was singled out as the archetypal figure. Brian Simon later produced the influential book *Intelligence Testing and the Comprehensive School*. In Deborah Thom's view the attack on testing wasn't so much ideologically driven as the result of popu-lar reaction among workers in the field—in this case teachers.[495] The same seems true in Kirman's case, given his experience of the system for 'mental deficiency'.

Tredgold, however, preferred to highlight Kirman's Marxism, accusing him of taking 'an authoritarian attitude' and telling people what must be done. According to Tredgold the proper role for psychiatrists 'must be to guide the community into more mature and understanding attitudes...rather than to use the authority of our professional status to insist on the acceptance of a politically coloured and pre-conceived type of social structure'.[496] Tredgold emphasized that this proper role depended largely on dynamic psychol-ogy's ability to solve 'inter-personal, inter-national and social aberrations' of human behaviour. Given such a mental hygienist statement it isn't surpris-ing that *This Matter of Mind* upset him. Kirman had no truck with dynamic psychology—Freudian, Jungian or otherwise. The title of his book made the point. Mind was material. In his view the mind was a product of biology and environment. Aspects of the environment were human and relational in content of course. But Kirman didn't invoke a Hegelian dialectic to show their importance, as Bosanquet had done some decades earlier. On the con-trary, the likes of Bosanquet and the COS no doubt represented examples of that element of society which Kirman termed 'parasitic'. He wrote that at one time the ruling classes had flaunted their idleness as a mark of their riches. But more recently, under the pressure of social change they had taken to

> ...concealing their idleness by rapidly spinning round in a multitude of small vortices designed to create the impression that, far from being idle, they are the most overworked people in the world, and, further, that they are continually burdened by a weight of care as to the welfare of the lower classes.[497]

For Kirman this was just a manifestation of their shaken confidence, and it was paralleled by their donning a belief in liberal democracy to hide their insecurity. His argument amounts to a neat reversal of COS-style claims about the threat of 'the idle', if you remember the second chapter of our story.

Kirman certainly didn't construe those aspects of the environment that were human and relational in terms of some dialectic of emotional development and parental authority either. In fact, he highlighted the way in which 'unorthodox political activity' was psychologized, especially by the schools of psychoanalysis, in such a way that communist, socialist and union activists were portrayed as psychopathic in order to 'merge them with the mentally ill'. Thus, political revolt was made equivalent to the alienated revolt in class society of those who embarked on a 'private war against society' for their own ends.[498] Kirman described how in this representation both forms were portrayed as a 'revolt against the "father figure," a continuation into adult life of a childhood revolt against a tyrannical and overbearing parent'.[499]

Anyway, Kirman had his own dialectic courtesy of Marx, Engels and Lenin. In fact, this provided him with his own version of history as progressive stages of increasing rationality, of social relations as part of a social organism and of each mind containing the stages of human evolution within itself.[500] Indeed, coupled to his own professional education, this allowed him to talk in terms of the health of society, as well as the health of the mind. But his diagnosis, prognosis and treatment were quite different to the mental hygiene movement.

As a materialist Kirman acknowledged evolutionary aspects of mental contents. But this had significance only over the long haul of human and pre-human history. The brain held vast resources for human advancement and therefore there was no requirement for hereditarian ideas to play any important part in the understanding and treatment of mental disorders, or inform the organization of society. The physician's province should be in the area of the 'three per cent of people whose minds are inadequately formed by reason of some defect in the formation or function of the brain, and that other, say two percent, of people who at some time in their lives suffer from some serious mental illness'.[501]

Regarding 'established mental disorder' Kirman listed three measures. Help to find a more suitable job or better housing and friends, occupation therapy, and physical treatments including electrical convulsion treatment (ECT) and lobotomy. Admitting the latter were crude he said that they should be used with great discretion and only by experienced psychiatrists.[502] Minor mental disorders, and the large group of behaviours that came under the vague term 'psychopathy', were, for Kirman, less ailments of the individual than 'symptoms of a disorder in the state of society'. They needed to be dealt with by changing societal organization. He rather grudgingly accepted, however, that the more marked of these should continue to receive medical attention.

Kirman wrote, towards the end of his book, that 'An attempt has been made in the preceding pages to show that it is the ability to perform creative work which constitutes the essential difference between man and animals, and that therefore labour is fundamental to human dignity and is the basis of all consciousness, science, knowledge, and culture'.[503] But surely this posed problems regarding the human dignity of some people categorized mentally deficient? This seems especially so as Kirman described people labelled 'idiots' as almost totally helpless like a small child, wholly dependent, unable to contribute to society in any way, and often with only some of the 'more animal functions' operating satisfactorily.[504] This does seem to come perilously close to Gunzburg and Stafford-Clark's denigrating comments.

Kirman could also be accused of failing to take seriously enough mental distress that fell short of psychosis. He had little to offer other than a full-scale revolutionary change of society. The divorce of work from its product produced a warped culture that threatened the 'social organism'.[505] Capitalist society with its parasitic ruling class had a morbid psychology; it promoted greed and acquisitiveness. Kirman acknowledged that in previous times people like 'William Morris, the Utopians, and the "Merry Englanders"' had recognized this, but they had only sought a reversion (and thus regression) to some previous age. They hadn't recognized the materialist historical dialectic.[506]

In fact, though, despite his radically different 'diagnosis' and 'cure', Kirman's description of the expressions of mental morbidity in society wasn't so far away from the mental hygiene movement's. Cinema (especially American films, of which, according to Kirman, 90% were based on sadism, greed, violence and pornography) was largely degenerative. So was contemporary art ('if he paints a sunset it looks like a fried egg'). Other expressions of social morbidity were the proliferation of gaming-houses, drug-pedlars and brothels.[507]

The mental hygiene movement had also been concerned about the effects of mass entertainment, such as the cinema and gambling. And, Kirman's assailant, R. F. Tredgold, had also expressed concerns about the effects of capitalism on the working classes. In his case though, Tredgold was specifically concerned about mass production. He worried about its connection to 'mass produced or spoon-fed leisure – listening to music rather than playing, watching games rather than taking part – and the growth of sensational reading or film going'.[508] He was, in effect, concerned about the moral effects of rationalizing and totalizing power. For him, mass production in industry didn't encourage initiative. But, in typical mental hygienist style, he worried that reformers might end up discouraging initiative by providing constructive forms of leisure. 'The masses' were all too vulnerable to becoming passive it seems.

Kirman, however, emphasized that capitalism couldn't provide a reliable environment for 'healthy emotional satisfaction'. It had the inherent and crazy kind of logic that saw efficiency as building 'bigger and better bombs'. He warned that without organized working class activity and resistance society was 'likely to take on to an increasing extent the appearance of a large mental hospital'.[509]

*   *   *

In fact, there was another level to the 1950s attack on the mental deficiency system. It wasn't just that communists and civil libertarians were attacking it, but that the mental hygiene movement had a nest of them in their midst. On 16 March 1950 a member of NAMH's mental deficiency subcommittee had reported to it that the NCCL was arranging a conference to discuss the operation of the Mental Deficiency and Lunacy Acts. This person added that the NCCL would be glad to discuss issues with NAMH representatives and went on to say that the Council did good work regarding public awareness of infringements of the liberty of the subject, including in the sphere of mental health. The member was, in fact, L. T. Hilliard. He continued by telling them that he was a member of the NCCL's Executive Committee. There is no record of the response of the other members other than that it was afterwards agreed to recommend that observers attend the NCCL's proposed conference and arrange for NAMH and NCCL representatives to meet.[510] It was out of this subcommittee meeting that NAMH had decided to send written enquiries to institutions in order to check some of the NCCL's allegations. We've seen that NAMH believed the responses largely refuted the NCCL's claims. These findings were reported to a later meeting of NAMH's Mental Deficiency Sub-Committee. Hilliard responded by saying that the NCCL had been instrumental in getting a number of patients in mental deficiency institutions released, and that it was becoming evident that the whole procedure of licensing and discharge needed investigation. Opposing views found common ground, however, in the ensuing discussion. Subcommittee members agreed that beds would be freed up in overcrowded institutions if there were residential hostels and more generous licensing. It was decided that NAMH's committee on legislation should consider possible revisions to the Mental Deficiency Acts' procedures regarding certification, licencing, discharge and appeal. This committee had already been created by NAMH because of the administrative and legal implications of the new National Health Service. Hilliard was its chairman.

As we've seen, a special meeting of NAMH's Mental Deficiency Sub-Committee concluded that the NCCL's allegations published in *50,000 Outside the Law* were 'limited and prejudiced'. It's unclear whether Hilliard attended the meeting but, either way, this became NAMH's official response. Despite it, Hilliard continued to have influence with NAMH. He remained as chairman of the Mental Deficiency Legislation Sub-Committee, remained also a mainstay member of the Mental Deficiency Sub-Committee, became a member of the Association's Honorary Medical Panel in 1953 and, on invitation, became a member of the Mental Deficiency Training Sub-Committee when it was created in 1954.[511] Brian Kirman also had contacts with NAMH. He spoke at NAMH conferences, co-wrote a book on 'mentally deficient' children for the organization and also had his *British Medical Journal* article on 'The Backward Child' released as an information pamphlet by NAMH.[512]

Much of this interchange can be attributed to the fact that since the end of the war Hilliard and Kirman had developed the Fountain hospital as a centre for multidisciplinary research and training.[513] The backdrop to this endeavour was the fact that through this same period the mental deficiency system in general was becoming an ever-growing backwater. The number of adults and children certified and detained continued to grow. NAMH was well aware of the serious issues of insufficient funding, overcrowding, poor staff ratios and poor staff training, as well as what it considered to be public apathy.[514] But it didn't publicize these deficiencies, still less seriously question the need for the existing mental deficiency system. It preferred, instead, to work behind the scenes to improve matters. This gave particular significance to work being carried out at the Fountain. In fact, NAMH themselves became directly involved in it. In the early 1950s a NAMH advisory service carried out experiments in 'training' women and children at the Fountain.[515] There was also some involvement between NAMH and two psychologists associated with the Fountain whose research would soon prove influential. These were Jack Tizard and Neil O'Connor.[516] By the end of 1957 Tizard had joined NAMH's Mental Deficiency Training Sub-Committee, remaining on it until 1960.[517]

* * *

There are metal bars in front and behind. They're the bars of a cot. In between them are sheets that look crisp and white. They've surely been washed many times. The hair on the top of the little girl's head has been pinned into one big curl. It runs along her head like a funnel: a very cute funnel. She lies half in the sheets, propping herself up with one arm. At the end of the other arm, peeping out of the woolly cardigan she's wearing, is a tiny hand gripping a comb. It's in the process of combing what looks like a toy bunny rabbit. From outside the bars a nurse in starched hat and puckered sleeves is gazing dotingly at her. The little girl looks back with compelling eyes and the sweetest smile you'll ever see. It's life, frozen in time in a photo: a photo in a technical, academic text. Beneath it runs the caption: 'A baby with Mongolism and congenital abnormality of the heart. Certified aged twelve months, died at eighteen months'.

Pictures speak louder than words. And this one has been placed slap bang in the middle of the very first section of a textbook called *Mental Deficiency*, published in 1957. It's the part that covers the historical and legal context, and which, in the appropriate formal prose, sets out this context in such a way as to expose the mental deficiency system's punitive and coercive fundamentals. The authors were Hilliard and Kirman, with the help of four contributors, including Tizard and O'Connor.[518]

Medical texts weren't noted for placing these kinds of emotive pictures between their covers, least of all those on 'mental deficiency'. *Tredgold's Mental Deficiency*, which had for decades been considered the authoritative text on the subject, contained only 'technical' photos. Even 13 years later in the 1970 edition edited by Soddy and R. F. Tredgold, the only informal picture included was one placed before the preface showing the latter's father standing among some women deemed 'feebleminded' and looking for all the world the great white male explorer of the Heart of Darkness.[519]

* * *

Tizard had no experience of 'mental deficiency' before he embarked on psychological research based at the recently established Medical Research Council Social Psychiatry Unit. He'd been recruited and directed to this area of work by the psychiatrist Aubrey Lewis. Lewis was a longstanding member of NAMH and had also been the President of the APSW from 1943–1944.[520] Tizard had, in fact, originally been attracted to studying psychology partly by Cyril Burt's application of science to human development and the elegant prose with which he relayed it. But, throughout his life, Tizard remained a 'passionate egalitarian' and retained 'an equally strong objection to authoritarianism'. While serving in the New Zealand Field Ambulance Unit during the war he had courted trouble through his habit of not saluting officers. He'd also refused a commission because he aligned himself with soldiers in the ranks. After the war Tizard had attempted to study industrial history at Oxford. Though admiring his tutor, the socialist G. D. H. Cole, he hated Oxford's elitism and couldn't handle what he saw as the arrogance of its undergraduates. He became a member of the Communist Party of Great Britain for a few years, believing at the time that they were 'the only party really trying to change the class basis of English society'. On Sunday afternoons during these same years, he could often be found on Clapham Common haranguing an audience from a soap box on behalf of the Ex-Serviceman's Movement for Peace (one of the forerunners of the Campaign for Nuclear Disarmament). Unfortunately, his audience very often only consisted of his wife (the psychologist Barbara Tizard), their two babies in a pram and few old-age pensioners.[521]

But Tizard found a wider audience with his work in the area of mental deficiency. And this work ultimately did a great deal towards combating the authoritarianism and gross inequalities experienced by people who had to endure that label. His work helped directly, through various research programmes, and also indirectly, as Tizard gave a substantial amount of advice to the NCCL.

During its campaign the NCCL was able to use his and Neil O'Connor's research, which appeared to show that the IQ of patients classified as feebleminded had been underestimated by earlier and less sophisticated tests. Tizard and O'Connor judged the average IQ of people detained as feeble-minded at more than 70 percent. This enabled the bulk of these people to be re-termed 'educationally backward' rather than mentally 'arrested'.[522] As a result, the NCCL was able to argue that if these people did require help it should be more in keeping with their difficulties. The NCCL also made use of Tizard and O'Connor's research into the traditional association of mental deficiency with crime. Its General Secretary pointed out in the press that their evidence showed the vast majority had no record of this or delinquency.[523] Tizard repeated these and other criticisms in a speech on 'Adult Defectives and their Employment' given at a NAMH conference in 1953.[524] Here he strongly questioned the traditional medical view of mental deficiency as relatively easily categorized and diagnosed. Tizard pointed out that the results of prevalence surveys of children and adults in the community showed that 'many individuals who, as children, were or would have been found to be mentally deficient, in later life became useful and well adjusted citizens who do not require special attention or supervision'.[525] He also cited evidence from a survey of 12,000 patients in mental deficiency institutions, which he and his colleagues had carried out. It showed, he said, that the 'great majority of patients appeared to be inoffensive, docile people who constituted no danger to society'.[526] Very few of those surveyed could be considered violent or dangerous. Only just over five percent of male patients had any history of indecent assault or exposure. And, 'contrary to general opinion', less than four percent of female patients had ever had venereal disease or been pregnant.[527] This last conclusion wasn't endorsed by some psychiatrists. One, responding to a published letter by the Secretary of the NCCL citing this research, suggested that the reason that the incidence of venereal disease was so low was because institutionalization had prevented feebleminded 'girls' having the 'opportunity'.[528] Thus, this psychiatrist was, in effect, arguing for preventive detention. Presumably, he didn't believe that this recourse should apply to the rest of the female population. Indeed, the fact that some detentions did amount to preventive detention in general appears to have gone largely unacknowledged by psychiatric staff. For example, a study of admissions to one hospital in the mid-1950s recorded without comment that 10 of its sample of 100 had been certified because social workers had failed to find employment for these people and they 'feared mischief if unemployment continued'.[529]

At the same time the NCCL highlighted the emotional effects on patients of the present system in order to show the detrimental effects of licensing and treatment. For example, it criticized prolonged indefinite licensing and the strict prohibition on forming attachments with anybody of the opposite sex, pointing out that these restrictions weren't just 'incredibly cruel' and humiliating, but that they placed an unfair burden on people already burdened with having to prove their worth and 'efficiency' in the community.[530] As the NCCL pointed out:

> Parliament has never knowingly approved the principle that association with the opposite sex must wait upon the attainment of a certain mental level, and the 200,000 or so mental defectives who have never been found to be in need of care and protection naturally suffer no such disability.[531]

In fact, several of these issues were raised and debated at the NAMH Mental Deficiency Sub-Committee. In 1952 the National Council of Women wrote to NAMH asking it to support a resolution on mental deficiency passed at its recent conference. This included a clause urging that all releases from an institution should be on 'extended licence with no decertification'. NAMH's subcommittee replied that if this meant, as it seemed to, that no people diagnosed as mentally defective should ever be discharged from order, then it couldn't give its support. In the same year the committee debated the issue of women having been certified to an institution because they had become pregnant and the related issue of these mothers being separated from their babies after birth.[532] Hilliard made clear his strong opposition to women being certified simply because on becoming pregnant they had been found to be 'feebleminded', as well as the subsequent separation of mother and baby. Most medical superintendents appear to have separated mothers from their babies soon after admission, but the trend of opinion at a conference arranged by NAMH in 1952 was against this.[533] The following year a social worker wrote to the subcommittee to complain about the special licensing precaution against forming attachments with the opposite sex. She wrote that this was 'unrealistic, frustrating and a hinderance to the process of adaptation'. The committee agreed and sought a redraft of the clause.[534]

Perhaps the most poignant comment in the NCCL's *50,000 Against the Law* was this: 'It should be emphasised that there is never a hearing for a mental defective. A charge, by itself, is sufficient'.[535] With this comment the NCCL summed up this professional power and authority, buttressed by coercive legislation, which asserted a domination that could not be questioned by those under its control. The NCCL relayed instances of arbitrary censorship of letters, revocation of licence without explanation and the denial of evidence of 'social competence', which included assurances from employers, trade unions and even confirmation of acceptance as a National Coal Board trainee.[536]

As we saw earlier, the growing power and authority of the state was a notable target of criticism for some post-war mental hygienists. They expressed concerns about centralized power and authority cascading down through society (despite its legitimization through popular mandate) in totalized fashion performing a detailed ordering and inspection of people's lives. The image was of an authority whose dominating power stultified individuality and freedom through an all-consuming rationalized power that paid no regard to emotional experience and its relationship to development. The similarities with the mental deficiency system and its hospitals seem clear enough. However, they lacked one crucial component, which meant that the mental hygiene movement didn't draw the comparisons. It's a component we've been highlighting. You could trust 'the primitive', but only if it had the capacity to develop into 'the civilized'. For those among the state's mental hygienist critics, the key issue was that adequate moralization and mental health rested on a concept of emotional maturity which was understood to be founded in the family unit. Blunt state power was unable to provide the intimate personal relations that tailored parental authority to the intellectual and emotional stages of development that each person needed to pass through in order to attain the mental maturity equated with mental health and adequate citizenship. The state needed to recognize the ultimate authority of the Family as organizing principle and methodology through which individual growth and maturity could be achieved. This was the underlying reason why the system for mental deficiency existed and why mental hygienists hadn't publicly criticized and challenged it. Through 'the primitive', through 'madness', through childhood, emerged the 'the mature individual'. But it didn't emerge through mental deficiency; this was already preconceived in terms of a permanent childhood. So the conceptual position from which state power and authority had come under criticism was, in fact, the very reason why the authoritarian system for mental deficiency hadn't.

For the psychodynamic understanding of mental growth and development that had formed the vanguard of the mental hygiene movement ever since its institutional establishment between the wars, mental deficiency represented the antithesis. So, in retrospect, maybe it isn't surprising that the psychiatrists and psychologists who were involved in the NCCL campaign worked in the mental deficiency system where any direct application of psychotherapeutic approaches was virtually non-existent.

But the 'permissive' therapeutic ethos disseminated by these psychodynamic approaches nevertheless reinforced the image of the mental deficiency system as out of date and custodial. Despite the psychotherapeutic prejudice against people designated mentally deficient the interactions in NAMH's Mental Deficiency Sub-Committee reveal that, under pressure, the mental hygiene movement could shift its position somewhat towards this more 'permissive' pole of its theorizing. The shift was inevitably ambiguous, but one of the consequences was that the descriptions of progressive history that mental hygienists had used as an authoritative foundation for their expertise in defining individual and societal health and citizenship became weakened. This would ultimately contribute to unravelling the logic on which the mental hygiene movement had founded its authority.

7
Alienation Revisited

In 1956 moral treatment announced its comeback in the asylum. Or rather T. P. Rees, the Medical Superintendent of Warlingham Park Hospital, announced it. In a presidential address delivered to the Royal Medico-Psychological Association (RMPA) he maintained that, although there had been real advances in scientific treatment, probably the most important change from the patients' point of view had been the return of moral treatment to the mental hospital.[537] For Rees, the modern progressive mental hospital had returned to the view of the insane epitomized by moral treatment. Patients were 'normal people who had lost their reason as a result of having been exposed to severe psychological and social stresses'. What mattered was the creation of the right 'atmosphere' for their care and treatment.

But this wasn't really a reappearance of moral treatment through its rediscovery so much as a continuation of the trajectory that we've been tracing. Given our story so far it isn't surprising that, having once more popped up in the asylum, this moral treatment bore all the hallmarks of its extended excursion into the community. In fact, Rees had long had connections with the mental hygiene movement.[538] And in the year that followed his presidential address to the RMPA, the National Association for Mental Health (NAMH) appointed him a member of its medical panel.[539] The ideas he expressed in his speech were clearly in keeping with the mental hygiene project and its emphasis on emotional relationships.

In fact, Rees noted in his speech that 'In recent years great attention has been paid to the importance of the emotional relationships existing between the members of any working group', and he maintained that this was also of the utmost importance for the staff of a mental hospital.[540] Rees had co-authored an influential World Health Organization (WHO) report on mental health in 1953.[541] This had also emphasized the crucial importance of the 'atmosphere' of the mental hospital. The hospital was a community constituted of emotional relationships, and it meant that the proper conceptualization of the mental hospital was as a 'therapeutic community'. Rees believed that:

Patients come to mental hospitals in order to learn how to live with other people, and to do that successfully they can reasonably be expected to contribute something to the welfare of the community in which they live... The role of the patient as an active member of the hospital team, promoting his own recovery through his contribution to the work of the hospital as a whole, brings us to the concept of the mental hospital as a therapeutic community, as an instrument of treatment in its own right. We, as doctors, are apt to flatter ourselves by attaching undue importance to specific methods of medical treatment. From the patient's point of view it is the total picture that counts, it is not the daily, weekly, monthly or six monthly hour he spends with his doctor that matters so much as what happens to him in between these periods.[542]

So, on this view, the therapeutic community is equivalent to moral treatment. It's a straight-forward equation that, as we've been seeing, can't be sustained. But, that aside, let's see how Rees' claim of moral treatment's rebirth follows the general trajectory that we've seen 'the family dialectic' take in our story.

We've seen that the nineteenth-century redeployment of moral treatment in the wider community made central the idea of the individual as a historical emergent. This was tied to the idea of human progress and the emergence of civilization. Originally described in terms of reason and rationality, the notion of emotional development was made a crucial constituent by the later mental hygiene movement. In the process, 'the primitive', 'madness' and childhood became equated. The WHO report made explicit reference to this:

In their gradual return to social effectiveness, patients often seem to need to recapitulate, not only the development of the interests and activities of the human being from childhood to adult life, but also the development of the human race itself. The group activities must therefore cover the scale from the archaic and primitive to the cultural and technical. In the demand which these group activities make upon the patient they must provide for the wide range of social response on the patient's part, ranging from a dependent and infantile attitude to one of initiative, responsibility, and self-sufficiency.[543]

Rees' image of the modern mental hospital amounted to a critique of mental welfare provision in post-war society and an attempt at promoting its renewal. He cast his image against two others that were common, those of a prison and a modern general hospital. For Rees the mental hospital couldn't become properly therapeutic until it got away from both. Here he was repeating the twin critiques of institutional care and the prevailing medical model in general medicine that we've seen emerge at sites of mental hygiene activity. Rees thought the mental nurse, in particular, had been caught between the role of a gaoler, manhandling patients and locking doors, and a general nurse, tucking patients into bed and keeping things clean and orderly. But neither role embodied what was properly therapeutic about mental treatment. The mental hospital should aim to create an environment in which patients could learn to re-relate. This view informed Rees' claim that 'the condition of the patients in mental hospitals is often the result of the conditions under which they are treated, rather than the symptoms of a disease process'.[544] He thus also emphasized a view that has become a classic of therapeutic community theorizing:

> When confronted with a disturbed patient, whether in the ward or in the home, we should ask ourselves not only what is wrong with the patient, but also what is wrong with the ward or the home. It is only too easy to deal with the patient rather than with the total situation, and sometimes even the wrong patient at that.[545]

But, actually, as we've seen, this type of thinking also informed the mental hygiene movement more widely. And though Rees challenged the traditional and entrenched hierarchical roles of the mental hospital, Warlingham Park's social organization retained significant authority relations of status and function. In fact, this structure echoed the traditional mental hygiene conceptualization of the Family as an organizing principle of authority over one's mental states and over the social order. Staff from the hospital had presented a paper at the 1948 *International Congress on Mental Hygiene*, which had been called 'An Examination of the Psychological Basis of Family Life and its Influence on Mental Health'. This modelled the treatment of mental patients on the family as this was considered the fundamental unit of society. One of the presenters, the psychiatrist R. A. Sandison, elsewhere described the social order of Warlingham as 'the authority of the hospital'. This authority emanated from the Medical Superintendent, through the medical staff, Matron, Chief Male Nurse and nursing staff, and imposed 'an ordered life and discipline on the patient'. Sandison considered this a valuable aspect of the patient's 'education':

> To get on with authority means, psychologically, to come to terms with the father. As the father image is one of the greatest of the primordial images, we have achieved a great deal if we can help a patient come to terms with it. It means that the patient can leave hospital with a better relationship possible towards his employing authority; and he becomes a better father himself in his own home. The woman, likewise, becomes

better adjusted towards what is admittedly a man's world. Likewise, a more subtle education in relation to the mother figure is secured through the female nursing staff, emphasising the need to have female nurses in both male and female wards.[546]

Similarly to the other critiques of institutional care associated with the mental hygiene movement Rees described how institutionalization was destructive of therapy and mental health. But he attributed this largely to the hospital being too comfortable and creating a kind of dependency. It's a view that expresses an apparent blindness to the deeply unpleasant experience of enforced and regimented living in drab and overcrowded surroundings with the ever present possibility of physical and emotional abuse perpetrated by those in a position of authority over a patient. At Warlingham Park, as well as some other hospitals promoting this 'therapeutic community' approach, patients were placed in a hierarchy of grades according to their perceived ability for social and emotional interaction. This included 'habit training' incontinent patients through rigid discipline, and tables of promotion and demotion.[547] Despite Rees' apparent aims, this emphasis on teaching the patient new ways of reacting could easily encourage the continuation of an attitude of moral guidance blind to its own prejudices. Describing the employment of Warlingham nurses in the after-care of discharged patients, for instance, T.P. Rees relayed the following 'success' story:

> I can well remember one man telling me his wife was no better and would have to come back into hospital. I had a suspicion that he had his eye on another woman and told him his wife was perfectly all right. Next week he came back and said she was dreadful at home, she would not do a thing in the house. So I talked to the ward sister who was working outside the hospital and she said 'That woman was on the ward with me; she had a leucotomy [lobotomy] and helped look after the ward kitchen; she was magnificent and kept the kitchen spotlessly'. When the ward sister went to the house about 9.30 a.m. next day she found nothing had been done since the husband had left; there were half-used milk bottles on the table and the beds were not made. So she sat down as a good mental nurse does and told the patient to get on with the work, which of course she did. After that the ward sister looked in about once a week; the patient never knew when she was coming. There were no more complaints from the husband![548]

Even so, Rees' concept of the therapeutic community at Warlingham Park was both a critique of prevailing institutional care and an attempt at its renewal. Others also developed this critique, building on both the themes of institutionalization and the general idea of a therapeutic community. The first term became commonly used to denote the perceived detrimental effects of prevailing mental hospital care, and the second became for a time a symbol of the possibilities for a renewal of therapeutic function. But other developers increasingly went further in their critique and attempted renewal.

One was the deputy superintendent of Claybury Hospital, Denis V. Martin. In 1955 he published an influential article on 'institutionalisation' in the mental hospital. He noted that phrases such as 'well institutionalised' implied being well-behaved, giving no trouble and ceasing to question one's position as a patient. Echoing Rees, Martin maintained that this process was detrimental to the therapeutic aims of the hospital and that it couldn't simply be attributed to the end result of an internal mental disease as it was so common across diagnostic classes.[549]

Martin's initial attempts to understand and combat institutionalization were made in the early 1950s through experiments carried out partly in association with Warlingham Park hospital.[550] Like Rees, he cited the unchallenging and relatively comfortable hospital routine as a cause. But he also emphasized the detrimental effects of the organization of power and authority in the mental hospital. Martin argued that institutionalization was produced through two forces:

> ... those that tend in many minor or subtle ways to relieve the patient of all responsibility for himself, so that he ceases to be aware of the need to tackle his problems seriously, and secondly, the authoritarian basis of staff and staff-patient relationships which, however benign, requires an attitude of submission on the part of the patient which readily leads to a loss of initiative, and to institutionalisation.[551]

The aim, for the nurse in Martin's experiments, was to reduce the usual authoritarian role to a minimum. Control and any issuing of directions was avoided in the interests of building up a relationship based on friendship, acceptance and understanding.[552] Martin reported that 'our experience amply confirms that where the role of the staff is authoritarian, based upon a relationship of submission to rule and the suppression or punishment of "bad" feeling, a truly therapeutic relationship is impossible'.[553]

Martin soon went on to introduce a therapeutic community-style approach at Claybury Hospital that was based on this understanding. Meanwhile, others were moving towards similar approaches. D.H. Clark, the Superintendent of Fulbourn Mental Hospital near Cambridge, had been enthused by the WHO report on 'The Community Mental Hospital', which Rees had co-authored. He had been moving towards similar ideas about locks and keys being largely unnecessary, and the hospital day needing to be characterized by organized work and activity, with the Superintendent ensuring an atmosphere of good relations.[554] Through the 1950s and 1960s he developed a close association with NAMH.[555]

In a retrospective account Clark admitted, however, that originally he and the staff saw themselves as the active initiators and the patients as passive. He added that:

> I had hitherto accepted the prevailing medical view of patients as pathetic beings, only kept from recovery by the failure of their illness to respond to medical treatment or their willful inability to do what doctors prescribed for them.[556]

Gradually, through years of reform and experiment at the hospital, Clark and his colleagues came to drastically revise these views. From the later 1950s forms of self-government were increasingly introduced into the hospital. Clark also came to criticize the therapeutic community model that had been advocated by Rees and the WHO because it didn't institute self-government properly and left the emphasis 'still on treatment flowing from the doctor downward'.[557]

The psychiatrist, Russell Barton, also had a notable influence with his 1959 booklet *Institutional Neurosis*.[558] This comprised a series of lectures he had given to nurses. Barton was a prominent member of NAMH through the 1960s, serving on its Executive Committee and Council of Management.[559] He claimed that long-term mental patients acquired an additional mental disorder caused by the institution itself. Barton described this as a disease process, but it was constituted of external relations—the behaviour of doctors and nurses, and the social order of the mental hospital.

These critiques of relationships of authority in the mental hospital weren't always fully endorsed within the mental hygiene movement. For example, Barton's *Institutional Neurosis* was praised by R. F. Tredgold in a review for NAMH's journal *Mental Health*, yet the processes of institutionalization Barton had detailed were reduced to 'abuses and shortcomings' occasionally ignored by staff and the remedy to reversing the gross under-staffing in hospitals.[560]

But, as we've seen, critiques of institutionalization and experiments under the banner of 'self-government' and 'therapeutic community' had been both embraced and promoted by the mental hygiene movement. These critiques run as an important thread through most of its history. The mental hygiene movement had been able to contain such seemingly radical approaches within its overall vision of a meritocratic social order founded on mental 'ability' and emotional 'maturity'. For the likes of Wills, though, these experiments were demonstrations of social organization and human relationships that had applicability to society as a whole. They were radical libertarian and egalitarian approaches that challenged the prevailing social order and its authority structures. They made psychological health and social justice two sides of the same coin. Earlier we noted the affinity of the work at Hawkspur with R. H. Tawney's 'qualitative' socialism. But we also noted that, during the 1930s, the emphases of 'qualitative' socialism were largely eclipsed by technological and Keynesian Socialist approaches. The 1950s, however, held possibilities for the therapeutic community to become connected with wider elements of social critique. Was the dialectic of the family finally bringing forth its fruit?

* * *

According to Stuart Hall, 'The "first" New Left was born in "1956"'. This is the same year that moral treatment was apparently re-born in the asylum. But almost inevitably for an intellectual, Hall's 1956 turns out to be 'a conjuncture (not just a year)'. That conjuncture was made up, on one side, with the Soviet Union's suppression of the Hungarian revolution and, on the other, with the joint British and French invasion of the Suez Canal. The first one marked a final disillusion for many British Communists with Stalinism and so-called State Socialism in the USSR; the other bluntly displayed the continuing imperialism and exploitation retained under Social Democracy. In a political environment dominated by the 'Cold War', the New Left emerged as a challenge to both.

What's the relationship between the New Left and the trajectory of moral treatment that this story has been tracing? The key word for us can be 'alienation'. At the start of our tale we noted that the original moral therapists had considered madness to be a failure of authority over one's mental states. It was a failure of self-government and a form of mental alienation. They attempted to treat it through the organizing principle of the Family. But we've traced Foucault's 'dialectic of the family' further through time. There had been a long existing analogy of authority relations in the family with those of the political order. Now psychologized, these were re-imported into the wider community. We traced this, first through the Charity Organisation Society (COS), then through the mental hygiene movement. For these reformers, faith in the march of reason and progress was tempered by concerns about an apparently accompanying social and moral estrangement.

Both the COS and the mental hygiene movement saw state power and authority as something too distant and blunt to be able to adequately moralize individuals. This view was linked to concerns about popular power leading, paradoxically, to a reinforcement of centralized state power. Through this the totalizing and rationalizing elements of power—its extension to the whole of people's life and being, and its relentless measuring, ordering and systemizing—would undermine the moralizing role of the Family. For mental hygienists the state must accept the Family's ultimate authority as the organizing principle through which individual growth and maturity could be achieved and a healthy developing society maintained.

But, as we've been seeing, the emergence of a psychotherapeutic emphasis on tailoring familial personal authority relations with perceived needs for emotional security and self-expression reinterpreted this organizing principle. Social and moral estrangement became no longer a refusal of authority, but a result of subjection to inappropriate authority. At radical experiments such as Hawkspur these elements became interpreted through a social, as well as psychological, understanding of 'self-government'. Within these therapeutic sites the Family changed. It remained an organizing principle for the self and the social order. But it partially inverted itself. Its moral order retained the close relations of personal intimacy and emotional depth, along with the image of the individual as inherently relational. But the Family's hierarchical authority, its presumption of an easy distinction between reason and unreason, and of the need for rigid orders of status, was undermined. Egalitarian relations were the common denominator of therapeutic self-government. A consequence was that, within such experiments, concerns about centralized power, along with its rationalizing and totalizing manifestations, were transformed. Popular power and authority became the means to offset centralized power and authority, while offering a measure through which the rationalizing and totalizing aspects of power might be weighed.

And so we return to the sociologist Robert Nisbet. As we saw in Chapter 2, Nisbet argued in the mid-1960s that although the *word* alienation took hold in sociological thought (and we may add, far beyond) through post-Stalinist Marxism in the 1950s, much of its *content* derived from a more longstanding and non-Marxist current of sociological and political thought. The New Left, in particular, drew together these Marxist and more longstanding understandings of alienation. It emerged in opposition to authoritarian centralized power. On the one hand it attempted to forge a socialist position opposed to forms of revolutionary leadership expressed in the idea of a seizure of power by a small elite vanguard and of 'democratic centralism' whereby open debate was followed by strong party discipline once a decision had been made. On the other hand it was against what it saw as the bureaucratic state control that characterized social democratic government in the West. The New Left acknowledged the substantive gains enshrined in the post-war Labour government's welfare state. But it repudiated the commonly associated faith that this achievement, along with increasing material affluence built on the post-war economic boom, either marked the end of exploitation and inequality, or the beginnings of a shift to a fully socialist society. The welfare state, as established so far, was a marker in the ground: it was a beginning, not an end. It was something to be critiqued in order that it might be renewed. To that extent the New Left's critique was similar to the internal critiques of the mental hospital that we've seen so far in this chapter.

This desire for socialist renewal employed the terminology of alienation. In part, this signalled the New Left's concentration on wider sites of class conflict and possibilities for change beyond the traditional left's concentration on 'the point of production'.[561]

Brian Kirman had warned that capitalism couldn't provide a reliable environment for 'healthy emotional satisfaction' because it had the crazy kind of internal logic that saw efficiency as building 'bigger and better bombs'. Kenneth Soddy had urged the necessity of a comprehensive post-war mental hygiene system in the face of 'the atomic bomb bogey' hanging over everyone. But while Kirman thought the alienation inherent in capitalism would likely turn society into something like a large mental hospital, mental hygienists such as R. F. Tredgold worried that mass-production and consumption undermined a mentally healthy community producing workers lacking in initiative and self-responsibility.

The New Left were very concerned about the possibilities of nuclear war and clearly concerned about the political effects of mass-consumption. But the proposed solution of a social and moral order mediated by mental hygienist professionals was just the sort of thing that they would have considered part of the problem. They clearly held similar views to Kirman, but they aimed for a new Marxist-inspired understanding of capitalism and class relations. Looking backwards, as well as forwards, they took inspiration from an English radical tradition and the associated interwar 'qualitative' socialist ideas of Tawney and Cole. An emphasis on egalitarianism, and suspicion of centralized authority and authoritarianism was clearly at the heart of the New Left.

One expression of this was the new Left's promotion of Left Clubs around Britain. These aimed to open up wide socialist debate and self-organizing participatory action. They represented the New Left's emphasis on popular power: the creation of socialism in the here and now of common cooperative experience and struggle.

> People have to be confronted with experience, called to the 'society of equals', not because they have never had it so bad, but because the 'society of equals' is better than the best soft-selling consumer capitalist society, and life is something *lived*, not something one passes through like tea through a strainer.[562]

Here was a stance similar to that expressed at Hawkspur in its emphasis on popular power as the theoretical antithesis to authoritarianism and centralization instead of its unwitting source. Similarly, rationalization and totalization might be ever-present in modern society, but popular participation would mediate these aspects of power. This egalitarian emphasis informed the New Left's attack on bureaucratic state control, and the idea that socialism could be created for the working classes by professional middle-class technocrats and experts.

The terminology of alienation also signalled the New Left's alignment with people who felt outsiders in current society. Although closely affiliated with the position of the working classes in general, this extended much wider to issues such as mass-consumption and generational conflict. The New Left's association of its stance with the younger generation expressed itself, for example, in its championing of more child-oriented education and its close affiliation with the Campaign for Nuclear Disarmament (CND) founded in 1958.[563] It was also associated with a more radical off-shoot of the CND called the Committee of 100. Formed in 1960, this contained radical social-ists, anarchists and libertarians, and advocated non-violent, direct action. A host of artists and writers ranging from Shelagh Delaney, the teenage working class author of *A Taste of Honey*, to Gustav Metzger, the creator of auto-destructive art, were among its original members, but there were a cou-ple more closely associated with our story. One was Dorothy Glaister, wife of Norman Glaister, one of the founders of Hawkspur. The other was a found-ing member of the Committee, Tony Smythe. Then in his early 20s, by 1966 Smythe would be the Director of the National Council for Civil Liberties (NCCL) and, by 1974, Director of NAMH.

There were other manifestations of 'qualitative' socialism breaking out around this time. A group of researchers associated with the economist Richard Titmuss were working closely with the Labour party.[564] Together, they challenged two dominant and related political views. One was that economic growth would eliminate poverty;[565] the other was that redistri-bution under Beveridge's welfare approach had become excessive. As we've noted, this second assumption was commonly alluded to by members of the mental hygiene movement with their claims that presenting problems of economic need often revealed themselves to be issues of 'emotional imma-turity'. The Titmuss group, however, argued that the better-off benefited disproportionately from welfare measures because welfare benefits extended beyond the payment of specific social benefits.[566] At the same time, they questioned whether equality and social justice ever could, or should, depend on economic growth. Peter Townsend and Brian Abel-Smith were especially influential with their 'rediscovery' of poverty. The focus on measures of 'absolute' poverty which had informed levels at which National Assistance was set directed attention away from the fact that, although affluence was increasing and numbers below the poverty line declining, deprivation hadn't been eliminated nor had society become more egalitarian. Townsend urged the Labour Party to radically improve the 'income and living conditions of the poor and handicapped'.[567]

* * *

A Royal Commission on the Law Relating to Mental Illness and Mental Defi-ciency was appointed by the government in 1954. Its membership must have seemed a significant success to the mental hygiene movement. Along with Hester Adrian and T. P. Rees it included two other NAMH supporters, Bessie Braddock MP and the psychiatrist D. H. H. Thomas. Meanwhile, as Clive Unsworth's analysis of its deliberations has noted, although there were three legal representatives none were strong proponents of enhanced legal safeguards.[568] The assumption informing the Royal Commission was that psychiatric medicine should be dealt with as equivalent to general medicine. Its terms of reference were to make recommendations on the possibility of treating mental and mental deficiency patients on a voluntary basis without certification. As we've seen, this paralleled the long-held beliefs of the mental hygiene movement. The development of apparently effective new physical treatments in the previous two decades had helped their cause. So, too, did the introduction of the new tranquilizer Largactil into English mental hospitals during 1954. It was argued that these tranquilizers would enable more patients to enter therapeutic relationships and this encouraged a further liberalization of the social order of mental hospitals.[569] The promise was that hospitals need no longer be warehouses, but should only be places for active treatment, just like general hospitals.

So the Royal Commission attempted to complete the assimilation of psychiatry with general medicine that had been pioneered with the dismantling of formal legal safeguards by the 1930 Mental Treatment Act. But it did so in the context of the post-war Labour government's introduction of legislation that finally swept away the Poor Law and established the comprehensive and freely-accessed services of the Welfare State.[570] This very different context for the Royal Commission's extension of voluntary status meant that it was partly a belated fulfilment in psychiatric care of the post-war political promise to make welfare services universally available, and provided without stigma and segregation. In accordance with this the Commission's proposals for voluntary status were formulated into a policy of community care that envisioned a major shift towards community integration of large numbers of patients previously detained in mental hospitals. The Commission's Report and the ensuing 1959 Mental Health Act was therefore presented as a progressive liberalizing extension of post-war welfare measures built on faith in the medical model. The removal of legislative differences in treatment of physical and mental illness would help avoid stigma at the same time as it allowed medical and psychological professionals the necessary freedom to apply their benevolent expertise on those people who truly needed and could benefit from treatment. Hospitals were to be specialist treatment locations, which fitted with the simultaneous promotion of the community as the primary arena for welfare. Voluntary status for mental patients now became aligned with other welfare state services that acted to maintain citizenship by their universality and non-stigmatizing, non-segregatory nature. Thus, the Royal Commission's Report, and the Act that followed it, was

intended as a bold extension of greater tolerance and liberty to mentally disordered people through the extension of benevolent professional
expertise.

NAMH welcomed the Royal Commission's Report, noting that it had
received 'general acclamation' and that it was a measure of the public's
sympathy for mental disorder that the Report's liberal recommendations
regarding admission and discharge had been widely accepted.[571] But there
were tensions at the heart of the Report and ensuing Mental Health Act
associated with the dialectical story we've been tracking.

As we've seen, an influential view in the mental hygiene movement held
that a failure of relatedness to others was a central and unifying component
of both minor mental disorders and mental illness. This view also informed
the approach of the Royal Commission. It announced that:

> Social care and treatment are of particular importance in the treatment of
> all forms of mental disorder... Indeed, mental health in its widest sense
> embraces the whole field of human relationships and human behaviour,
> and many forms of mental disorder are evidenced by, and often arise
> from, disturbance in a person's relationship with other individual human
> beings or with the society in which he lives.[572]

One direct effect of this view seems to have been that the Commission
accepted that there were debilitating effects caused by institutionalization.
It maintained that this should be avoided by making community care the
preferred option with hospitals taking patients only where this was necessary in the interests of treatment.[573] Consequently, the mental hygiene
movement performed something of a fudge in its mid-1950s endeavour
to promote voluntary status for patients in mental deficiency, as well as
mental illness hospitals, and in its promotion of community care for both
groups.[574]

Beyond this, Unsworth makes two observations that are also pertinent.
As we've already noted, the Royal Commission's deliberations ought to be
understood in the context of the post-war Labour government's sweeping health and welfare legislation. Unsworth notes that, along with public ownership of important sectors of the economy, regulation of private
economic activity and tax reform, these measures were envisaged by their
supporters as the road to socialism. They were the means to bring about a
transformation in the quality of social relations. But, as Unsworth adds,

> The Labour Party's rejection of syndicalism and belief in the effectiveness
> of the mechanical transfer of private assets into public ownership ensured
> that the working-class perception of the changes was of a shift in the com-
> position of management toward the state and trade union bureaucracy
> rather than a democratization of industrial and social decision making.[575]

It's a point that recall's the New Left's critique and also Hawkspur's affinity with R. H. Tawney's 'qualitative' socialism. But, as Unsworth later remarks,

> Social reconstruction was ... not only a political impulse which impacted on mental health through the reorganization of health and welfare services, but one which operated within the mental hospital itself as a challenge to its traditionally authoritarian principles of organization. The application of radical democratic values in the resocialization of psychiatric patients by the most advanced disciples of Social Psychiatry was to cultivate a critique of the relationship of orthodox psychiatry to the political order and contribute to the genesis of Anti-Psychiatry.[576]

As we've been observing in our story, both the aspects of post-war social reconstruction that Unsworth describes were entwined, along with their apparent contradictions, with the mental hygiene project.

* * *

The year that saw the passing of the Mental Health Act also saw the Buttle Trust offer NAMH a grant to open a new hostel. This was to be a 'prototype hostel for boys leaving schools for maladjusted children who had no settled homes and needed a bridge to independent life in the community'. Reynolds House opened in 1963. It was a large detached Victorian redbrick house with a good sized garden that backed onto Bromley cricket club. NAMH's General Secretary, Mary Appleby, commented that 'if an experiment is really needed and timely there is the right man waiting in the wings to step forward'. She added, 'How lucky [it] was that in 1961 David Wills was that man'.[577] Wills reckoned NAMH were falling over themselves to get him. And he was extremely chuffed because he hadn't forgotten how Evelyn Fox, the Central Association for Mental Welfare's (CAMW) General Secretary had treated him all those years ago.[578]

* * *

It's a posh voice, a male voice. Also gentle, kindly, nurturing...

> Ward one is the home of forty young children. But a home with the inevitable routine of a big hospital. The children must be got up in the morning, pottied, washed, dressed and fed. An endless repetition of routine: toileting, washing, feeding and so on, and on ... and on. The nurses work hard but they are ruled by the clock and they have adopted methods which save time. For instance, it is quicker to carry a child than to help him to walk. It is quicker to dress him than to help him try to do it himself. Thus children have little chance of developing their abilities,

of gaining even personal independence. Instead of activity and learning being encouraged the children are often left to sit motionless, waiting for the next thing to happen. No one talks to them.[579]

It's part of the voice-over for a film called *Mentally Handicapped Children Growing Up*. Made in 1961, and sponsored by the National Association for Mentally Handicapped Children, this film promoted a late 1950s experiment that became known as the Brooklands study.[580] The work came too late to directly influence the Royal Commission on mental illness and mental deficiency, but it would become highly influential in the following decades. Headed by Jack Tizard, this study marks the moment when childcare theorizing developed by mental hygienists during the war was finally extended to encompass the majority of children categorized as mentally deficient. (By 1959 this term had been replaced by the terms 'subnormal' or 'mentally handicapped'.) In our story it marks the point when the bulk of the members of the anti-Family finally began to become accepted into the fold. But—ever full of contradictions—it also marks the point when the organizing principle of the Family that we've been tracing took a blow that fundamentally destabilized it.

We've seen that the Curtis Committee was formed in response to criticism of child care provision during the war, and that the mental hygiene movement substantially influenced the thinking on which its judgements were based. There was a prioritization of the home environment and its accompanying emotional atmosphere, and a related critique of existing institutional provision. Large-scale group provision and relationships that were seen as emotionally illiterate were condemned; these left many children either withdrawn, destructive or expressing a 'pathological clamouring for attention'. But so-called mentally deficient children had been excluded from this concern and the resulting legislation for child care. Institutional care for these children remained with the mental deficiency system that also dealt with adults. Through the post-war decades this system had become an ever-growing backwater.

Public attention to poor conditions in mental deficiency hospitals had been encouraged by the NCCL. Along with pressure from the National Association for Mentally Handicapped Children and the Spastics Society, this had made it a concern at government level.[581] Tizard and his colleagues took the opportunity to argue that the upbringing of mentally handicapped children in mental deficiency institutions should be measured against the accepted principles of care for other children deprived of a normal home. Noting that the Curtis Committee's 1946 report had set the foundations of post-war policy for the latter children, they applied it as a basis for the care of children at Brooklands.[582] The thesis was simple: it was that 'severely retarded children are entitled to the same opportunities and quality of care that we give to normal children'.[583]

Brooklands was a three-storey late-Victorian house with a large lawn and gardens. Between 1958 and 1960 it became home to a group of children then categorized as 'imbecile'. In keeping with Curtis, the project attempted to create a small 'family style' atmosphere. It provided for a group of children of mixed sex and an age range of 4 to 10 years. (The children's mental ages were given as half between 3 and 4 years, and a quarter younger than 2 years). They were divided into family groups under a housemother. Contrary to the theories that informed the mental hygiene movement, Tizard emphasized that these children had similar 'intellectual, social and emotional needs' to other children. In fact, they were often more complex and therefore needed increased understanding.[584] Just as recommended by Curtis, the aim was for a continuity of close affectionate care with each child by specific adults, emphasizing, in particular, the children's present emotional needs. This was contrasted with existing approaches that emphasized training through exercises and drill.[585]

Tizard believed the children to be suffering from profound emotional maladjustment, as well as mental handicap. Some of the children's difficulties were attributed to the 'sudden and profound' change in their lives brought about by their move to Brooklands. Most of the staff at their new home were unknown to them and, at first, changed as much as at hospital. Familiar routines disappeared and new experiences were thrust upon them. Echoing John Bowlby's attachment theory, Tizard remarked that 'The children were thus in the kind of situation that children face when they go unprepared to hospital'.[586] But Tizard believed the children's institutional upbringing to be the more important factor. They had suffered greatly from emotional maladjustment due to institutional deprivation and lack of close and continuous affectionate relationships. As the wartime critique had done, Tizard condemned large-scale institutional provision that was ordered for organizational efficiency and staff interests. Discussing the initial problems of settling children in at Brooklands, Tizard noted that:

> Not only were they severely subnormal intellectually, but they were institutional children, used to a constant routine and the uniformity of experience of ward life in a hospital. Their lives were, inevitably, governed by ward practices, rather than by emotional links with particular adults with whom they identified themselves.[587]

All the children had been selected from the Fountain hospital where conditions were poor and typical of those in many institutions for mentally handicapped children at the time. Here they lived on wards of around 60 beds with harassed nurses unable to provide individual attention amid constant noise. Incontinence was a constant problem, and the smell of faeces and urine often impossible to eradicate. Children were grouped together by sex, age and level of handicap. Tizard noted that the effects of emotional deprivation were clear among the children. His description echoes that of the Curtis Committee whose members had visited institutions for 'normal' children: 'Rocking and head-banging were commonly observed; they crowded around strangers, clutching and pawing them. The children were apathetic and given to tantrums. They rarely played'.[588]

Tizard reported that at Brooklands children's emotional maladjustment became considerably lessened, their ability to play socially and constructively improved, and they developed close attachments to staff and other children. All became more independent in caring for themselves and their ability to use and understand language improved drastically.[589]

So, in terms of our story, the Brooklands study represents a further elaboration of the organizing principle and methodology of the Family. Brooklands also marks an extension of the mental hygiene movement's associated attack on elements of institutional care. A decade and a half after a battle was waged on behalf of other children, those now called mentally handicapped were, for the first time, openly admitted to suffer the same psychological problems and to benefit from the same recommended changes in care.

But Brooklands amounts to more than this. It also represents a fundamental blow against the Family in the role that it had traditionally informed the mental hygiene movement. 'The self' emerged under the rubric of the Family; through 'the primitive', through 'madness', through childhood, it grew, until it reached adequacy as a citizen. But one thing it didn't grow through. It didn't grow through mental deficiency. These people had been defined out of this process. Brooklands didn't so much snap this supposed progression, as slow it down to the extent that the intended ultimate end became almost irrelevant. Under the mental hygiene movement the means had clearly been subordinate to the ultimate ends. But with Brooklands the means became closer to ends in themselves. These children would never become the kind of rationally marshalled, self-sustaining, 'responsible' citizens that the mental hygiene movement had traditionally considered the aim of 'mental health' to comprise. But why did that matter? In a revealing passage of his account of the Brooklands experiment, Jack Tizard confessed that, if it had only resulted in the children becoming happy and content then this would have been vindication in itself.[590]

*   *   *

By far the majority of the young men at Reynolds House were from working class families or families on the poverty line. They all displayed difficulty in forming relationships. Common behaviours before and during life at Reynolds House were frequent stealing, 'abnormal aggression', constant lying, enuresis and insomnia. Wills described the 'general picture' of the residents as of 'rootless young men from disturbed broken, and sometimes chaotic backgrounds, virtually homeless, without normal family relationships or feelings, without experience of a stable and loving home'. He characterized the way Reynolds House was organized as 'in some respects like a family home, and in others simply a boarding house in which an experiment in communal living was being carried out'. It followed all the principles for which Wills had by now become well-known in child-care circles. The general aim of care was to provide three essential elements: 'warmth, security, and freedom'.[591] The goal was conceived as helping the young men to learn how to relate.

* * *

Around the time that Barton's *Institutional Neurosis* was published, the sociologist Irving Goffman produced a series of articles based on his field work at a large public mental hospital in Washington DC, which, at that time, housed more than 6000 patients.[592] These were subsequently published as *Asylums: Essays on the Social Situation of Mental Patients and Other Inmates* in 1961. Goffman's analysis wasn't part of those stemming from the mental hygiene movement, but it had closer associations than has been recognized.

Goffman's social interactionist study presented an intimate description of the 'career' of the mental patient. Here the transition from 'pre-patient' to 'in-patient' was relayed as a process of contingencies. Only in retrospective reconstruction, he claimed, could these contingencies become case histories of a progressive mental illness.[593] Goffman's work clearly championed the patient's perspective and was openly critical of the psychiatric system. He also had some sharp things to say about psychiatrists' contemporary assumptions that faulty human relations lay at the core of much mental illness. Mental hygienists commonly touted their expertise in 'human relations'. But Goffman reckoned that providing a patient with an environment within which he or she might re-learn an ability to relate couldn't really be described as a technical skill. Goffman added that, anyway, such skills that staff might have in this area couldn't be 'broken down into the skill status hierarchy' attributed to the mental hospital. His research led him to point out that a ward orderly could often seem 'as well equipped to give a "good" relation to a patient as a highly trained psychiatrist'. He added that, in any case, as the psychiatrist's contact with any particular patient was usually so tiny, the orderly's relations, whether good or bad, would affect the patient far more. Goffman remarked that despite this 'hospital administrations, operating within the medical model, give to psychiatrists the right to make crucial decisions concerning the disposition of the patient'.[594]

But, in spite of these pretty scathing criticisms, several influential figures involved with the mental hygiene movement who were also engaged in attempts to reform institutional provision, were clearly receptive to Goffman's work. This receptivity may have been facilitated by the fact that Goffman believed 'total institutions' such as the mental hospital were, as he put it, 'incompatible with a crucial element of our society, the family'. They were incompatible because the total institution was 'a social hybrid, part residential community, part formal organization'. In Goffman's view the household was a bulwark against the modern social power of the total institution.[595] Goffman's claim certainly directly informed the research and theorizing in mental handicap that followed from Jack Tizard's Brooklands study. In 1963 Tizard, Norma Raynes and Roy King set up comparative studies of institutions, their organization and the effect on childcare.[596] Building on the Brooklands work, these studies sought to differentiate between 'institutionally oriented' care, as described by Goffman's total institution ideal type, and 'inmate oriented' care, based on the 'family' model of care.[597] This approach was explicitly justified by reference to Goffman's opposition of family relations to those of the total institution.[598] Tizard also helped to set up the Wessex project. This set up a series of small community residential homes for children and adults categorized as severely mentally handicapped.[599] Its director, the psychiatrist Albert Kushlick, employed the same differentiation between institutional and inmate-oriented care.

Russell Barton also employed Goffman's concept of 'total institutions' in the application of his 'institutional neurosis' concept to hospitals for people diagnosed mentally handicapped.[600] David Clark was influenced by Goffman's description of the 'moral career' of the mental patient.[601] Others involved with the mental hygiene movement responded favourably to his description of the social construction of stigma in society.[602] An article in NAMH's journal *Mental Health* reckoned, however, that Goffman's rendition of total institutions ought to have been differentiated in its consequences by gender. Gerda Cohen claimed that mass treatment and loss of individual identity was felt harder by women. 'Women resent regimentation', she wrote, 'whereas some men, at least, glory in it'.[603]

Because of the arguments made in *Asylums* (as well as their extensive influence) Goffman has subsequently been described as part of the anti-psychiatry movement that emerged during the 1960s.[604] But, if Goffman cast the mental institution negatively against the family institution, British psychiatrists associated with the so-called anti-psychiatry movement drew them both together for sustained criticism. R. D. Laing's early work criticized the medical model for treating the mental patient as a bundle of 'its' and 'bits' that were then read as signs of disease. His attempt to listen to patients commonly related their condition to the family interactions within which they had grown up.[605] Later works depicted family relationships in general as largely adversarial, confusing to their members and, ultimately, mystifying. With the Marxist-influenced psychiatrist David Cooper and others, Laing established several radical experiments in therapeutic community-style living. This work became increasingly associated with the New Left and its cultural critique.[606]

*　*　*

In April 1964, Stuart Hall could be found addressing the *Twentieth Inter-Clinic Child Guidance Conference*. We can get a sense of the depth of the New Left's radical critique compared to the mental hygiene movement by taking a look at his talk.[607]

The conference had opened with a screening of one of a series of films made for the BBC by the young director John Boorman. Together they had been called *Citizen '63*. This particular episode had depicted the life of an adolescent girl called Marion Knight. Marion provided the voice-over for scenes taken from her life at home, school and beyond. As the child psychiatrist Jack Kahn described in his report of the conference, it was hoped that the film would give a 'living reality' to their discussions and enrich their observations and 'theoretical formulations'.[608]

Addressing himself to this film, and sprinkling his argument with quotations from the psychoanalyst Erik Erikson, Hall advocated what appeared to be the long-held child guidance aim of getting a picture of 'the *whole child*'. But he pointed out that you couldn't get it by just adding up 'factors' affecting something internal called adolescence, such as family, school, peer groups or 'teenage culture'.[609] Meanwhile, he asked his audience to try to keep in mind the ways in which they themselves were influenced by a 'general climate and by mediated values' beyond their own family relationships and personal life-histories.[610]

Echoing the longstanding fears about power in the modern world as rationalizing and totalizing, Hall argued that current 'technologically complicated' society promoted 'lives of an artificial kind'. Consequently, it was increasingly important to look at the complicated interaction between, 'what people are, what people feel themselves to be, and what they feel others expect them to be'. Hall added that

> This tendency to dramatise or 'invent' with our own persons states of feeling which are alien to 'us', or which we can only half authenticate in our own experience, to use ourselves to 'act out' feelings which are 'true' only because they are true for others like ourselves, or representative of people-like-us, is a real denaturing of identity, a loss of the self, a cultural alienation.[611]

Hall argued that it was during adolescence that this was most clearly taking place. Attention to this required something more than investigating separate compartments of people's lives at home, school or in peer groups. It required an attempt to 'read this story from the inside, from where the child *is*, and to see it as a total picture'.[612] Many working in child guidance might well have responded that this was what they had been trying to grapple with for some time. But Hall was saying that teenage culture (as opposed to 'adolescence') was something apparently new and penetrating. As something cultural it acted at emotive and imaginative levels that he called 'primitive' in the psychic sense, where it engaged with the development of 'values and life-orientations' before they were socialized. Adolescence, in the sense of something happening within and to an individual no longer made sense; something *cultural* was cross-hatched with it.

Along lines similar to the New Left's depiction of the working classes, Hall emphasized that young people weren't just the passive recipients of 'teenage culture'; they were also its active constructors. They made use of this culture in different ways for their own ends. What might appear simplistic or 'false, second rate and shoddy' was, in fact, genuine and expressed valuable communication.[613] 'Teenage culture', unlike the home or the cultures of the school or community, was something that spoke to one generation. It was this 'culture of other-people-of-their-own-age' that meant that it engaged directly with youth. It had a 'direct emotional claim' because, unlike home, school or community, it lacked those elements of authority associated with age and tradition. As such, this culture was a rebellion against authority, but also a genuine questioning of its forms in contemporary society. Hall argued that underlying it was a profound questioning of the nature of identity and freedom, in an increasingly organized, conformist and artificial 'mass culture'. All social work with youth must understand, he maintained, that this 'fundamental and philosophical problem' lay at the heart of the work. The condition of building a 'whole life' was freedom, and this meant 'change, and the choice between meaningful goals'.[614]

Inevitably, there was a battle of authority here. 'Where then does one draw the line? Where does one yield the integrity of one's own position to the integrity of the adolescent?' asked Hall. In attempting to answer it he turned to the film about 'Marion'. Referring to the poem that she had recited, he admitted that it was both moving and naive. Her utopian hopes were longings which, he said, 'adults know will never be there in quite the way in which she wants'.[615] But he also pointed to her argument with a woman about nuclear weapons, 'spies for peace' and anarchism:

> ... where does one cease to be the professional social worker, interpreting the stresses on her face or her tone of voice, and begin to take her opinions and her attitudes as 'real' in themselves? Where do we authenticate her life for her [?] ... I was struck in that exchange not by the naivety but by the maturity of her views. If I reacted strongly against anything, it was against the paternalism, the quietism, the gently, embracing conformism of her opponent ... If we want, every ideology can be explained simply as the search for inner security ... [616]

Hall was, in his own way, struggling with the inveterate and unending questions of power and authority between people in the social world; and the inevitable fact that authority can't just be spoofed away by being 'against it' or, conversely, by acting as if it doesn't exist or isn't worth analysing. But he used the specific case of Marion and the general case of youth to throw into relief generational, class and professional prejudices. For good measure, he threw at his audience an extract of an exchange between Norman Mailer and Ned Polsky on the question of hipsterism and psychoanalysis, where Mailer had written:

> Still, the impolite question remains to be asked: does the direct experience of the analyst's own life prepare him to judge the inner states of hipsterism? Sedentary, middle class, in fief to fifteen years training, living among the absurd magpie scrutinies of wife, children, colleagues, patients and hostile strangers... The analyst is Gibraltar in a pathless middle class sea... For what would the analyst do, and what would become of his tidy, narrow, other-directed little world, if he were to discover, and may God help him, that the hipster way out of the lip of danger may conceivably know more of the savour and swing of the damn dialectic of the orgasm than he... the educated ball-shrinker who diagnoses all joys not his own as too puny?[617]

Well, the New Left hadn't yet recognized its own gender prejudice had it? But, that aside, it's difficult not to imagine that Hall's audience might have felt a bit got at.

* * *

Reynolds House's puppy had annoyed the neighbouring Bromley Cricket Club. They hadn't enjoyed it digging holes in their pristine cricket 'table'. Apparently, instead of sending someone round to see the Warden, club members simply swore at odd residents. The result was that Wills didn't know a thing about it until the solicitor's letter arrived. But the club officials were actually very tolerant. They had a large garage near Reynolds House fronted by tall lamps. Only after one had been broken for the third time did a director of the club come round and very politely ask if they could stop doing it. Wills wrote that he was so apologetic and kind you'd have thought that he was the guilty party. In fact, he only sent a bill for the cost of the third lamp rather than all of them as Wills had insisted. The Housemeeting, which was Reynolds' version of Hawkspur's Camp Council, isolated the culprits in minutes and made them pay up.[618]

* * *

There's no record in the published conference paper of how Stuart Hall's talk went down with his audience. But it wasn't as if some of them hadn't asked for it. Take the 1959 inter-clinic conference, for example. This had been entitled *Truancy or School Phobia?*. 'School phobia' was rapidly becoming a vogue term in child guidance, and the conference generally agreed that although it was a condition related to truancy, it was, nevertheless, quite distinct from it. The class and intellectual prejudices at the heart of this distinction are obvious. One leading child psychiatrist embodied the general consensus by describing truancy as 'on the whole a *social* problem...arising in bad homes, with inadequate care; hence from somewhat lower strata than the school phobic'.[619] A child guidance psychologist remarked similarly that, 'truants mostly come from poor homes, and have on the whole, rather less than average intelligence'.[620] School phobics were, in contrast, generally from materially good home backgrounds, with either average or above-average intelligence, and of a sensitive nature. But attitudes towards these latter children were no less normative and unreflective than those towards the apparently distinct and lower class truants. Most truants and 'phobics' were boys. The 'phobics' were described as having 'feminine' characteristics and often fixated on their mothers, who were considered to be frequently over-protective.[621] Physical characteristics were correlated with these 'feminine' psychological attributes. One child guider remarked that, in his experience, the boys tended to be 'characteristically aesthenic with pale, translucent skin, large protruding ears and a tendency to rabbit teeth'.[622] The mothers were apparently 'as might be expected...all "neurotic" in some sense or another'. As another speaker put it, 'you all know the kind of mothers that one meets in these cases'.[623] The fathers were variously, aloof, distant, 'almost maternal, too involved in the family', submissive, inadequate, 'rather interfering, assertive or irritable', or 'unstable'. This is such a litany of 'characteristics' that it leaves you wondering who wouldn't represent one of these descriptions sometimes. But, apparently to the psychiatrist who supplied these details, out of a sample of 64, this left 'only about 6' that he 'could honestly describe as anything like normal'.[624]

As to the reason why there was apparently such a number of school refusers, one posited that the general population had become 'softer and more neurotic'.[625] Another argued that 'parents have discarded an authoritarian role without always reaching the maturity...needed for replacing the traditional family pattern with something as strong'. According to him, children were not learning the self-discipline necessary to save them from self-indulgence.[626] But neither the speakers nor the reports of the group discussions (which comprised 20 groups) actually seem to have questioned whether the situation at the school itself could be the primary factor. Nor was it questioned whether any particular child might have perfectly reasonable grounds to want to avoid school. And apparently beyond these particular child guiders' comprehension was any idea that the 'norm' of compulsory schooling might itself be questioned. It was taken for granted that schools were a functional pre-requisite of full 'socialization' in society.

To avoid school was, almost by definition, not only to show inadequate socialization in the family, but also to compound this by attempting to avoid the final significant institution in society that might bring about appropriate socialization. The norm was to go to school and thus the 'illness' was to seek to avoid it and the 'cure' to get a child back there. These child guiders, therefore, focussed almost exclusively on the 'disturbance' of the child and his emotional interactions at home. 'School phobia' was simply seen as a symptom of these deeper problems.

* * *

This wasn't the whole story though. We saw in the previous chapter how important elements of child guidance theorizing had disrupted the prevailing medical model. We saw also how this had spilled over into a questioning of professional authority in child guidance and associated work. There were other examples of this. For instance, at a conference on mental health just after the war, the psychiatric social worker Sybil Clement Brown had worried that the promotion of child guidance as a preventive service risked selection of children only on the basis of problem behaviour defined by other institutions. 'We must', she argued, 'guard against a definition of problems only by those who stand to gain by their prevention'.[627] In 1953 another psychiatric social worker warned of the danger of a psychiatric 'technical authoritarianism' whereby experts kept themselves in work by creating problems and 'anxiety' in the community.[628] Similarly, the leading psychiatric social worker E.M. Goldberg published an article in the *British Journal of Psychiatric Social Work* in which she admitted that psychiatric social workers were beginning to realize that the early family relationships, which they had been studying intensively, were highly complex. She went on to stress that cultural practices and norms couldn't be separated from human relationships. It followed from this, she argued, that psychiatric social workers needed to question what, in fact, they were asking their patients to adjust to:

> Do we always remember that the deviants of to-day may be the 'normal' of tomorrow... [?] Do we ever stop to question the nature of our cultural norms that compel us to label an ever increasing number of people as neurotic or abnormal [?] Do we in fact make enough use of the accumulating knowledge which throws light on these questions [?][629]

In 1957 the same journal published an article that brought this questioning of professional authority roles home to the professional groups who made up the classic, and much touted, child guidance team. The article was called 'The Psychopathology of Inter-Clinic Conferences'.[630] Its author was Christopher Beedell, at that time a psychologist at the Bristol Child Guidance Clinic.

Beedell rendered the inter-professional teamwork approach a conflict-ridden affair. He noted that at conferences the disciplines mixed very little. When, at regional conferences, each discipline had a separate sectional meeting, those who attended an inappropriate group by mistake were met, he remarked, with a 'gleeful and adolescent attitude of exclusion'.[631] Equally, formal satisfaction with inter-clinic conferences was expressed during their progress, but underlying disputes and discontent were only revealed privately later. Beedell attributed these issues to the emotional aspects of professional relationships militating against full, open communication. He related this phenomenon to the relative satisfactions of power derived by membership of each profession and speculated upon what these might entail:

> The psychiatrists have the most obvious satisfaction from power at an administrative and decision-making level. The psychologists feel power and draw satisfaction from their ability (real or supposed) to predict certain aspects of behaviour. At first sight the psychiatric social workers appear to withdraw deliberately from power satisfactions; but it seems fair to say that the assumed role of unsurprised, and by implication omniscient, listener has the same sort of power satisfaction attached to it as that of the most bigoted behaviourist.[632]

Maybe it isn't surprising that one of the most forthright critiques of professional conceit among child guidance workers came from a man like Christopher Beedell. During the war, while a young man in his late teens, he had registered as a conscientious objector and for a time been resident in Wormwood Scrubs. Later in the war, while studying chemistry at university, he'd come across a copy of Wills' book *The Hawkspur Experiment*. It had so impressed him that by the end of the war he'd joined the camp (by then beginning its second phase as Hawkspur camp for Boys—without Wills) as a student helper.[633] He was later to become a leading figure in theorizing residential childcare.

* * *

The critique of mental hospital organization and function by psychiatrists associated with the mental hygiene movement had repudiated the traditional image of the mental patient as an 'isolated individual, inside whom things are happening'.[634] In doing so, it linked the questioning of authority roles with an emphasis on patients' expressed experience and the relational milieu within which this took place. This entailed a therapeutic emphasis on 'open communication'.[635] If attempted earnestly on both sides, this held the possibility that the nature of psychiatric expertise and the role that it should play might become even more radically questioned both within the therapeutic encounter and outside it. An example of this is provided by David Clark in a later reflection on his experiences at Winston house. This was a 'half-way' hostel linked to Fulbourn hospital and developed with the support of the Cambridgeshire Mental Welfare Association.[636] Winston house was run independently of the hospital and didn't operate as a therapeutic community. And this, in fact, enabled another level of open communication. Direct experience at the home made Clark begin to understand more clearly, he believed, how the power of the psychiatrist in the hospital permeated the social relations within it. Clark's experience of Winston House reinforced to him that most patients were very well aware, and wary, of the doctor's power, for example to order electro-convulsive therapy (ECT), confinement and seclusion, or to deny discharge. Clark began to accept more fully that the psychiatrist's role in any therapeutic relationship could never be neutral nor could it simply and always be therapeutic.

> Gradually I learned ... more subtle lessons and began to respect patients' judgement of their own needs. Some of them said they felt no need to see a psychiatrist again and I learned to accept that. Others were guardedly polite; their referring psychiatrist had spoken enthusiastically about how their psychoses had been cured and their state stabilised on Largactil; gradually, as they came to trust me, they revealed that for many months they had been putting the pills down the lavatory.[637]

This kind of appreciation of the negative role that therapeutic authority could unintentionally perform was also expressed in child guidance. The psychiatrist Martin James expressed some of the consequences of such thinking at the *Child Guidance Inter-Clinic Conference* in 1965. He had worked at Northfield during the war and served on NAMH's consultant medical panel in the 1950s and 1960s.[638] Summing up the conference's deliberations, James announced that what he feared was the multiplication of experts. For him, psychiatric training, in particular, portrayed people as 'machines', rather than people, desensitized doctors to emotionality, and engendered an authoritarian, paternalistic attitude.[639] It was a view shared by some among the psychiatric social work profession. A few years later Barbara Butler told the Psychotherapy Section of the RMPA that psychiatrists' training had not helped them to 'deepen their humanity or their understanding of people in the community'. 'The question is', she said, 'how far the training of psychiatrists is moving to include content concerned with the meaning of life and feelings and relationships in society'.[640]

In child guidance, the idea that the child with its presenting symptoms was most often a lightning rod, or scapegoat, for disturbed relations in the parents or the wider family became commonplace.[641] So did an associated term, 'collusion'. Were family members colluding in channelling maladjusted relations into an embodied maladjustment in a particular child?

Was the child guidance team being drawn into colluding with them in this manoeuvre? We've just noted that, in the post-war years, E. M. Goldberg, among others, had broadened this out to suggest that psychiatric social workers might be colluding in the labelling of increasing numbers of people neurotic or abnormal. In 1965, a child psychiatrist at the Tavistock Clinic claimed similarly that child guidance was colluding with the community in labelling more and more children 'ill' or 'maladjusted'.[642] By the mid-1960s, however, these contentions held wider currency.

* * *

The staff at Reynolds House tried to subtly reproduce the warm affective circumstances of personal homes. They sat in 'the quiet room' and had a cup of tea ready for when those who had jobs came home from work. It was an attempt to let the young person 'assure himself of the parental presence'. The 'boys' did it in their own way. They 'burst into the house through the kitchen entrance' and asked what was for dinner first.

Some of the many visitors to the house thought that the boys were over indulged. The boys, on their part, resented the visitors' intrusions into their home. In any case, Wills defended the practices. He said it was true that 'in the matter of small personal services such as getting a chair, or an extra cup from the kitchen and that sort of thing, the position tended to be precisely the opposite of that in many residential establishments, in the sense that the staff of Reynolds House usually did for the boys the kind of things which in many such establishments the boys do for the staff'. These practices, and the fact that the boys had their beds made for them on the days they weren't working, were defended on the basis that they were common in the ordinary working class home. 'Rightly or wrongly', wrote Wills, 'the working class adolescent boy does very little in the home and it was thought right, by and large, for the hostel boys to enjoy the same privileges as the friends they made outside'. That said, they were expected to help with the washing up after the evening meal along with the male members of staff. The Housemeeting arranged this and Wills claimed there was little trouble over it. He reckoned it was a notable achievement given that such demands, in his view, 'didn't meet with much success in most working class families'.[643]

* * *

Laing and Cooper's emphasis on destructive group relations gradually widened through the 1960s to include, not only the family, but other social institutions up to and including nation states. All these were theorized in terms of interpersonal protection rackets that were destructive and mystifying.

It is around this point in our story that Foucault became adopted (to use an inappropriate word) as an honorary anti-psychiatrist. An abridged translation of his book *Madness and Civilization* came out in English in 1967 with an introduction by David Cooper. He hailed Foucault's stylistic power in showing 'the nature of the violence that mental patients meet'. For him, Foucault made it clear that 'the invention of madness as a disease' was actually a disease of civilization: 'we choose to conjure up this disease in order to evade a certain moment of our existence – the moment of disturbance, of penetrating vision into the depths of ourselves, that we prefer to externalize into others'. Cooper then brought this interpretation together with the anti-psychiatric focus on the family.

> Recent psychiatric – or perhaps anti-psychiatric – research into the origins of the major form of madness in our society, schizophrenia, has moved round to the position that people do not in fact go mad, but are driven mad by others who are driven into the position of driving them mad by a peculiar convergence of social pressures. These social pressures, hinted at by Foucault, are mediated to certain selected individuals by their families – themselves selected by processes that are intelligible – through various mystifying and confusing manoeuvres.[644]

It was hardly a surprise that anti-psychiatrists saw an affinity with Foucault. He had claimed that 'What we call psychiatric practice is a certain moral tactic contemporary with the end of the eighteenth century'.[645] This stemmed from moral treatment, and Foucault's negative account of it at the Retreat had described how 'Tuke, precisely, reconstitutes around madness a simulated family, which is an institutional parody but a real psychological situation'.[646] Tuke's modern psychiatric and psychoanalytic successor was therefore 'an alienating figure'. All of this must have seemed in close alliance with much of what Laing and Cooper had been claiming. And Foucault's 'alienation in guilt' sat nicely with the prevailing counter-cultural critique, too. In fact, Foucault's rendition of moral treatment surely wouldn't have seemed so very far away from Goffman's description of how the mental patient was treated as a child, and of the mental hospital as promoting a 'self-alienating moral servitude' in which any expression of revolt was read as justification for the regime.[647]

David Cooper was the most sustained and radical exponent of anti-psychiatry's anti-family view. Fusing it with his Marxist perspective, he described the family as performing the same function in all exploitative societies; it was 'an ideological conditioning device'. The present Western bourgeois family was, however, the most perfected example.[648] Cooper described its relevance to the social order this way:

> The power of the family resides in its social mediating function. It reinforces the effective power of the ruling class in any exploitative society by

providing a highly controllable paradigmatic form for every social institution. So we find the family form replicated through the social structures
of the factory, the union branch, the school (primary and secondary), the
university, the business corporation, the church, political parties and governmental apparatus, the armed forces, general and mental hospitals, and
so on.[649]

And so we have reached a complete reversal. The Family that we have been
tracing, with its function as an organizing principle and methodology, has
turned upside down. For the Bosanquets, and for the mental hygiene movement, the Family's relations of authority were both the means through
which the self emerged and became moralized, and a more general model
for the authority relations of the wider community. The Family created
and sustained secure personhood and mentally healthy relations. Cooper
acknowledged this Family as organizing principle and methodology. But, for
him its real effects were the absolute reverse. The Family, the mental hospital
and the state were key components of the same exploitative and alienating
system. We were all in an institution now, and we were all institutionalized.
   In terms of our story, what's happened here is that the Family has become
totally inverted. The traditional image of the Family as a hierarchical authority constituted of rigid levels of status, claiming an easy distinction between
reason and irrationality, is refuted. But so, too, are its relationships of emotional depth and intimacy. These are apparently always adversarial and
merely reinforce the institutions of an exploitative society.
   But hold on, aren't the contradictions of a dialectic supposed to resolve
themselves into some sort of 'negation of the negation'? Can you have an
outright reversal?

*   *   *

Towards the end of 1966 NAMH's General Secretary, Mary Appleby, com-
piled a report for its Council of Management titled 'National Trends and
the Mental Health Services'.[650] It detailed elements of rapid social change
that NAMH needed to respond to. The report highlighted the phenomenon
of 'the dropout from society (more perceptible so far in America) resulting
in higher rates of drug-taking, alcoholism and suicide'. But it cast against
this 'the growing movement towards consumer and voluntary groups'. The
relationship of the family to mental health remained a crucial issue.

Earlier maturity and community tolerance of sex experiment and irregu-
larity is leading to earlier and more fragile marriages. The effect of this
must be a potentially damaging insecurity for the next generation: a
different view of the sanctity of life. The pill, the change in abortion
and suicide law, may herald a situation in which not only the control

of birth but the control of death may become socially acceptable ... we have a generation ... who know more than the older generations, and this accentuates the usual generation gap ... It has been brought up with the television set, and therefore is the first to have been exposed ... to the desensitizing influence of the box.

To these descriptions of social change in relation to the family, the report added:

> We are indeed faced with the dilemma that, while the roots of good mental health seem to be laid in childhood, and therefore a secure family has been a plank of our preventive arguments, there is increasing evidence that the roots of psychiatric illness are also laid in the family.

The doubts and uncertainties expressed by the report were summed up in its rendition of the question raised by a recent meeting with television producers: 'What, they ask, must they avoid so as not to undermine the mental health of the population? They are not sure, and nor are we'. The report concluded: 'We on the staff suggest these points as background to our consideration of the future'.

* * *

The 1970 NAMH Annual Conference was notable for the number of patients who had managed to gain admittance to the auditorium. As the first morning's discussion progressed many spontaneously rose from their seats to graphically express home-truths, fears and criticisms. Other members of the medical and ancillary professions joined suit and the normally staid and polite NAMH conference was thrown into shocked confusion.

But it was just pretend. The views that the 'patients' and professionals 'spontaneously' aired were taken from various opinions that had been expressed publicly through the previous year. 'Happenings' had been all the rage in the 1960s, and now even NAMH was having one. The 'cast' were made up of various doctors, nurses, hospital administrators, social workers and members of the public. The patients were played by members of the Casualties Union. It was intended as a 'dramatic charade' that would highlight in an immediate and personal way the 'critical state of the mental health services' and stimulate more outspoken discussion among the audience. Unfortunately, it seemed to have had the reverse effect. There was no continuation of the debate in 'the same vehement vein' by the rest of the audience. NAMH suspected that 'delegates were rather numbed by the outpouring of home-truths, complaints, and fears that must have been common to them in their everyday work'. In fact, several delegates were so taken in by the 'patients' that they became outraged and denounced what seemed to them the obvious exploitation of patients for publicity purposes. NAMH representatives had to reassure them repeatedly that 'the listless, shambling, pathetic figures really were actors'.[651]

The advocacy of 'open communication' that had become increasingly prominent across sites of mental hygiene activity had largely been contained at the level of the therapeutic encounter. Psychiatric patients, and relatives of mentally ill or handicapped people only began to be heard at NAMH conferences from the mid-1960s, and these appearances were extremely limited to begin with. A couple of earlier annual conferences had aired and discussed 'lay' opinions, but these were carefully controlled. The 1961 annual conference had played tape recordings made up of extracts of 'typical opinions and prejudices' of the public, which were then discussed by an expert panel.[652] At the 1964 annual conference another expert panel, which included D. V. Martin, commented on abbreviated transcripts relaying the opinions of mental patients' relatives.[653] But, in 1969, NAMH had finally arranged for several real relatives and patients to be given the opportunity to speak from the platform, as members of a 'consumer panel'.[654] It had been encouraged to do so partly by its acknowledgement that there had been a 'rapid development of consumer participation in all areas of public life, not least in the realm of the medical and social services'.[655] Clearly, as well, it was dragged towards this wider 'open communication' by the growth of a more radical view of what it meant to make a sympathetic attempt to appreciate a patient's experience. Additionally, NAMH's more open engagement with the views of patients and relatives coincided with the series of hospital scandals that broke around this time. These seemed to underline the need, not just for increased funding and better facilities, but also a reorganization of the institutional order to enable the kind of reflexive care and treatment that had become advocated through attention to the primacy of emotional relationships.

A letter published in *The Times* in November 1965 provided one important trigger for media and political debate on the condition and treatment of patients in mental hospitals.[656] A book stemming from the letter appeared in 1967. *Sans Everything: A Case to Answer* included responses and additional evidence mostly from nurses and social workers, and charged that serious abuse and neglect had taken place in specific hospitals.[657] Although Kenneth Robinson arranged for Regional Hospital Boards to appoint committees of enquiry, the subsequent reports largely dismissed the allegations as inaccurate and often the product of distorted and 'overly-emotional' accounts. Robinson publicly called the allegations 'totally unfounded or grossly exaggerated' and lamented the effect on public attitudes to mental hospital care.[658] Long involved with NAMH, Robinson was essentially repeating his response to the NCCL's criticisms of mental deficiency institutions in the 1950s. NAMH had publicly supported this view.[659] But this time, despite disagreements on its Council of Management about the issues raised in the book, NAMH publicly refused to accept the Minister's interpretation and pointed out that in only three of the cases were the allegations found to be disproved. It argued that management and administrative organization, as well as staff training and attitudes, were all deficient.[660] Both Brian Abel-Smith and Peter Townsend criticized Robinson's handling of the *Sans Everything* allegations at separate NAMH annual conferences at the end of the 1960s. Townsend asked 'May it not be time for patients to be consulted, if not participate, in the management of hospitals, and more of those in the community organise and administer their own community care? In the last analysis good social planning promotes the ends of democracy'.[661]

In fact, more than one person closely associated with NAMH had contributed to *Sans Everything*.[662] Russell Barton provided a foreword in which he criticized mental institutions as developing powerful defence mechanisms and producing 'neurotic' misplaced loyalty amongst staff. The result, he argued, was that justified criticisms of practice and allegations of abuse were dismissed, outright denied, or their authors discredited as mentally unstable, over-zealous or motivated by malice.[663] A year later, he was cited in the *Observer* as saying that psychiatric hospitals 'ran the risk of growing more and more like prison camps'. He claimed that drugs were being used to accommodate patients to intolerable surroundings instead of helping to treat emotional disorders.[664]

In 1968 Ann Shearer, a journalist for the *Guardian*, published a damning article about her visit to a sub-normality hospital. She wrote of having found children locked in a room covered in urine and excrement, which the children walked in and ate. Though one researcher replied that she'd come across no such squalor, Roy King replied that he and his colleagues Jack Tizard and Norma Raynes' four-year comparative study of institutions revealed 'abundant evidence' that such circumstances were frequent.[665] In the same year, Peter Mittler, head of the Hester Adrian Centre, set-up with NAMH's help, wrote in its journal that inmates of mental and subnormality hospitals didn't need 'medical or nursing attention or supervision'. He argued that some of the worst symptoms of chronic disorders were the result of 'impersonal institutional organisation'. Early treatment and a therapeutic community approach should, he claimed, be the main means to combat this.[666]

In the same year the report of the committee of enquiry into allegations of ill treatment and misconduct at Ely hospital for mentally handicapped people was published. As John Martin's analysis of it shows, the report revealed a 'comprehensive failure of care'. Given the fact that any nursing staff who aired concerns about conditions at Ely were bullied and victimized, the patients' position can only be assumed to have been all the worse. According to Richard Crossman, the then Secretary of State for Social Services, a Nursing Division of the Ministry of Health, had reported appalling conditions and poor nursing after a visit to Ely several years beforehand. This had simply been filed. Crossman concluded that the Ministry probably had a good idea there were a lot of long-stay hospitals in which conditions weren't much different to those at Ely.[667]

The Ely report reinforced criticisms of the hierarchical authoritarian order of the mental and mental handicap hospitals. Those patients and relatives who spoke on the 'consumer panel' at the 1969 NAMH Annual Conference expressed experiences that reinforced these images of hospital care. At the same time, many of their views also mirrored the mental hygiene movement's expressed emphasis on tailoring authority commands and relations to the perceived need for emotional security and self-expression.

Two speakers spoke as parents of children diagnosed with mental handicap. Both of them reported on their children's experience of relatively short stays in hospital or other residential care.[668] These parents had found leaving their children traumatic and the children themselves had displayed seriously disturbed behaviour during and after their stays. The medical professionals disregarded factors obvious to one parent: her son's separation from his parents clearly caused him to suffer emotionally, and this was made worse by the inadequate care and poverty of the relational environment.[669]

A father of a child diagnosed as a Mongol (now termed Down's syndrome) told of how a young doctor broke hospital policy by informing him of the diagnosis soon after his wife gave birth. The father was left to tell his wife alone and he noted that the doctor had only risked telling him straight away because he considered the father 'intelligent'. The father was left wondering how long senior doctors would have been prepared to leave him and his wife in ignorance of the situation and without support. He remarked that the motives were unclear, but suspected that they were 'devices to protect the medical profession from emotional stress and involvement'.[670]

All of those speakers who had been patients in psychiatric hospitals denounced the rigid hierarchy and control.[671] Stifling organizational procedures and rules of behaviour were criticized, along with what was seen as unnecessarily close surveillance by nurses. But these weren't just expressions of a desire simply to be left to their own private liberty. These criticisms of power and authority in the hospital were made because the speakers saw them as detrimental to their recovery and well-being.

One patient, diagnosed with schizophrenia, contrasted this prevailing hospital organization with his experience of living on Villa 21, the therapeutic community-style experimental ward that David Cooper had set up at Shenley mental hospital. He contrasted what he saw as the more 'natural, human atmosphere of the place', with the restrained and regimented orthodox ward from which he had come; there, the patients had been dominated by the 'watchful eyes of white-coated nurses'.[672] He felt it a great relief to meet with staff on what felt like equal human terms. This had been especially helped by staff giving up their uniforms, taking their meals with patients and dispensing with a separate staff room. A complementary aspect had been the unrestricted visits by people from outside the hospital and from other wards, which had been encouraged as a vital aspect of therapy. He also valued the open discussions in which visitors could take part and believed that the visitors themselves also benefited from the experience.[673]

Another speaker diagnosed as schizophrenic was also a psychiatric nurse. She had been hospitalized several times, and described how many of the elements of hierarchy and control in mental hospitals were exemplified in the use of ECT. She had received the treatment 80 times. Some of these had been given 'straight', without anaesthetic. She reported how this was sometimes justified as a means to 'save time' and, at others, considered by staff as a method of punishment.[674] She, and another of the speakers, emphasized that patients' emotional experience of ECT was often one of extreme fear and a sense of enforced indignity. Clare Wallace maintained that patients were well aware that it could sometimes be used as punishment.[675] All of these factors were considered, by patients themselves, to be detrimental to their well-being and therapy.

The speakers at this conference wanted to be listened to as people whose experience was a valid aspect of knowledge about the value and effect of treatment. Those professionals reflecting on the therapeutic importance of dynamic emotional relationships supported this to varying degrees. The psychiatrist J.H. Kahn, for example, introduced the speakers by emphasizing the commonality of feelings and experiences between those diagnosed as mentally ill and other people. He remarked: 'Mentally ill persons have scarcely been allowed to give expression to the normal part of their personalities. When they give descriptions of their own experiences, they often meet with what Goffman has called the "institutional smirk", which means, "Yes, that's what you think you mean, but we know better"'.[676]

* * *

Dialectic run ragged? The mental hygiene movement pulled apart at the seams for sure. It had been weakened by developments in leading psychotherapeutic strands of its own theorizing, and associated 'radical' therapeutic experiments. These theories and practices had themselves been influenced by political stances from across the political spectrum. Critiques inspired by various political positions had pulled at the movement from both within and without. Under pressure, the movement had shifted and developed as it attempted to accommodate various standpoints within a coherent whole. Meanwhile, it had been buffeted by wider social changes that, increasingly, it struggled to incorporate into its mental hygiene vision.

The post-war critique of the mental deficiency system was significant. Resisting and then partially embracing the critiques of civil libertarians, communists and other socialists the movement ultimately allowed psychological claims to be able to measure intellectual capacity to be divorced from the movement's use of them as a mediating factor of citizenship. This took place gradually and was obscured by the fact that the separation was largely achieved while retaining and, in fact, extending the use of the Family as an organizing principle and methodology. But, even so, by the end of the 1960s the mental hygiene movement had reached the point at which it could no longer confidently assert that the diagnosis of a level of intellectual capacity, which was accepted as signifying 'mental handicap', either denoted a failure of mental health or in itself precluded the rights of citizenship.

Meanwhile, the critiques of the traditional medical model and of prevailing institutional care, derived from the movements' vanguard psychotherapeutic theory of emotional relations had become more influential through the post-war decades. The theory penetrated beyond the areas of childcare and the treatment of mental 'maladjustments' to the realm of mental illness in the mental hospital and the community at large. As it did so it brought with it and extended the emphasis, developed in experiments before and during World War II, on the therapeutic primacy of 'emotional security' and self-expression, and its associated questioning of the relations and commands of authority.

Classically, the mental hygiene movement had sought a social order of function and fit mediated by their professional determinations of intellectual capacity and the contents of personality. But the therapeutic imperatives that became increasingly influential left this vision threadbare. Torn and rent the movement appears to have collapsed in on itself. Maybe the dialectical fruit was emerging from the dialectical blossom and becoming the 'truth' of the plant instead?

*   *   *

The Editors,

    British Journal of Criminology.

Dear Sirs,

I was much relieved when I read Dr. H A Prins's letter in your July issue on the subject of the shift he sees in your editorial policy. I had assumed that I must be suffering the effects of approaching senility when I found so much of recent *B.J.'s C.* incomprehensible, but hurrah, Dr Prins says it is your fault not mine!

I would not mind so much if the *Howard Journal* did not seem to be clutching at your coat tails and getting dragged in the same direction. I recognise of course that you are not a 'popular' journal, but you seem now to be getting beyond the reach of the average decently informed non-academic. We all know what a specialist is; must the *Journal* become an example of *specialisation* – that process in which fewer and fewer know more and more about less and less?

Yours Faithfully,

W David Wills

Reynolds House,
Bromley,
Kent.[677]

8
Dialectic Dismembered

'Smash the dictatorship of the head!'. It was 1969 and the Schools Action Union had targeted the selective school Dulwich College in south London. The Union had decided to go and find out whether the College's open day really was an 'open day'. With active members from the Young Communist League, along with a variety of Maoists and anarchists, the Schools Action Union was a short-lived and increasingly sectarian revolutionary group of students. Its central aim was the dismantling of what it saw as the school Head's sovereign power and a democratization of school control among teachers and pupils. Associated demands were freedom of speech, the outlawing of corporal punishment, abolition of school uniforms, co-education in universal comprehensive schools and more pay for teachers.[678]

But Wills was unimpressed with the invasion of Dulwich College. He wrote a letter to *Freedom: Anarchist Weekly* in response.

Dear Friend,

I have devoted my life to the furtherance of freedom in education. If anyone had burst into any of the libertarian establishments in which I have striven to express that ideal, and had daubed the walls 'Discipline:Punishment', I should have considered them mindless hooligans, and the effect upon me would have been to confirm my prejudice against discipline and punishment.

Can anyone tell me what in the name of freedom is gained by bursting into Dulwich College and daubing on its walls "Anarchy"? My opinion is that it does incalculable harm to our cause, and I would be glad to see a reasoned defence of such action. I suggest that those who talk about freedom should consult those of us who have tried to practice it before they indulge in this kind of hooliganism.

Yours Sincerely...

Someone wrote back describing his claims about having devoted his life to 'the furtherance of freedom in education' as 'a pompous Speech Day phrase that means probably no more than that he has been earning his living around the education factories and tried to be a bit tolerant'. Wills composed a response. It was a long letter. But these sentences can stand as its summary: 'Freedom is an idea. Ideas are communicated by discussion, by argument, by debate...Freedom will be spread by the intelligent portrayal of its benefits, by the clash of mind upon mind, not of fist upon flesh - or of paint upon wall!'.[679]

* * *

Meanwhile, the walls of the mental hygiene movement had actually been falling in. So, in that case, what had happened to the dialectic of the Family? Under mental hygiene the Family had remained an organizing principle of the self and the social order, even if increasingly influential elements had reinterpreted its content. But if these elements' engagement with wider forces of social critique contributed to the demise of the mental hygiene movement, did they also contribute to the demise of the Family dialectic?

At the turn of the 1970s the National Association for Mental Health (NAMH) announced a radical change of emphasis. It adopted the 'brand name' MIND and took on the role of a pressure group campaigning for patients' rights. This transformation has been portrayed by supporters and critics alike as a radical break with its past.[680] This is connected with the perception that the strategy was classically civil libertarian. In fact, some have drawn direct parallels between it and nineteenth-century civil libertarian challenges to psychiatry.[681] The 1890 Lunacy Act, for example, is generally regarded as a 'triumph of legalism', in prioritizing and protecting civil rights over psychiatric care and detention.[682] This apparent parallel, along with connections to legal advocacy for mental patients in the USA, has commonly been invoked to characterize the approach as framing the treatment of mentally ill people in terms of the deprivation of liberty.[683] There is, it has been argued, a fundamental dichotomy between the interests of civil libertarianism and of psychiatry.[684] Critics have maintained, for example, that the rights approach represented an unnecessary obstacle to therapeutic endeavour.[685] Supporters have argued that it was a welcome return to legalism in as much as it accentuated control by statute and prioritized the legal rights of the patient.[686]

In terms of our story such views seem to suggest that the dialectic has been broken; mental hygiene dies, so does the Family as an organizing principle, and individual civil rights usurp its role.

* * *

One vehement attack on MIND's rights strategy came from a sociologist who had adopted Foucault's theorizing as the foundation for his studies of psychology and psychiatry. In the mid-1980s Nikolas Rose wrote an article in which he cast the strategy similarly to other critics as a classic civil libertarian defence of individual liberty.[687] Like other depictions this implied a radical break with MIND's past. But, in terms of our story, Rose's stance was curious. He claimed that MIND was deliberately placing itself in opposition to the psychiatric profession but that, ironically, this only served to extend and 'modernize' psychiatry. He reckoned that both civil libertarianism and psychiatry had developed 'in the same transformation of social and intellectual rationality that gave birth to the concept of the individual free to choose'.[688] Consequently, they were both enmeshed within a liberal moral humanism and a bourgeois ideology of individualism.[689] Fundamentally, they both shared a contractual notion of 'the self'. Their objectives were the same: to produce and maintain privatized autonomous selves. So, Rose simultaneously gave the impression that a mental hygiene organization had been colonized by civil libertarianism, but that both, in fact, shared the same fundamental imperatives.

Regarding our tale, Rose's stance is odd for several reasons. As our story shows, the Family and civil liberties can be brought to overlap and engage in various ways but, even so, they cannot be said to be equivalent. In any case, Rose made no attempt here to show how the Family was, or could be, the driving organizing principle and methodology of civil liberties. Beyond this, Rose's critique made little attempt to disentangle the varying notions in psychiatry of what constituted healthy individuality or to show how these varying notions have interacted historically with those he considered representative of civil libertarianism.[690] Neither discourse can be considered static over time, as the shift of NAMH to MIND, in fact, exemplifies.

It is clear, in fact, that Foucault's 1967 description of the Family's fundamental role in modern psychiatry and its origin in moral treatment had no part in Rose's argument. Modern psychiatry, including mental hygiene, had been redefined. Something else lay at its core. The reason lies in the fact that around the time NAMH was transforming itself into MIND, Foucault was transforming his own theoretical position from the one that he'd expressed in *Madness and Civilization*. Here is the moment referred to in the introduction to our story. Hadn't Foucault used the very Reason that had apparently silenced madness in order to write a history of that madness? And didn't his tale suggest a romantic image of madness as some sort of pre-rational 'truth' that had been suppressed? Foucault had been forced to respond. The turn of the 1970s saw him admitting that he had been 'positing the existence of a sort of living, voluble and anxious madness which the mechanisms of power and psychiatry were supposed to have come to repress and reduce to silence'.[691] His rendition of madness as 'alienated in guilt' under the organizing principle of the Family must now have appeared to him an inadequate representation of power. So what happened to the 'dialectic of the family' that in 1967 Foucault had claimed was still grinding on, to a termination date impossible to predict?

Damned dialectic. Smash it. Smash it with a philosophical hammer. It would no longer even be recognized as illegitimately injected into history. It would be denied entirely. Inspired by Nietzsche, Foucault developed a genealogical approach that conceived history as a 'profusion of entangled events' whose 'essence' is not intrinsic, but fabricated in a haphazard fashion from 'alien forms'.[692] With this conceptual understanding, he now argued that 'Nothing in man – not even his body – is sufficiently stable to serve as a basis for self-recognition or for understanding other men'.[693]

* * *

Isaiah Berlin wasn't a fan of Rousseau, that radical, but dangerously ambiguous, proponent of self-government. Berlin included him as one of his *Six Enemies of Freedom*. Rousseau was a 'tramp', 'a guttersnipe', and, worst of all, he was silver-tongued with it. He had a way of writing, said Berlin, that was spellbinding. With eloquent prose he wielded words and concepts that appeared to be entirely familiar to his readers. But these were immediately re-spun in ways that produced an 'electrifying effect' and seduction into an entirely foreign world.[694]

You couldn't describe Foucault as a lower-class street urchin, but the rest of this could as well be said of him. Having caused a stir with his poetically written *Madness and Civilization*, and appearing in accord with both the burgeoning counter-culture and anti-psychiatry, Foucault's work transformed into a Nietzschean-inspired theory of 'power/knowledge'. In the process, concepts like 'power', 'power relations', 'politics', 'government' and 'truth' itself were taken up and transmuted in Foucault's hands.

In fact, truth became the principal target of attack. The traditional couplet knowledge/truth became knowledge/power, or power/knowledge as Foucault was to order it. He announced that 'The political question... is not error, illusion, alienated consciousness, or ideology; it is truth itself. Hence the importance of Nietzsche'.[695]

In Foucault's revised approach the organizing principle of the Family no longer figured, except in subsidiary fashion as one of the many constellations of the primary power of reality—power/knowledge. The shape of humanity was no longer, as it had so often been touted, that of a household or family. So what was it? Well it seems more like a city of torment. The comforts of the present now became revealed as mere skin stretched over mechanisms of torture. Municipal gardens apparently representing enlightenment, reform, love and even humanity itself were all uncovered to expose only the sites of exquisite pain. Political or personal self-government was a sham, and the sovereign hierarchy of knowledge/truth just an apparition that concealed the inverted and dispersed hierarchies of power/knowledge.

In 1975 Foucault published *Surveiller et Punir: Naissance de la Prison*, translated into English in 1977 under the title *Discipline and Punish*. The book focused on prison and punishment, but elaborated a general schema for understanding how power had transformed in modern society and how it now operated as a productive rather than repressive force. He attempted to show that 'from the seventeenth and eighteenth centuries onwards, there was a veritable technological take-off in the productivity of power'.[696] An often cited passage reads:

> Perhaps, too, we should abandon a whole tradition that allows us to imagine that knowledge can exist only where the power relations are suspended and that knowledge can develop only outside its injunctions, its demands and its interests. Perhaps we should abandon the belief that power makes mad and that, by the same token, the renunciation of power is one of the conditions of knowledge. We should admit rather that power produces knowledge...that power and knowledge directly imply one another; that there is no power relation without the correlative constitution of a field of knowledge, nor any knowledge that does not presuppose and constitute at the same time a field of power relations.[697]

It's a statement that, at first glance, doesn't seem so far away from the extrapolation of the notion of institutionalization in the mental hospital and prison to wider society. Were power and authority being further examined in the attempt to throw off an authoritarian hierarchy of power relations that trapped and distorted people? You could have been forgiven for asking. Certainly, to say that there could be no position from which power relations could be suspended and true knowledge apprehended might have struck a chord with those who saw all knowledge as mediated by social relations. But the crucial difference was that Foucault was re-spinning the definition of 'power relations'. These didn't exist between people. They were relations between power and knowledge themselves. Foucault said that they couldn't be analysed on the basis of 'a subject of knowledge who is – or is not – free in relation to the power system'.[698] Elsewhere, he made this point clearer by stating that 'In short, it is a matter of depriving the subject (or its substitute) of its role as originator, and of analyzing the subject as a variable and complex function of discourse'.[699] So power and knowledge worked in relation and inseparably. And they pre-existed the individual. Thus was the famous 'knowledge is productive' claim. It's become like a mantra to some. Foucault has told us that we must stop, once and for all, thinking of power as repressing, excluding, censoring, masking or concealing.[700] No more illusion, no more alienated consciousness, no more ideology. The crucial element in a nutshell is that the 'urground' of everything for Foucault is power/knowledge. Reality is the ongoing result of power/knowledge combat. Power/knowledge is everywhere. It makes things up. It makes people up. We are power/knowledge fabrications. Power produces reality.

This was the theorization that informed Rose's studies of the 'psy' disciplines and his critique of the rights approach in mental health. Indeed, these days it's difficult for any academic to hear terms like 'government' and 'power' without also hearing Foucault in their heads. But in our dialectical tale of moral treatment and the organizing principle of the Family we've used understandings of power and government that all the actors would have broadly understood. They were in current usage at the time and, of course, largely remain so. So, instead of applying Foucault's theory of power/knowledge to our story, let's do it the other way around. As Foucault is clearly entwined in our historical tale, let's try a reading of his approach in terms of power as centralization, power as having a popular basis, and power as rationalized and totalized. How might it look?

Famously, Foucault cut off the King's head. He maintained that traditional theories of sovereignty considered it fundamentally linked to power over a territory and that the way this power was expressed was 'juridical and negative'. He claimed:

> Sovereign, law and prohibition formed a system of representation of power which was extended during the subsequent era by theories of right: political theory has never ceased to be obsessed with the person of the sovereign.[701]

But Foucault asserted that the essential problem was to show how power had, instead, become 'technical and positive'. This transformation began, he said, around the sixteenth century when government by the sovereign became defined around the perceived requirement to order and make efficient the population, the family and the economy. What emerged was a modern regime of 'bio-power'. Its twin points of engagement were the human species and the human body. Interventions at the level of population were matched by technologies that acted on the individual body, making it controllable, teachable and transformable. In *Discipline and Punish* Foucault used prisons and punishment to show some of the ways this took place. Jeremy Bentham's architectural panopticon had been designed for prisons and other applications, and it provided an ideal paradigm for Foucault. In his hands, what he called its 'diabolical power' could convey the fact that power was actually multiple, technical and productive. The device was concerned with the efficiency of power and, for Foucault, this was linked both with its role of normalization (the gradation of individuals against a norm) and the interiorization of such power by those who were subject. *Discipline and Punish*, along with Foucault's later work, detailed various ways in which technologies of power invested individuals with a subjectivity. Panoptical power was thus a technical and productive 'gaze'. But, in Foucault's description, it also denied any one person the command of this gaze. Power couldn't be held at one central point, be it the sovereign, the state or any particular individual; there was no centre or pinnacle to power outside of us, or within any of us. And, by the same token, everyone was caught in this power. The tactics and technologies of power/knowledge originated and organized around 'local conditions and particular needs'. The mechanisms of power didn't form a homogenous whole, but a complex interconnection of engagements and assemblages.

So how might we understand this power/knowledge in terms of the 'old' co-ordinates of power that our story's engaged with? Without doubt it's a rationalizing and totalizing power. In fact, it's rationalizing and totalizing power with knobs on. It's as if calculating, ordering and systemizing have become the one and only focus for theorists who have followed in Foucault's wake. And there's good reason for this impression. It's closely related to the fact that these components of power are separated from any notion of power as centralized or popularized. As we've noted, the stripping of power from the sovereign applies to the personal sovereign, a monarchy, a state, and also any and every individual. So any notion of popular power or centralized power is subverted under the subterranean regime of power/knowledge.

But it seems strange, on the face of it, that Foucault chose not to focus analytically on authority when he attended to shifts in the nature of sovereignty. It is a 'modern notion of political authority' and its core meaning has, after all, generally been accepted to be 'supreme authority within a territory'.[702] Why is the issue of authority so absent in Foucault's power/knowledge? Sure, Foucault used the term every now and then, but its dynamic association with power had, essentially, been severed. In effect, 'sovereignty' was reduced to power at a centralized point. Foucault referred, at times, to sovereignty as an 'apex' or 'summit', but this meant little as he'd shorn it of any hint of the notion of authority that the use of such terms might just contain. Authority had become merely an epiphenomenon of the machinations of power/knowledge.

So, for Foucault, where there is power there is knowledge and where there is knowledge there is power. But surely where there is knowledge there is also authority? How could authority not therefore be bound up dynamically with power/knowledge? And a crucial consequence of placing an accent on authority is that it draws attention to the fact of imbalances of power between people. These imbalances of power may be accepted by groups and individuals, or they may be questioned, resented or rejected in whole or part. The general relational co-ordinates of authority can vary wildly over time and place, but the issue of human beings accepting or rejecting authority is clearly an inherent aspect of authority as a concept. Foucault's power/knowledge subverts this. Somewhere, at some prior point, power/knowledge, it is tacitly claimed, existed before—and is the basis of—authority in its various manifestations.

* * *

It was this theory of power/knowledge that informed Rose's assertion that civil libertarianism and psychiatry actually shared the same fundamental imperatives of producing and maintaining the autonomous, privatized individual. The key word was 'produce'. The theory of power/knowledge said that these activities produced the individual. Their theories and activities emerged with the era of the bourgeois individual and liberal moral humanism. So, first, Rose claimed that the rights strategy amounted to the classic civil libertarian defence of individual liberty, and the assertion of a 'right to be different' outside of psychiatric control and coercion; second, that this strategy was cast in complete opposition to the 'truth-claims' of psychiatry; third, that this was paradoxical as they both, in fact, worked to 'produce' the same bourgeois individual; fourth, that the rights approach couldn't, therefore, either transform 'social provision' or 'provide authoritative solutions' for the 'contemporary fragmented moral order'; and, finally, that the 'psy disciplines' and the rights approach actually only operated to produce selfish self-absorbed people who had invested in an idea of freedom that was entirely subjective.

As we'll see, in terms of MIND's activity, none of these statements is very accurate, and the inability of the theory of power/knowledge to integrate an analysis of authority appears to underlie much of the inaccuracy. In fact, the effect of what I have called the partially inverted Family and its connections with contemporary elements of social critique is clearly apparent in MIND's work. These connections were unstable and can be seen as a further development of the continuities and contradictions we've been following. So, who knows, perhaps if that's the case then maybe our 'dialectic of the Family' continued to play itself out as well? Perhaps the pressure group and rights strategy adopted by MIND was the 'negation of the negation', the fruit transforming from, and refuting, the bud?

* * *

If MIND's rights approach was classic civil libertarianism how come it aimed to promote the civil and social status of people whose diagnosis had traditionally led to their separation from civil society? Our story has suggested an important reason. We've seen that through theorizing about the relationship between family authority relations and a duality of instinctive emotionality in the child, mental hygienists linked childhood with 'the primitive' and 'madness'. These had a common emotional core and this endured in diluted form in the rationally marshalled adult. There was, therefore, a fundamental continuity of normal and mentally disturbed experience. We saw, for instance, that the psychiatric social worker Sybil Clement Brown declared in the 1930s, 'most of us are to some extent, or sometimes, "insane"'. But that contention was accompanied by another. This was that emotionality was an important and shared language only in so much as it was progressive, or potentially progressive. This limited attempts at communication to therapy with selected groups of people. In particular, people labelled mentally deficient were ruled out. Despite this, the range of people encompassed by this therapeutic attempt at communication widened through the life of the mental hygiene movement. As it did so, the questioning of authority commands and relations that accompanied the movement's psychotherapeutic theorizing encouraged an emphasis on greater freedom and egalitarian relations in care and therapy. This, along with a suspicion of doing things to passive or unwilling patients, encouraged critiques of institutional treatment and the prevailing medical model. Advocacy of 'open communication' became influential across sites of mental hygiene activity. As it did so an emphasis on similarities of human experience and response, rather than deficiency and difference, also became more pervasive.

The apparent compatibility of these elements with wider forces of social critique during the post-war decades, contributed to their partial release from containment solely to therapy. At NAMH, by the end of the 1960s, patients' voices were beginning to be listened to as sources of knowledge that could challenge and inform the organization of care and treatment.

The continuity of normal and mentally disturbed experience outlined by the mental hygiene movement also informed MIND and its desire to raise the social and civil status of mental patients. For instance, the MIND manifesto, produced in 1971 as the policy basis for its early rights campaigning, noted that:

> No boundaries mark out mental illness from mental health. Just as most mentally ill people have periods of stability and insight, so do 'normal' people experience feelings of irrational anxiety and depression. Mental illness may begin as a distortion and exaggeration of moods and emotions which we all share. So the mentally ill are not a separate race divorced from our world and our experience: they are 'we' and we are 'they'.[703]

In 1974, as MIND was embarking on a more radical interpretation of its strategy, it produced a report on psychotherapy that appears to have appreciated some of this provenance regarding its promotion of patients' rights. It stated that

> ... psychological treatments offer our best hope of increasing understanding of mental illness. The psychotherapeutic approach is to the patient

as a whole person (including his physical make-up, for which physical treatment may be prescribed). It is this approach that has advanced our treatment of mental illness in the past – it was when psychoanalysts began to shed light on the individual's personality that the mentally ill ceased to be lunatics to be controlled and became people to be understood. Increased opportunities to control symptoms by drugs should not mean that psychotherapeutic skills are allowed to atrophy – or our patients will become lunatics again.[704]

* * *

The photo shows some young people who lived at Fairhaven and Fairlop, MIND's hostels for school-leavers labelled Educationally Subnormal. They are standing in front of what is described as a 'Ford 12-seater coach'. In the driving seat is a man with suit and tie. His name is Tony Smythe and he has just been appointed the new director of MIND. He is receiving the keys from two other men in suit and tie. On the side of the door is emblazoned 'Sponsored by Ovaltine'. It makes you wonder what was going through Tony Smythe's mind as he accepted the keys. 'We are the Ovaltineys'?[705]

* * *

On 1 January 1974 Tony Smythe was appointed MIND's Director and given the task of integrating the MIND campaign with the whole work of the organization. We noted in the last chapter that Smythe had been a member of the Campaign for Nuclear Disarmament's off-shoot direct action group, the Committee of 100. This had a strong contingent of anarchists, and Smythe himself was, in fact, a pacifist anarchist. His radical activism had been sparked by conscription. Resistance and refusal to take on alternative civilian service led to a short time in prison. Soon afterwards he had become the assistant Secretary of War Resisters International. One of his major contributions was to help organize its 1962 conference in Beirut with the aim of establishing a world peace brigade that would offer nonviolent intervention at places of conflict and tension. He was appointed head of the National Council for Civil Liberties in 1966, broadening and popularizing its activities to include defending the rights of children and minorities.

Under Smythe MIND's pressure group and rights approach became more radical. One expression of this was MIND's appointment of Larry Gostin, a young lawyer from the USA. In 1975 the first volume of his book *A Human Condition* set out proposals for reforming the 1959 Mental Health Act. But even under Smythe's more radical approach strong continuities between this and those we've traced emerging in the mental hygiene movement remain clear. For example, Smythe also endorsed the continuity of normal and mentally disturbed experience. Describing the psychiatric system he noted that:

> Distinctions between individuals, their needs and behaviour, and between groups of people who share diagnostic labels or common problems are blurred to give the impression that mental patients are a uniform class somewhat separate from the mainstream of humanity.

> The misapprehension that there is a firm dividing line between normality and abnormality diverts attention from individual needs and capacities and the dangers of discrimination, paternalism and oppression. In fact there is no such line – more a grey area which is not entered by many of us, occupied permanently by some or crossed regularly or spasmodically by others.[706]

Both the MIND of the MIND manifesto and of Smythe's directorship openly shared wider objectives established by the mental hygiene movement. From the interwar period the mental hygiene movement had advocated a comprehensive preventive mental health service for the community at large. Increasingly, this vision had included rehabilitation and treatment services whose location was to be the community. Post-war developments, such as the Social After-Care Scheme, tended to emphasize that organizational control should also be based in the community. In the lead up to the 1959 Act NAMH had pressed for legislation to ensure the development of services based in the community. After enactment NAMH criticized the failure to make mandatory the community care provisions that it had been envisaged would be carried out by local authorities. MIND's rights strategy continued this emphasis. In fact, it was growing discontent with the failure of successive governments to fund and statutorily enforce the extension of community care and treatment that partly spurred NAMH's adoption of a rights emphasis. In 1974 MIND released a document criticizing the slow development of community provision by local authorities. The 1946 National Health Service Act had only provided local authorities with the discretionary power to provide this care. But MIND argued that the 1959 Mental Health Act (along with the subsequent Circular 22/59) and the 1968 Health Services and Public Health Act had reinforced this power to the extent that it could now be considered mandatory. On the basis of this view MIND signalled its decision to pursue test cases to firmly establish the legal right to community care.[707]

Post-war, the psychotherapeutic emphasis on emotional relationships increasingly emphasized that a failure of such relationships was a core unifying component of both minor mental disorders and mental illness. This entailed a destabilization of the prevailing medical model, and a critique of institutional care and treatment. We've seen that the latter informed the 1954 Royal Commission's aim of shifting the basis of welfare towards the community. The social isolation and passivity associated with institutionalization was the opposite of an engaged relational integration as close to normal family and community life as possible that was associated with mental health. MIND shared and developed this stance. Its approach can't be reduced to the simple assertion of liberty and choice against psychiatric control and restraint.

The very first page of *A Human Condition* asserted that 'there is a consensus of opinion among mental health professionals that community-based care should, wherever possible, be preferred to confinement in traditional institutions'. The 'poverty of the social environment' in traditional hospitals, it said, created a 'functional pathology' termed 'institutionalism'.[708] This was correlated with isolation, withdrawal from relationships, apathy, dependence and conformity. Community provision was correlated with social integration and participation.[709]

MIND's critique covered both mental illness and mental handicap hospitals. A pamphlet it published in 1972 was written by Ann Shearer, the former *Guardian* journalist. She helped initiate the Campaign for the Mentally Handicapped, which worked closely with MIND.[710] This pamphlet maintained that the institutional system had not only fed a myth that mentally handicapped people were 'uncontrolled and perverted in their sexual appetites', but fabricated the 'odd behaviour' that such stereotypes depicted by its separation of these people from 'normal emotional patterns of life'.[711] The same issues were stressed in MIND's campaigning throughout the 1970s and 1980s.[712]

For people termed mentally handicapped, MIND continued to advocate the kinds of approaches to residential provision proposed by Jack Tizard at Brooklands, and associated ventures, such as Albert Kushlick's work at Wessex.[713] These, in turn, were linked with Goffman and Barton's earlier work on the effects of institutionalization.[714] MIND emphasized that Kushlick's work showed care and integration within the community could be provided for all, including the most severely mentally handicapped people.[715] When his research unit came under funding threat MIND publicly defended the importance of its work.[716]

Regarding hospitals for people considered mentally ill, MIND emphasized the need for integrated community care sustaining supportive human relationships with its 1976 'Home from Hospital' campaign. Along with the government, MIND estimated that between a third and a half of patients in mental hospitals could live in the community if they had somewhere suitable to live.[717] The aim was to increase this accommodation for these, and already discharged, patients. Many of these patients required continuing support in the community.[718] MIND maintained that:

> Many of these patients will be severely institutionalised and while recognizing the reasons for this it is important not to assume that to place

people in the community is to modify their institutionalised behaviour – 'Institutionalisation does not disappear when asylums walls fall down'.[719]

This was no assertion of a 'right to be different' outside of psychiatric control and coercion, as Rose asserted. In fact, MIND's 1974 report *Co-ordination or Chaos?* remarked that the Department of Health and Social Security 'appears to have shaped its policies on the assumption that chronic mental illness no longer exists'.[720]

MIND promoted a variety of accommodation and support options. These included day centres, boarding out schemes and group homes.[721] It emphasized that community services must avoid setting up 'mini-institutions', and that hostels should be 'homely' and integrated with the general community.[722] NAMH's local associations had pioneered 'group homes' in the 1960s. Similar principles informed the establishment of group homes for people termed mentally handicapped, as well as people from mental illness hospitals. Through these, selected patients could leave hospital and live together in small groups in the community. A MIND working party on residential care set up in the early 1970s described the 'principal ingredients':

> The experience of group life and the opportunity to learn to live harmoniously in a small community in a somewhat sheltered situation... As in a hostel, residents have the opportunity to learn from one another, to mix, and to make allowance for other people, but without resident staff to intervene. The individual must be ready to exercise his own choice within the daily routine and to manage with a degree of independence. The amount of supervision afforded by visiting social workers varies considerably but the opportunity for group interaction differentiates this form of care [from] that offered in flatlets or other sheltered accommodation. Generally the number of residents in a group home does not exceed eight.[723]

This statement points us again to the way in which the Family, as organizing principle and methodology, had become partially inverted. A specific aim of group homes was to achieve rehabilitation through enabling a shared life as similar as possible to a family environment.[724] Mixed sex groups of no more than five or six people were recommended to retain a 'family atmosphere' and avoid the creation of a mini-institution.[725] But, at the same time, a distinctive feature of these ventures was considered to be the absence of staff, and residents' responsibility for their own affairs. Commenting on its promotion of group homes for people labelled mentally handicapped, MIND noted that living together without direct supervision had resulted in improved emotional relationships and communication skills. One result of this had been improved self-confidence and a greater willingness to stand up to what residents sometimes experienced as unnecessary authority.

It added, however, that researchers had shown that this had sometimes caused hostility from people who still expected mentally handicapped people to be 'docile and accepting'.[726] MIND responded that 'the label "difficult" applied to mentally handicapped people may be a reflection of staff attitudes – and it is time that this fact was recognised'.[727] Ann Shearer argued that, instead of an outmoded insensitive and hierarchical system, there needed to be shared and egalitarian relationships that enriched the lives of clients and staff together. The entire impetus of the professional ethic, however, had been the creation of distance between carer and client. Weren't we being too pompous about professional qualifications? she asked.[728] A Campaign for the Mentally Handicapped report, publicized in MIND's magazine *MIND OUT,* argued that the medical model of care was increasingly considered to be obsolete. It reiterated that community care should be based on a democratic multi-disciplinary model of educational and social models of care. Care for mentally handicapped people, it was maintained, entailed a 'hard critical look at the roles of powerful professionals'.[729]

These views were echoed in two of MIND's experimental day-care and drop-in centres for people with mental disorders, which were run with the Tavistock Institute. The Brecknock Community Centre in North London aimed at providing non-institutional support in the community for people with different needs. Located in an area with a large bed-sitter population, it was intended to provide support for isolated people (as well as those 'who may have been institutionalised for years') in a non-categorizing, non-labelling, informal environment.[730] Also based in North London, the Junction Road Project aimed to provide a community mental health project at a 'neighbourhood' level. It hoped to alter the nature of the relationship between caregivers and clients, through an advice and information service, and supported self-help groups.[731] A multi-disciplinary team provided the backbone of the project, with support from local volunteers.

* * *

Wills wasn't just coming under fire from certain anarchists. One of his oldest colleagues from the Hawkspur days was having a pop at him now. In the early 1970s Arthur 'Bunny' Barron attacked him for sanctifying Homer Lane and claiming that 'self-government' was central to their therapeutic work. In response, Wills admitted that he'd over-egged Lane's importance, but he still revered him. Wills didn't accept Barron's strictures on Lanes work because he reckoned that they were based only on Lane's early approach at the Ford Republic in Detroit. According to Wills, Lane, at that time, thought he'd found a 'panacea for youthful misconduct'. Wills gave a crude description of it:

> You simply collect all your offenders together, impose upon them a mode of democratic government based on the American constitution, and an

economic system based on capitalistic free enterprise and tell them to run it themselves. Then presto, in a couple of years they are cured.

Wills added, 'Obviously it is all nonsense to assume that is a panacea for anything' and he noted that Lane had come to realize many of its shortcomings before he got to England. Wills didn't see the 'self-government' he had just described as therapeutic. But he did see the kind of 'self-government' or, more specifically, 'shared responsibility' that had been applied and developed at Hawkspur as having 'considerable therapeutic value'. However, his main defence of these methods was that he'd have used them even if it didn't.

> The word *right* in the English language, and its equivalent in several languages...is extremely fascinating because it carries such an enormous range of meanings...I should like to draw your attention to two of those meanings – on the one hand we have the sense of correctness as in 'You've got your sums right', and on the other we have moral rectitude as in 'God defend the right'.[732]

This statement drives to the heart of our dialectical story and its contradictions related to authority. Was this therapeutically inspired version of liberty, egality and fraternity only a means to an end? Wills didn't think so. For him the means and the ends must remain as close as possible.

Wills acknowledged that all the key areas of Hawkspur's operation were adopted by the Q Camps committee because they were held to be in the interests of psychological investigation and treatment. 'Regimented discipline' was believed to create artificial behaviour, whereas an egalitarian order would allow observation and diagnosis of behaviour that was uncoerced and free of facades. The version of self-government termed 'shared-responsibility' adopted at Hawkspur reinforced this and also allowed cathartic debate for the members. Meanwhile, the emphasis on productive work within this environment encouraged purposive mental activity and social reengagement. But, while Wills accepted that these imperatives were derived from psychological 'science', he insisted that, for him, there was a pre-eminent moral 'right' informing them too. He emphasized that even without the psychological reasons he would have run the place in the same way. This was for the moral reason that

> ...all men are intrinsically of equal worth. I do not mean of equal intelligence, much less of equal virtue, but that all are by reason of their human birthright entitled to equal consideration from their fellows and an equal freedom in pursuit of their own destiny. I accepted the old anarchist tag – no man is good enough to be another man's master.

Wills added that all people shared an equal responsibility to try to ensure that their freedom to pursue their own ends didn't frustrate their neighbour's freedom to do the same. People therefore needed to meet together in order 'to examine and control and preserve those boundaries'.[733] Work was necessary to any community, but it should be organized democratically and as freely entered into as could be made possible. Similarly, the educational aspects of Hawkspur had, for Wills, been arranged on egalitarian lines in an attempt to restore 'a right that had been denied' in a 'class-ridden and inequitable society'.[734]

* * *

We've seen that Nikolas Rose reckoned both psychiatry and civil rights had developed historically 'in the same transformation of... rationality that gave birth to the concept of the individual free to choose'. They were thoroughly implicated in the bourgeois market governed by private contracts between atomistic autonomous individuals. Therefore, said Rose, they shared the same objectives. They aimed to produce and maintain the privatized, autonomous individual.[735] Rose made use of socialist arguments to bolster his position. Socialists, he reminded his readers, had long struggled over the utility of rights as a strategy capable of promoting fundamental social change. This was certainly true. Peter Sedgwick, for instance, argued in the 1980s that rights-based moves to reform medicine peddled an illusion that liberty and equality could be achieved through focusing on the misuse of particular medical practices, rather than challenging the structural aspects of that power.[736] Others, however, while recognizing these contradictions, nevertheless considered rights campaigns to be useful, though limited and pragmatic, strategies of empowerment for minority groups and the working class in general.[737] This seems to have been the position of MIND. However, subsequent evidence appears to support some of the socialist inspired criticisms. Despite the evident fact that, as one contemporary rights protagonist emphasized, the 'discourse of rights' cannot be reduced to a 'homogeneous and undifferentiated [school] of thought',[738] it remains the case that rights-based strategies were easily appropriated by the Conservative governments of the 1980s, and put to the service of 'consumer choice' in its policies of privatization and marketization in the National Health Service (NHS).[739]

But, regarding Rose, it needs to be emphasized that while he made use of Marxist and other socialist criticisms of rights concepts, he elsewhere ruled out historical research and political strategies based on these philosophies. For him, they illegitimately reduced the means of social regulation to the state in the interests of capitalism and viewed power in terms of its repressive nature on subjectivity.[740] Rose's theoretical position was, as we've seen,

Foucault's power/knowledge, so no more thinking of power as repressing, excluding, censoring or concealing. No more alienated consciousness, no more ideology.

Rose maintained that radical critics of the psychiatric system ought to look elsewhere for methods and approaches. But surely Phil Fennell had a point when he argued in response to Rose that

> ...neither the origins nor the character of these rights should allow us to lose sight of their intrinsic value. Whilst they may not often affect substantive outcomes, they do open up areas of the psychiatric system to scrutiny which might otherwise remain hidden, and they require those who operate the system to reflect on and justify what they are doing.

And he was surely correct when he added:

> When reminded by Rose 'of the limited nature of law and legal mechanisms *vis-a-vis* other mechanisms of organizing, monitoring and transforming social provisions' we are surely entitled to ask what these 'other mechanisms' are before baby and bathwater disappear together.[741]

In fact, it's here that the key to understanding Rose's 'radical' critique seems to lie. Rose, in fact, proposed no potential alternative means for transforming social provision. Instead, he offered only vague and tentative suggestions. According to him, rights approaches and the psychiatric system shared the goal of producing and maintaining the atomistic, autonomous individual. So what was it about a rights approach that was most problematic for Rose? Well, it doesn't seem that it was because, at root, it failed to fundamentally challenge, and, indeed, might appear to support, the unequal and exploitative social relations of the bourgeois liberal market economy. This is what many socialists would argue. Instead, Rose chose to emphasize that the 'rights discourse is incapable of providing authoritative solutions to the problems of our contemporary fragmented moral order'.[742] He asserted that:

> It is a sign of our times that one has to remind oneself that it is possible to think of an ethics without rights, perhaps framed in a language of duties and obligations, of social support given not because it is a right, but because it would be virtuous to give it, or politically correct to give it, or because it would make the giver a better person. It is worth remembering that other grounds for morality exist than those in which humans are to be valued only in so far as they get what is due to them.[743]

The criticisms implicit in this statement were hardly applicable to MIND's work, as we'll see. What comes across from this statement is that Rose seemed to believe people were being encouraged to think only of themselves and their desires. It is a curious thing that such an amoral theory as power/knowledge was turned here into something so moralistic. Rose continued this fundamental theme in his later writings. On the effect of psychotherapy in the 'fabrication' of the self, for example, he wrote:

> It is the self freed from all moral obligations but the obligation to construct a life of its own choosing, a life in which it realizes itself.[744]

Under the theory of power/knowledge, of course, this thinking just of yourself and your own desires is 'real' in the sense that it is created by external 'technologies' and 'disciplines' (in this case the 'psy' disciplines). Doesn't that make it an illusion though? And weren't we to have no more truck with the idea of illusions and concealment? Well, the Foucauldian answer to that is that what 'technologies' and 'disciplines' fabricate *is* real. As we saw, power/knowledge makes things up. It makes us up. Power/knowledge produces reality. The extent to which this 'fabrication' is claimed to create 'the real' has been attested to in a recent paper advocating that a Foucauldian approach be adopted by the profession of medical history. Referring to Foucault's concept of bio-power, Roger Cooter states that it is 'knowledge producing processes through which institutional practices come to define, measure, categorize and construct the body and somatically shape all experience, meaning and understanding of life'.[745]

In the introduction to a 1990's compilation of some of his articles, Rose actually went so far as to claim that *'the very meaning of life'* in modern liberal democracies had been 'made possible by, and shaped by, the modes of thinking and acting' of the psy disciplines. These, he claimed, had 'infused the shape and character of what we take to be liberty, autonomy, and choice in our politics and our ethics'. Through this process, he said, 'freedom has assumed an inescapably subjective form'.[746] Elsewhere, Rose gave an example of how this operated regarding psychotherapy. This technology induced 'the self' to speak and confess, and, through this, he argued, 'the self' itself became fabricated. He stated, for example, that:

> In the subtle communicative interaction of the confessional scene, the expert gently brings the subject into a relation with a new image, an image that appears more compelling because it is their own... They become, in the passage through therapy, attached to the version of themselves they have been led to produce.[747]

Regarding such processes, Rose concluded that 'The irony is that we believe, in making our subjectivity the principle of our personal lives, our ethical systems, and our political evaluations, that we are, freely, choosing our freedom'.[748] But doesn't this rendition of psychotherapy present an extraordinarily passive depiction of people? Isn't it also highly unreflective given that academics themselves spend years in an induction process? What is the version of themselves (and others) that *they* have been led to produce?

In any case, Rose was clearly against the kind of 'real' (world and 'self') that he argued was fundamentally created by the 'psy disciplines' and, by extension, legal rights activity. He portrayed the rights approach as a classic civil libertarian strategy 'grounded in a right to privacy, to control over one's own internal thoughts and feeling, and to protection from assaults, no matter how well-meaning their motivation'.[749] In response, he wrote that

> It is clearly inadequate to appeal to some fundamental opposition of coercion to liberty, freedom and privacy. For even the demarcation of space of personal autonomy which is 'not the law's business' does not constitute an absence of regulation so much as a change in its modality.[750]

In other words, under the theory of power/knowledge there's nothing to be set free. Social regulation doesn't ever coerce and repress, it only produces. But, more to the point for us, not only does the theory maintain this, it also rules out any attempt to appreciate the role of authority in knowledge and power. Authority suggests power imbalances between people in the social world, and power/knowledge is simply unable to grapple with the fact that they take place and what the consequences may be. What could Rose offer instead? Not a lot. And this isn't surprising as the theory of power/knowledge doesn't leave you a lot to offer. If you're going to argue that the 'essence' of individuals is that they are, in their entirety, fabricated by power in the form of power/knowledge then you're going to be hard-put to find a moral justification for the type power/knowledge fabrication that you might prefer, let alone come up with some way of showing how it could be brought about. Still, Rose *did* tentatively offer some alternatives from the past. Instead of what he saw as the misconceived and self-defeating application of civil rights, Rose suggested that other historical discourses which propounded alternative values might be remembered:

> The ideals of community promoted both by the moral treatment of the last century and by the contemporary therapeutic-community movement certainly led to the development of profoundly moralizing institutions, but they also make thinkable a mode of support and care for distressed people that locates them within a matrix of emotional and practical affiliations, and that sees autonomy as a problem and not a solution. In a somewhat similar manner, the communitarian, mutual aid approach of Geel in Belgium, maintains the mentally distressed in a system of family placements linked together by structured collective support mechanisms.[751]

Rose was quick to say that he wasn't, in fact, advocating either, just pointing out that they were 'difficult to conceptualize within the horizon of the
contractualization and autonomization of both rights discourse and contemporary psychiatry'.[752] But this inevitable equivocation aside, there are
some obvious problems with the passage. In the first place it's hard to see
how Rose could claim that nineteenth-century moral treatment wasn't about
the resurrection of an autonomous and encapsulated individual. Rose himself had argued that it was. In an earlier work he had maintained of moral
treatment that

> This continual play of judgement had the objective of forcing the inmate
> to take into himself the role of judge, to internalize the moral order which
> constituted the asylum, to incorporate the rules and principles of that
> institutional space of morals into the moral space of his own character.[753]

What's that, if it isn't a representation of moral treatment as consider-
ing 'autonomy as a solution'? Likewise, regarding therapeutic communities,
Rose later wrote that they were 'a profound strategy of normalization of
maladjusted selves'. They should be understood, he has said, as coextensive
with 'the reconstitution of the patient as a person'.[754] Ambiguous as these
statements are, given Rose's account of the constitutive role of the 'psy disciplines' in the fabrication of 'the modern self', they surely can't offer much,
on his view, beyond what he had described as the individual free to choose
a 'subjective' freedom.

As we'll see shortly, Rose was quite wrong to separate therapeutic community approaches from MIND's rights strategy. But, for the moment, we
should note that in drawing moral treatment and therapeutic communities
together as alternatives to rights discourse, he overlooked the fact that there
are important differences between them that derive from the issue of authority and its relationship to mental health and therapy. This appears to be at
the root of his misrepresentation of MIND's approach.

We return, again, to the partial inversion of the Family that had
emerged within the mental hygiene movement. This had derived from a
psychotherapeutic emphasis on tailoring familial authority relations to a
need for emotional security and self-expression. Questioning of authority
commands and relations encouraged an emphasis on greater freedom and
egalitarian relations, advocacy of 'open communication', and concern about
passive or unwilling people having things done to them in the name of
care or treatment. This reinterpretation of the Family had emerged before
the war, as we've seen, with the reconsideration of the relationship between
authority, freedom and 'love'.

The direct connection with MIND's approach is made distinct in its 1978
written evidence to the Secretary of State for Education and Science on corporal punishment in Schools.[755] It was written by the psychiatrist Anthony

Whitehead and Maurice H. Rosen, a general practitioner. Whitehead had been a member of the council of MIND for several years. As an assistant psychiatrist in the 1960s he had worked with Russell Barton in transforming Severalls hospital.[756] Despite some disagreement during discussion at MIND's Council of Management, the evidence on corporal punishment was cleared for submission through the support of Tony Smythe and MIND's chairman, Charles Clark.[757] It began with what might appear to be a typically civil libertarian statement. Corporal punishment had been abolished for everyone in Britain except school-age children, it stated. The result was that 'Everyone has been protected against assault and battery by the state, except the child attending school'.[758] But the paper's opposition to corporal punishment could be described as coming straight out of Bowlby and Durbin via Wills (minus Bowlby's maternalist emphasis). 'All children need love and security', it emphasized. Ideally, this should come from the child's parents or substitute parent figures.[759] As so much time was spent at school, it continued, teachers were, inevitably, an extension of the home and their attitudes significantly affected children.

> Corporal punishment in schools is used for a number of reasons, including low educational attainment, revolt against authority, persistent lying, stealing and particularly for offences which may be regarded as aggressive, i.e. disobedience, insolence, open provocation, vandalism and assaults upon other children. Children who behave in these ways have not developed effective control over their impulses, and in the main are found to come from homes where there is little love and security. Where there is no love and security the child feels unloved and rejected and often reacts in anger, hate and lack of concern for others. Children who have a loving relationship with their parents learn to control their anger and to use it constructively.[760]

In homes where love and security were lacking, parental authority varied unpredictably, with acts punished sometimes, and ignored or condoned at others:

> Consequently, the child becomes confused, may react by passive submission to authority, or an exaggerated swaggering independence, which reveals itself in aggressive acts against property, authority or other children. It is often found that beneath the bravado of an aggressive child, there are real feelings of fear and anxiety. Often, unloved children have aggressive parents and as a consequence may not only act aggressively because of his or her feelings of insecurity, but may also act aggressively because of the model presented by the parents. If the teacher also acts in an aggressive manner, the child's feelings are reinforced.[761]

The paper added that it was worth noting that, except where it was very severe, corporal punishment seemed to be feared only by children who had never, or very rarely, received such punishment. Children who were often punished were usually 'singularly unmoved by the prospect'.[762] It was vital that the teacher either adopted the parent's loving role or compensated in some way for those parents who didn't do this. The paper noted that 'Sadly, the very child who needs warmth, acceptance and love from the teacher in the class is very often the least likeable member of that class'.[763] Praise and recognition shouldn't simply be given for achievement. It should be 'a recognition of the individual'. MIND's paper noted that the 'intelligent, healthy, well-adjusted, attractive child tends to receive recognition, but in contrast the intellectually slow, emotionally deprived, or disturbed child gets little, if any, in spite of his or her greater need'.[764]

Given this stance, it isn't surprising that MIND continued to promote the therapeutic community-style ventures that NAMH had long supported. In the 1960s NAMH had been involved in conferences and training with the Richmond Fellowship, the primary organization providing small-scale therapeutic community care.[765] MIND continued this during the 1970s and 1980s. After the Association of Therapeutic Communities was founded in 1972 it held some of its regular conferences in conjunction with MIND.[766] MIND also offered a consultancy service for setting up a therapeutic community approach to residential care.[767] Meanwhile, in the late 1970s, MIND was vocal in opposition to the threatened closure of the Henderson Hospital, a prominent therapeutic community in the NHS.[768]

After Tony Smythe became Director, MIND closed a number of its residential homes, some of them on the grounds that they were too institutional and provided the wrong form of care. Years later he recounted that the two residential institutions MIND provided for young women included a seclusion suite in which they could be locked up 'for days on end' while '"privileges" like clothing, visits and personal possessions, could be withheld'. He recalled that 'Gradually we got out of the incarceration business', but he also emphasized that 'to be fair, we did run one exceptionally good school for highly emotionally disturbed young children'.[769] This was Feversham School near Newcastle. It had opened in 1969, just before NAMH adopted the name MIND. Taking explicit influence from Wills' work, it operated as a therapeutic community.[770] Under Smythe, MIND also transformed Fairhaven, its hostel for leavers from schools for children labelled 'educationally subnormal'. This also took inspiration from some therapeutic community ideas regarding open communication, flattened hierarchies and self-government.[771]

A 1975 edition of *MIND OUT* alludes to some of the connections between community therapy experiments that we've traced and MIND's approach.

The pioneers of child oriented education, like A S Neil, David Wills and the proponents of free schooling, have introduced fresh approaches but perhaps have not shown whether the small unit approach is capable of surviving without charismatic educationalists or that it can be adapted to meet the needs of wider society. Still, ventures like the White Lion Street Free School, Islington, are refreshing in their attempt to find alternatives to pre-package curriculae, examinations, petty rules and regulations, uniforms, punishments, compulsory time tables and religious instruction, staff hierarchy, limited age-groups and enforced attendance. There are no instant solutions in the field of education but the free school movement cannot be ignored.[772]

In fact, by 1977 the White Lion Street Free School was under threat of immediate closure. Smythe agreed with his Chairman of Council, Charles Clark, that MIND should make an urgent grant of £4000. One result of this was a written protest signed by 21 members of MIND's staff and sent to the Council of Management. MIND was under great financial constraint at the time, and staff were annoyed that Smythe and Clark hadn't discussed it with them beforehand. Some staff felt the money would have been better spent on internal projects, others that the Free School wasn't an outstanding case for support. Smythe told them that the money couldn't have been used for internal projects and that a decision had to be made within 24 hours. Clark expressed 'puzzlement that some members of staff appeared not to see the central importance of this type of preventive measure to MIND's work. It was directly in line with the decision to pull MIND out of Duncroft [A community home for young women] and to look for pioneering projects in community-based care for emotionally disturbed adolescents'.[773]

* * *

We've seen throughout our story of moral treatment extended to the community that the organizing principle of the Family held political, as well as psychological, imperatives. The authority relations of the Family imbued the social order, as well as each developing mind. These imperatives had emerged in relation to concerns about political and social changes affecting the social body. Fears about centralized power and authority that were derived from mass popular mandate clearly informed the use of the Family as the organizing principle. Associated with this were concerns about the effects of the rationalizing and totalizing aspects of power. Under the mental hygiene movement worries about the latter continued to be expressed, even while promotion of psychological and psychiatric expertise extended these processes deeper into the family. At the heart of these issues was the view that any political power and authority shorn of the moralizing structure of the Family risked creating demoralization, and passivity among people in a social environment made increasingly isolating by the modern rationalizing and totalizing elements of power. This Family principle had expressed itself across the wider community as a hierarchy of status, informed by a presumed hierarchy of rational ability and personality structure. Each successfully raised individual took a mentally ingested Family authority with them into this wider community and so it didn't require the nurturing affectional aspects of each personal family. But, as we've seen, the concentration on emotional relationships produced a partial inversion expressed in a drive towards more egalitarian relationships. As radical experiments associated with the mental hygiene movement these became the centrifugal element that returned popular political participation to its original political position as, theoretically, the antithesis to centralization. If egalitarian relations could be the common denominator of therapeutic self-government, then couldn't this apply to self-government in wider society? An emphasis on popular power and authority might yet help dissipate some of the more rigid institutional centralizations of power and authority in society, while at the same time providing a better way to mediate the negative effects on people of rationalized and totalized power.

This partial inversion in understandings of the co-ordinates of power in modern society is clearly evident in MIND's approach from the turn of the 1970s. The 1971 MIND manifesto, for example, challenged its readers to question whether material progress had been 'matched by advances in human well-being'.[774] Increases in gross national product and 'hard won improvements in living standards' hadn't done much to increase people's sense of well-being. It argued that 'We all need to "belong" in the family, the neighbourhood and the wider community'. But this was damaged constantly 'in the name of technological advance'. High-rise blocks of flats, for example, isolated people from family and neighbourhood contacts. If these contacts were to be preserved, planners and policy-makers needed to take account of the psychological consequences of their decisions and actions.[775] Smythe reiterated such concerns in more trenchantly radical style. For example, he made these comments to a Social Services Conference in 1979:

> How can we create an environment at home, in the community, at work or at play in which mental health would flourish? Every bad planning decision from high-rise flats to the suburbanisation of the fittest, manufactures stress and intensifies vulnerability... Even if we could get tolerably civilised standards for institutions and community support for the social casualties of Western industrial life, we may still have to make conscious choices about the kind of society we want. Will we all be made consumers in a therapeutic society which puts us down and peps us up? Will we go for authoritarian control and be told how to live? Or

will we share our wealth and our problems without being directed and manipulated by a benevolent, know-all state to retain only an illusion of freedom, and a reality which has much in common with the back-wards of our mental hospitals?[776]

Here, rationalizing and totalizing elements of power are brought together with politically centralized power and authority as detrimental to both mental well-being and social justice. Popular power and authority, expressed within egalitarian relations, is seen as a means to dissolve rigid institutional centralizations of power and authority in society. At the same time, the implication is that more personal egalitarian relations within the community will help control modern rationalizing and totalizing power.

MIND's approach clearly promoted popular participation. Its journal *MIND OUT* stated in April 1974 that 'MIND exists to give a voice and a forum to the neglected and the under-privileged and also to co-ordinate and promote your efforts'. Through research, education and service provision it hoped to offer a framework for individual action, creating an 'irresistible, irrepressible force for change'. This would involve, not only 'a crash programme' for comprehensive mental health facilities, but 'radical improvements in the quality of care and the creation of a climate of sympathy in which care can flourish'. It urged people to start an 'action committee' or a local association, or find out what was going on by vetting services or going into hospitals as visitors or volunteers so that they could 'see things through the patients' eyes'.[777]

In June 1974 MIND announced that it would devote the October edition of *MIND OUT* to the views of people on the receiving end of mental health services.[778] Smythe extended this orientation in October 1974 by suggesting to MIND's Council of Management that, as the interests of professional groups, administrators and other staff in the mental health services were well represented, 'Should not MIND place particular emphasis on expressing the views and needs of those at whom the mental health services are directed, and also their relatives and friends?'. He argued that this wouldn't prevent MIND from acting as a forum for debate for all who were concerned with mental health, and that MIND could act as a bridge between 'users and professionals'.[779]

MIND OUT received hundreds of letters in response to its call. They came from patients, ex-patients and relatives. The October *MIND OUT* editorial remarked that, although one patient's experience couldn't simply be generalized, it was important that patients' experiences were heard. 'We do not think psychiatrists will like being criticized by their patients', it said, 'but [we] would suggest that criticism is a necessary function of this particular relationship'.[780]

MIND OUT reported that most people who wrote in 'seemed to find hospital a depressing or terrifying experience and for a multitude of reasons'.

For many, poor food, constant noise from radio or television, or just sheer boredom were major issues. Others emphasized the 'alienation they felt, and the indiscriminate herding together'. *MIND OUT* added that 'Sometimes the problem was that indescribable substance – atmosphere'.[781]

In the next issue several letters in response were published. One, from a consultant psychiatrist, dismissed the 'paltry standards of the October edition' which, he wrote, appeared 'to consist of nothing more than anecdotal alarmist accounts from disgruntled and querulent people of their experiences in mental hospitals'. A woman wrote in warning that the issue was 'dangerously slanted', emphasizing complaints 'which are those of a sick mind anyway'.[782] Meanwhile, a principal social worker from Hertfordshire wrote in congratulating MIND and remarking that 'Psychiatrists and social workers can fall into the trap of "daddy knows best" and it is refreshing to hear the views and thoughts of those who do know what mental illness is really about'. The General Secretary of the Samaritans also wrote in to congratulate MIND on the issue.[783]

But the incitement of individual initiative and popular power clearly had its moralizing components. We saw in previous chapters that critiques of the detrimental effects of institutionalization had often mixed two views. One reckoned that it could be attributed to the hospital being too comfortable and unchallenging. This encouraged patients to wilfully adopt a position of dependency and inactivity. The other view placed attention on the relations of authority and hierarchy that appeared destructive to mental well-being. The first view was in keeping, as we've seen, with the more traditional and longstanding views of the mental hygiene movement. On this view 'the masses' were all too vulnerable to a loss of initiative. Often this was related to state policies that, it was claimed, encouraged dependence, and therefore mental 'immaturity' and mental ill-health.

There's a similar mix in some of MIND's pronouncements. In the issue of *MIND OUT* devoted to service users' views, the editorial noted that common themes were calls for better hospitals, more informed responses from general practitioners, more psychotherapy on the NHS and the need for community care to be more than just a slogan. The editorial commented that 'We can at least assert that neither the NHS nor social services can offer a solution for the loneliness which is so graphically described by some of our contributors. Only we, the community, can do that'. But it also argued that another, less familiar theme, was suggested:

> Could it be that many who feel they need psychiatric help or social support have an exaggerated view of what available services have to, or ever could, offer? To what extent do patients, professionals and public have, in varying degrees, a vested interest in encouraging dependency and, in consequence, discouraging self-help? Are we as a society becoming dazzled by professional jargon and machines?[784]

An article by Tony Smythe and Denise Winn in the April 1975 edition of *MIND OUT* appears to flesh out the views underlying these questions. Noting that capital expenditure on hospital and community services was to be reduced in the coming years, it argued that 'perhaps the time is ripe to look again at our expectations of statutory services and to start to take responsibility for our welfare back into our hands'. It reckoned that 'intricate administrative machinery, as provided in the statutory sector' could only deal with 'quantities' and not individuals. It argued that 'over and above this provision', 'everyone, professionals and non-professionals together' could help establish small community projects and support groups. These could help encourage an awareness of each individual's potential to deal with the stresses of daily life. It contended that:

> As a society, we have succumbed to the indulgence of being ill. If being ill does not resolve problems, at least it removes the responsibility of having to face them.... Many patients try to escape from their problems by taking tranquilisers. Rather than admitting that in some way they need to change their feelings or attitudes they assume what is known as a "sick" role. They try to justify their actions by blaming them on their illnesses which may not exist, in fact.[785]

Smythe and Winn maintained that it was 'all too easy to become dependent' and 'push problems onto someone else's plate'.[786] It's a view that seems to have something in common with the traditional mental hygienist assumption that rationalizing and totalizing power centralized in the state, was a blunt and abstracting force that, if left unmediated by mental hygienist expertise and services, encouraged a child-like dependence in the populace. What needed to be recognized instead was the central importance of the Family as an organizing principle and methodology for the establishment of mentally healthy development in individuals, and the maintenance of a mentally healthy social order. But Smythe and Winn's assertion that people seemed all too easily made passive and dependent, rested on what I have called the partially inverted version of the Family. This invoked egalitarian relations as both the common denominator of therapeutic self-government *and* of self-government in wider society. It advocated relations of personal intimacy and commitment, along with the promotion of open communication and a voice for patients and consumers. What it disavowed was paternalistic authoritarian hierarchies characterized by status and function. This disavowal included a deep suspicion of doing things too, and for, unwilling or passive people.

Smythe and Winn argued that everyone should look at the deeper causes of stress and mental troubles, such as poor housing, badly paid jobs, inadequate pensions and poverty. For them, self-help was community self-help. One side of it emphasized 'social help and change', the other 'individual growth and thus better functioning at a personal level'. They pointed to the emergence of law centres and legal advice agencies that could help poorer people with things like social security, employment issues and rent problems. These they argued could 'bring the law down to the level of the people', helping them to appreciate their rights and exercise them. They also highlighted claimants unions, relatives' support groups ('who suffer, say, the agony of having a schizophrenic member in the family whilst being unable to find any real help for him'), the Mental Patients' Union, Depressives Anonymous and organizations for gay rights. But Smythe and Winn also warned that these groups needed to 'see themselves and be seen as an integral part of a wider social movement'. Emphasis on 'victimisation' might bring unity and strength in the short term, but it encouraged isolationism and entrapment in an identity politics that could be exclusive and condemning of others.[787]

* * *

So MIND's approach can't be reduced to a straightforward defence of individual liberty from psychiatric coercion and restraint. It wasn't ignorant of the commonalities between its approach and elements of the mental hygiene movement's post-war strategy. On the contrary, it highlighted many of them. Whether MIND's approach ever offered the 'authoritative solutions' to 'our contemporary fragmented moral order' that Rose had claimed was required is open to serious debate (not least perhaps in its over-concentration on critiquing the state and relative lack of attention to marketization). But what *is* obvious is that the theory of power/knowledge clearly cannot fulfil such a task.

Foucault's theory emerges within our story as a reconfiguration of the concerns about power in modern society that accompanied the extension of moral treatment into the community. Under his power/knowledge, concerns about centralized and popular power appear severed from those elements of power seen as rationalizing and totalizing. The former expressions of power are obliterated, leaving only the latter any significance—but this significance is absolute. The anthropologist David Graeber has commented recently on the way in which he believed the theories of Max Weber and Foucault had been used for conservative political ends in US academia. He associated this with their interpretations of the power of bureaucracy:

> Foucault was far more subversive [than Weber], but in a way that made bureaucratic power more effective, not less. Bodies, subjects, truth itself, all became the products of administrative discourses; through concepts like governmentality and biopower, government bureaucracies end up shaping the terms of human existence in ways far more intimate than Weber would possibly have imagined.[788]

We needn't just limit this to bureaucratic regulation. Definitions, measurements and categorizations, along with the practices purportedly associated with them, are the very stuff from which, according to Rose, the 'psy disciplines' create individuals' knowledge, experience and desires. But, as I've said, where there's knowledge there must also surely be authority. How can the theory of power/knowledge grapple with this? In fact, in the 1990s Rose wrote an article in which he argued that attention must, indeed, be directed to the issue of authority (among other areas) in the interests of performing a 'genealogy of subjectification'. But, as the aim of this proposed enterprise implies, the concept of authority remained subordinated to the analytical power of power/knowledge. All we were going to get was more of a focus on the terminology of authority, while the concept itself remained theoretically severed from any dynamic relation to power and knowledge. Its conceptual utility would continue to be denied. Instead 'disciplines', 'technologies' and 'assemblages' of power/knowledge would be revealed as the producers of its varying manifestations.[789]

The theory of power/knowledge dissolves every concept it touches. Roger Cooter has recently highlighted the fact that under the theory of power/knowledge not only has 'the social', 'the political' and 'the cultural' lost 'discrete analytical power', so also have 'vision', 'rationality' and 'reality' submitted to 'intellectual disembowelling'.[790] But, in the process, of course any idea of the dynamic interaction of power, authority and knowledge in a social world is lost. The theory of power/knowledge can't analyse such dynamics and it can't openly engage with them either. Power/knowledge creates reality apparently and so the theory of power/knowledge must inevitably subordinate everything to itself. Everything is dissolved in the corrosive solution provided by the theory.

* * *

The dialectic does not exist of course. There is no 'dialectic of the Family'. We have seen that in idealized abstraction the modern Western family has provided an influential organizing principle and methodology in modern psychiatry. Its significance has been both psychological and political. But this abstraction, with its variations and amendments, does not constitute elements of a dialectic elaborating itself through the forward motion created by its contradictions. The idea of a dialectic may provide a useful way to think about things, but the past is too subtle and too complicated for such a concept to be credible as an understanding of it and its relation to the present.

Foucault shed any idea of 'the dialectic of the Family' whose end-point it was apparently, in the late 1960s, impossible to predict—and so must we. Yet, ironically, his subsequent theory of power/knowledge recapitulates one of the glaring omissions of the dialectic: an engagement with the relationship of authority that exists between the past and its interpretation. Power/knowledge theory sucks out the fact that such interpretations are always sustained and produced within social relations of power and authority. It achieves this by redefining power, linking it with intellectual knowledge and making them together the creators of everything else. The lodestone of this imperialist theory is 'the self' as a zero-sum fabrication of power/knowledge. The wielder of the theory sits in sovereign intellectual authority above professional texts and pronounces on the ways in which they, and the practices presumed to have accompanied them, have created this self. There are no vying authorities here, no competing knowledges, nor, indeed, is there any possibility of a shared language through which varying views might communicate and debate. Whose language is this? Is the reader of such texts given the space to ask? Or is everything all sown up by the theory?

And how does this theory as dissolvent render the subjects it claims for its kingdom? Inevitably, they appear drained of life. Mere power/knowledge carriers. Ironically, Dewey's criticism of an older abstract psychology in the name of 'the new psychology' is apt:

> They emasculated experience till their logical conceptions could deal with it; they sheared it down till it would fit their logical boxes; they pruned it till it presented a trimmed tameness which would shock none of their laws; they preyed upon its vitality till it would go into the coffin of their abstractions.

'It wasn't the cough that carried 'im off, it was the coffin they carried 'im off in'. Lifeless, flat, a monochrome world. Reality with the world sucked out of it. Not even a pale imitation.

Afterword

In Foucault's eye, moral treatment's use of the organizing principle and methodology of the Family made madness forever childhood. In his view, psychoanalysis rediscovered this Family only in mystified form and in so doing became its quintessence, distilling the Family order into the personage of the therapist. But, in fact, it was the simultaneous psychoanalytic focus on emotionality and authority relations that proved pregnant. And here the mental hygiene movement appears pivotal.

If the mental hygiene movement inherited from moral treatment a primary emphasis on the Family as an organizing principle and methodology for conceptualizing and treating madness, it also inherited an amended version that extended across the wider community. Stretched in this way, the Family brought together notions of individual mental self-government with community self-government.

Here, the Family was a hierarchy of rationality and a structure for the distribution of status and function. This psychological and political order was fundamentally inegalitarian. Moral treatment's attribution of the status of 'childhood' was extended to encompass people labelled 'mentally deficient' (most notable numerically were people diagnosed 'feebleminded') and an array of people deemed 'socially inefficient'. These people were overwhelmingly targeted among the 'lower orders' of society. Indeed, the status of 'childhood' was cast as a shadow across the 'lower orders' in general.

Mental hygiene inherited and continued this general schema, interpreting it under the strong influence of a theory of instincts and versions of psychoanalysis. To this extent psychoanalysis can, indeed, be understood as a recapitulation of the Family instituted by moral treatment. But it was here that the enduring psychoanalytic concern with the significance of authority relations contributed to other effects.

If mental hygienists fused madness, childhood and 'the primitive', they also placed them at the core of the rationally marshalled adult. And this implied a shared language. Resting on assumptions about progress from 'the primitive' to 'the civilized', this endeavour retained the Family's hierarchical social order with its prioritization of intellectual capacity and distribution of status and function. Yet, there was a simultaneous assertion that the exercise of authority needed to take account of the emotional contents of actions and relationships. A prioritization of emotional security and emotional expression emphasized greater freedom, choice and emotional expression, along with a reduction of hierarchy and status, and more equal relations. One result was an attempt to mitigate authority commands and relations in the personal family, and in therapy. In the process, exercise of this authority was opened up to deliberation and debate.

These emphases can be traced with varying weight and consistency across the sites of mental hygienist activity. Yet, while such emphases were held at the level of therapy, and thus incorporated into the inegalitarian order of status and function, their effect was attenuated and, at times, bogus.

Two areas emerge as of particular significance here; they lie at polar ends of the spectrum of care and treatment, and have been commonly portrayed by historians as marginal to psychiatric strategy.

One area is the conceptualization of people categorized as 'mentally deficient'. If the rationally marshalled, mentally healthy individual emerged through 'madness', 'the primitive' and childhood, it didn't emerge through mental deficiency. These people were deemed incapable of benefiting from the mental hygienist interpretation of the Family. But they were, nevertheless, intimately linked to it: they were its flip-side. The Family disowned them.

The psychoanalytic versions of this gaze couldn't even look these people in the face. In fact, if ever there was a total silencing, as Foucault maintained in *Madness and Civilization*, perhaps its quintessential enactment was against those people who laboured under the categorization 'mental deficiency' and its successor terms. In institutions for mental deficiency these people were truly left with no means of communication other than in terms of the definition that had been provided for them. Here was the most glaring attribution of permanent 'childhood'.

In their role as the flip-side of the Family and mental health, people designated 'mentally deficient' acted, in fact, as a cornerstone of the mental hygiene project. We have seen that from the 1960s the mental hygiene movement was dragged, in principle, into a positive application of the Family to these people. But, for this very reason, despite its extension of the territory of the Family, this manoeuvre pulled the rug from under the mental hygiene movement. It dislocated the discourse of emotionality embedded in the Family from histories of individual and societal development. By the end of the 1960s the mental hygiene movement was unable to sustain the idea that a diagnosis of 'mental handicap' (previously 'mental deficiency') precluded rights of citizenship or necessarily indicated a failure of mental health.

It seems no coincidence that the point at which this happened was the point at which the mental hygiene movement collapsed and also the point at which the other significant area of our narrative—community therapy experiments—were (under the terminology of 'therapeutic communities') at their most socially prominent.

Experiments in community therapy are highly significant because they pushed the reconfiguration of the Family into a partial inversion. They brought together mental self-government and community self-government in one therapeutic endeavour. Here, the Family's relations of personal intimacy and depth were retained, but its hierarchical authority, easy distinction between rationality and irrationality, and distribution of status were undermined. Scepticism towards authority relations encouraged an emphasis on greater freedom, advocacy of open communication, suspicion of doing things to and for passive or unwilling people, and a central emphasis on more egalitarian relationships.

This partial inversion of the Family, brought about by making mental and community self-government operate simultaneously as part of the same endeavour, was inconsistent with the overall mental hygiene conceptualization of mental health and its relationship to 'progress' and the social order. Held at the level of therapy, however, it could be contained as a specific treatment model. But, from the later 1950s the emphases of this partial inversion increasingly linked up with wider elements of social critique. The emergence of MIND from the rubble of the mental hygiene movement reveals a clear, yet unstable, expression of this partial inversion of the Family, and this is related to the latter's release from total containment within therapy.

* * *

You could interpret Foucault's development of a theory of power/knowledge as an attempt, once and for all, to throw off the power and accompanying moral authority of the Family. Yet, if this is so, it has been a forlorn attempt. As this book has shown, ironically, the theory of power/knowledge only serves to cut us further adrift from appreciating the importance, trajectory and transformation of the Family as organizing principle and methodology. Alongside this loss are others.

The theory of power/knowledge disavows self-government. Yet, paradoxically, power/knowledge theory operates like a kind of non-dialectical rendition of Bernard Bosanquet's resolution to his paradox of self-government. When it comes to 'the self' *any* personal authority is rooted out and cast from the unsuspecting individual. But, of course, authority always lies somewhere. In reality it has not been dissevered in discourse. It has been usurped by others, by the wielders of the theory of power/knowledge.

Habitual claims that such power/knowledge analyses are merely intended as 'disruptions' and 'destabilizations' are not cogent. And, in any case, what do these 'disruptions' not disrupt? What institutional and professional hierarchies? What enduring practices? Which tutelages do they re-impose?

The application of power/knowledge to the history of psychiatry over the last few decades has taken the form of a voice that is never surprised, never at a loss in the face of any question it seeks to answer. It is a sober and intellectual voice. Its discourse takes the form of a lecture. The parental gaze may have succumbed here to the faceless gaze, but a fundamental element of the Family, the hierarchy of heads—of rationality and status situated in a social hierarchy—seems to remain.

Notes

1. W. D. Wills, Personal and Professional Papers, Planned Environment Therapy Trust (PETT) Archives, PP/WDW 1.5, September 1929 (full date unclear).
2. For example: A. Digby (1985) 'Moral treatment at the retreat, 1796–1846' in W. F. Bynum, R. Porter and M. Shepherd (eds) *The Anatomy of Madness: Essays in the History of Psychiatry, Volume 2.* (London: Tavistock Publications) pp. 52–72; A. Digby (1985) *Madness, Morality and Medicine: A Study of The York Retreat 1796–1914* (Cambridge: Cambridge University Press); A. T. Scull (1979) *Museums of Madness: The Social Organization of Madness in 19th Century England* (London: Allen Lane); R. Porter (1987) *Mind Forg'd Manacles: A History of Madness in England from the Restoration to the Regency* (London: Athlone Press).
3. M. Foucault (1989) [1967] *Madness and Civilization: A History of Insanity in the Age of Reason,* trans. R. Howard (London: Routledge). The original French edition was published in 1962. There has recently been a new unabridged English translation of the book: M. Foucault (2006) *History of Madness,* trans. Jonathan Murphy and Jean Khalfa (London: Routledge). Citations are taken from the 1967 abridged translation, with the equivalent page number for the 2006 translation in square brackets afterwards.
4. M. Foucault, *Madness and Civilization,* p. 241 [463].
5. G. Gutting (1989) *Michel Foucault's Archaeology of Scientific Reason* (Cambridge: Cambridge University Press), p. 109.
6. M. Foucault, *Madness and Civilization,* p. 247 [484–5].
7. R. Porter (1996) *A Social History of Madness: Stories of the Insane* (London: Phoenix), p. 19.
8. S. Tuke (1813) *Description of the Retreat: An Institution near York for Insane Persons of the Society of* Friends (York: Alexander), p. 137.
9. M. Foucault, *Madness and Civilization,* p. 251 [488].
10. M. Foucault, *Madness and Civilization,* p. 253 [489].
11. S. Tuke, *Description of the Retreat,* p. 150.
12. M. Foucault, *Madness and Civilization,* pp. 247–8 [485–6].
13. M. Foucault, *Madness and Civilization,* pp. 182–3 and p. 254 [326, 490].
14. S. Tuke, *Description of the Retreat,* pp. 172–3.
15. M. Foucault, *Madness and Civilization,* quotes at p. 251 and pp. 251–2 respectively [488–489].
16. S. Tuke, S. *Description of the Retreat,* p. 187.
17. M. Foucault, *Madness and Civilization,* pp. 253–4 [490].
18. R. Castel (1983) 'Moral treatment: mental therapy and social control in the nineteenth century' in S. Cohen and A. Scull (eds) *Social Control and the State: Historical and Comparative Essays* (Oxford: Martin Robertson) pp. 248–66.
19. R. Castel. 'Moral treatment: mental therapy and social control in the nineteenth century', pp. 256–8 (quote at p. 258).
20. W. D. Wills. Unpublished autobiography. PETT Archives, p. 118.
21. P. Linebaugh (2003) *The London Hanged: Crime and Civil Society in the Eighteenth Century* (London: Verso) p. 9 (and throughout Chapter 1). See also, Raymond Williams' comments on the term under the section devoted to the

word 'Unemployment' in R. Williams (1988) *Keywords: A Vocabulary of Culture and Society* (London: Fontana).

22. K. W. Swart (1962) '"Individualism" in the Mid-Nineteenth Century (1826–1860)', *Journal of the History of Ideas*, 23, 1, 77–90.

23. Etymonline (2008) Alienation. http://www.etymonline.com [home page] (accessed 1 August 2008); R. Williams. *Keywords*.

24. R. A. Nisbet (1966) *The Sociological Tradition* (London: Basic Books).

25. E. J. Dionne Jr (1997) 'Authority, community, and a lost voice', *The Responsive Community*, 7, 4, 67–9, p. 67.

26. R. A. Nisbet. *The Sociological Tradition*, pp. 264–5, 284–92. This influence in Marxism itself was the subject of bitter debate between so-called humanist Marxists, who considered the concept central to Marx's thought, and structuralist Marxists, the most notable of whom was Louis Althusser. Althusser argued that Marx ditched the concept in his later, 'mature' work, and replaced it with 'exploitation'.

27. R. A. Nisbet. *The Sociological Tradition*, p. 48.

28. R. A. Nisbet. *The Sociological Tradition*, p. 108.

29. S. Lukes (1979) 'Power and authority' in T. B. Bottomore and R. Nisbet (eds) *A History of Sociological Analysis* (London: Heinemann) p. 646.

30. W. D. Wills. Unpublished autobiography, pp. 119–20.

31. G. J. Schochet (1988) *The Authoritarian Family and Political Attitudes in Seventeenth-Century England* (Oxford: Blackwell), p. xvi.

32. G. J. Schochet. *The Authoritarian Family*, p. xvi, pp. xxii–xxiii.

33. S. Collini (1976) 'Hobhouse, Bosanquet and the State: Philosophical Idealism and Political Argument in England 1880–1918', *Past and Present*, 72, 1, 86–111, p. 91.

34. Isaiah Berlin was one of the most vocal in this view. See, for instance, *Freedom and its Betrayal: Six Enemies of Human Liberty* (Princeton: Princeton University Press), 2002.

35. J. H. Randall Jr (1966) 'Idealistic Social Philosophy and Bernard Bosanquet.', *Philosophy and Phenomenological Research*, 26, 4, 473–502, p. 491.

36. B. Bosanquet (2001) [1899] *The Philosophical Theory of the State*, 2nd edn (Kitchener: Batoche Books).

37. B. Bosanquet. *The Philosophical Theory of the State*, p. 49.

38. B. Bosanquet. *The Philosophical Theory of the State*, p. 62.

39. B. Bosanquet. *The Philosophical Theory of the State*, p. 68.

40. J. J. Rousseau. (1968) [1762] *The Social Contract and Discourses*, trans. M. Cranston. (Harmondsworth: Penguin), p. 54.

41. B. Bosanquet. *The Philosophical Theory of the State*, pp. 79–80.

42. B. Bosanquet. *The Philosophical Theory of the State*, p. 102.

43. B. Bosanquet. *The Philosophical Theory of the State*, pp. 88–9.

44. B. Bosanquet. *The Philosophical Theory of the State*, pp. 90, 102, 131, 205.

45. B. Bosanquet. *The Philosophical Theory of the State*, pp. 101, 199.

46. B. Bosanquet. *The Philosophical Theory of the State*, p. 101.

47. W. D. Wills. Unpublished autobiography, p. 125.

48. W. D. Wills. Unpublished autobiography, p. 125.

49. W. D. Wills. PP/WDW 1.5 (Journal Record for 29.9.29).

50. G. Steadman Jones (1971) *Outcast London: A Study in the Relationship Between Classes in Victorian Society* (Oxford: Clarendon Press), p. 14.

51. W. D. Wills. PP/WDW 1.5 (Journal Record for 27.9.29).

52. W. D. Wills. PP/WDW 1.5 (Journal Record for 14.10.29).
53. W. D. Wills. PP/WDW 1.5 (Journal Record for 15.10.29).
54. W. D. Wills. PP/WDW 1.5 (Journal Record for 25.10.29).
55. H. Bosanquet. (1906) *The Family* (London: MacMillan) p. 242.
56. H. Bosanquet. *The Family*, p. vi.
57. H. Bosanquet. *The Family*, p. 267.
58. H. Bosanquet. *The Family*, p. 264.
59. H. Bosanquet. *The Family*, p. 265.
60. See, for example, H. Bosanquet (1968) *The Family*, Chapter 9; and B. Bosanquet (1968) [1895] 'Socialism and natural selection' in B. Bosanquet (ed.), *Aspects of the Social Problem* (London: MacMillan), pp. 289–307.
61. H. Bosanquet. *The Family*, p. 221, 222.
62. W. D. Wills. PP/WDW 1.5 (Journal Record for 10.12.29).
63. B. Bosanquet. *The Philosophical Theory of the State*, p. 108.
64. B. Bosanquet. *The Philosophical Theory of the State*, pp. 108–9.
65. B. Bosanquet. *The Philosophical Theory of the State*, pp. 192–5.
66. B. Bosanquet (1968) [1895] 'The reality of the general will' in B. Bosanquet (ed.) *Aspects of the Social Problem* (London: MacMillan) pp. 319–32, 322–3.
67. Bosanquet, in fact, claimed to be a 'moral socialist' at one time. See A. M. McBriar (1987) *An Edwardian Mixed Doubles, The Bosanquets Versus the Webbs: A study in British Social Policy 1890–1929* (Oxford: Clarendon Press), p. 114.
68. B. Bosanquet. *The Philosophical Theory of the State*, p. 125.
69. A. M. McBriar. *An Edwardian Mixed Doubles*, p. 112; H. Bosanquet 'Socialism and natural selection', p. 291.
70. See, for example, H. Bosanquet 'Socialism and natural selection' and B. Bosanquet (1968) [1895] 'The principle of private property' in B. Bosanquet (ed.) *Aspects of the Social Problem*, pp. 308–18.
71. B. Bosanquet 'The principle of private property', pp. 314–15. See also A. M. McBriar. *An Edwardian Mixed Doubles*, p. 130.
72. W. D. Wills. PP/WDW 1.5 (Journal Record for 27.9.29).
73. A. Thornton (2005) *Reading History Sideways, The Fallacy and Enduring Impact of the Developmental Paradigm on Family Life.* (Chicago: University of Chicago Press).
74. She elaborated a distinction and an advantage of the English version over the French version of this family type. See H. Bosanquet. *The Family*, e.g. p. 227, 237.
75. H. Bosanquet. *The Family*, p. 99.
76. H. Bosanquet. *The Family*, p. 96, 97.
77. H. Bosanquet. *The Family*, p. 97.
78. J. Lewis (1995) *The Voluntary Sector, The State and Social Work in Britain: The Charity Organisation Society/Family Welfare Association since 1869* (Aldershot: Edward Elgar), p. 33.
79. J. Lewis. *The Voluntary Sector*, pp. 30–1.
80. J. Lewis. *The Voluntary Sector*, p. 41.
81. J. Lewis. *The Voluntary Sector*, p. 41.
82. B. Bosanquet (1968) [1895] 'Character in its bearing on social causation' in B. Bosanquet (ed.) *Aspects of the Social Problem*, pp. 103–17, p. 105. See also preface pp. vi–vii.
83. B. Bosanquet (1968) [1895] 'The principle of private property', p. 314.

84. In the original COS terminology this division was between the 'deserving' and the 'undeserving'.
85. S. M. Den Otter (1996) *British Idealism and Social Explanation: A Study in Late Victorian Thought* (Oxford: Clarendon Press), pp. 69–70, 101, 182–3.
86. J. Lewis. *The Voluntary Sector*, p. 26.
87. H. Bosanquet (1973) [1914] *Social Work in London 1869–1912* (Brighton: The Harvester Press), p. 197 (footnote).
88. The most useful is M. Thomson (1998) *The Problem of Mental Deficiency: Eugenics, Democracy and Social Policy in Britain c. 1870–1959* (Oxford: Clarendon Press). Another valuable work that covers this history is C. Unsworth (1987) *The Politics of Mental Health* (Oxford: Clarendon Press).
89. M. Thomson. *The Problem of Mental Deficiency*, p. 217.
90. H. Dendy (1968) [1895] 'The industrial residuum' in B. Bosanquet (ed.) *Aspects of the Social Problem* (London: MacMillan) pp. 82–102, 83–4, quotes at p. 84.
91. H. Dendy. The industrial residuum, pp. 84–5.
92. B. Bosanquet. 'Socialism and natural selection', pp. 299–304.
93. C. Unsworth. *The Politics of Mental Health Legislation*, pp. 151–2.
94. Rt. Hon. Lord Justice Scott (1937) 'Miss Evelyn Fox, CBE: A Tribute', *Mental Welfare*, 18, 48.
95. The organization was officially recognized by the Board of Control in 1914. It was originally known as the Central Association for the Care of the Mentally Defective. With the expansion of its activity to adolescents and adults not certifiable under either the Mental Deficiency or Lunacy Acts, it changed its name to the Central Association for Mental Welfare. See A. Hargrove (1965) *Serving the Mentally Handicapped* (London: NAMH) p. 23, 31.
96. Hargrove, *Serving the Mentally Handicapped*, p. 23.
97. E. Walford (1878) *Old and New London: Volume 4* (London: Cassell, Peter and Galpin) pp. 14–26 (apparently citing an 1875 article in the *Builder*), accessed at British History Online: http://www.british-history.ac.uk/report.aspx?compid= 45179 (accessed 5 January 2009).
98. Mathew Thomson (1996) concentrates, in particular, on community control in 'Family, community and state: the micro-politics of mental deficiency' in D. Wright and A. Digby (eds) *From Idiocy to Mental Deficiency: Historical Perspectives on People with Learning Disabilities* (London: Routledge), pp. 207–30. Sheena Rolph sums up the approach succinctly: 'Families were effectively under surveillance: if it was thought that they were unable to either care for, control, or segregate their children, inspectors were empowered to certify the children or young people and remove them to an institution'. See S. Rolph (2002) *Reclaiming the Past: The Role of Local Mencap Societies in the Development of Community Care in East Anglia, 1946–1980* (Milton Keynes: Open University).
99. Rt. Hon. Sir Lesley Scott (1926) Chairman's introduction to discussion on, 'The Proper Care of Defectives Outside Institutions' in Central Association for Mental Welfare, *Report of a Conference on Mental Welfare Held in the Central Hall, Westminster, London, SW on Thursday and Friday, December the 2nd and 3rd, 1926* (London: CAMW), pp. 17–23, p. 21–2.
100. W. D. Wills, PP/WDW 1.5 (Journal Record for 15.4.30).
101. M. Thomson (1998) *The Problem of Mental Deficiency: Eugenics, Democracy and Social Policy in Britain c. 1870–1959* (Oxford: Clarendon Press), p. 175; G. Jones (1986) *Social Hygiene* (London: Croom Helm), pp. 27–8.
102. H. V. Dicks (1970) *Fifty Years of the Tavistock Clinic* (London: Routledge and Kegan Paul), p. 1.

103. H. V. Dicks, *Fifty Years of the Tavistock Clinic*, p. 41, 42, 59, 84, 312.
104. C. Thomson (1922) 'A National Council for Mental Hygiene', *BMJ*, 1, 3196, 538.
105. See, for instance, NCMH (1933) *Tenth Report of the National Council for Mental Hygiene 1932* (London: NCMH), p. 5.
106. MRC, MSS.378/APSW/P17/3:7, CGC (1936) Response to questionnaire circulated by the Feversham Committee.
107. On the mental hygiene movement in the USA, as well as some of its connections with the British movement see J. Pols (1997) *Managing the Mind: The Culture of American Mental Hygiene, 1910–1950*, PhD Dissertation, University of Pennsylvania; J. Pols (2010) '"Beyond the clinical frontiers": the American mental hygiene movement, 1910–1945' in V. Roelcke, P. Weindling and L. Westwood (eds) *International Relations in Psychiatry: Britain, Germany and the United States to World War Two* (Rochester: University of Rochester Press), pp. 111–33.
108. Wills, PP/WDW 1.5 (Journal Record for 15.11.29).
109. For example, N. Dain (1980) *Clifford Beers: Advocate for the Insane* (Pittsburgh: Pittsburgh University Press); R. Porter (1996) *A Social History of Madness: Stories of the Insane* (London: Phoenix).
110. For instance, see J. R. Lord (1930) 'American Psychiatry and its Practical Bearings on the Application of Recent Local Government and Mental Treatment Legislation, Including a Description of the Author's Participation in the First International Congress on Mental Hygiene, Washington, D.C., May 5–10, 1930', *Journal of Mental Science*, 76, 456–95; H. Boyle (1939). '"Watchman, What of the Night?": The Presidential Address Delivered at the Ninety-Eighth Annual Meeting of the Royal Medico-Psychological Association, Held at Brighton, July 12th 1939', *Journal of Mental Science*, 85, 858–870, p. 865.
111. R. Porter, *A Social History of Madness*, p. 194.
112. Wills, PP/WDW 1.5 (Journal Record for 27.10.29).
113. G. Stanley Hall (1885) 'The New Psychology', *Andover Review*, 120–35 and 239–48, accessed at Classics in the History of Psychology: http://psychclassics. yorku.ca/Hall/newpsych.htm (accessed 2 May 2012).
114. W. McDougall (2001) [1919] *An Introduction to Social Psychology*, 14th edn (Kitchener: Batoche Books), p. 23.
115. It is usually accepted that Dewey was only influenced by Hegel early in his career. There is now debate regarding this. See, for example, J. A. Good (2006) *A Search for Unity in Diversity: The 'Permanent Hegelian Deposit' in the Philosophy of John Dewey* (Oxford: Lexington Books).
116. For an analysis of the medicalization of British children's socially problematic behaviour using the terminology of 'maladjustment' see S. Hayes (2007) 'Rabbits and rebels: the medicalisation of maladjusted children in mid-twentieth-century Britain' in M. Jackson (ed.) *Health and the Modern Home* (Abingdon: Routledge), pp. 128–52.
117. R. R. Thomas (2004) 'Fox, Dame Evelyn Emily Marion (1874–1955)', *Oxford Dictionary of National Biography* (Oxford: Oxford University Press), http://www. oxforddnb.com/view/article/33231.
118. W. D. Wills, Unpublished Autobiography, (PETT Archives), pp. 121–2, p. 126.
119. See Cyril Burt's introduction to W. McDougall (1952) *Psychology: The Study of Behaviour*, 2nd edn (London: Oxford University Press).
120. A.F. Tredgold (1924) Review of A. Wohlgemuth, *A Critical Examination of Psycho-Analysis* (London: Allen & Unwin), in *Studies in Mental Inefficiency* 5, 3, p. 67.

121. See, for instance, NCMH (1930) 'First International Congress on Mental Hygiene', *News Bulletin* (London: NCMH) pp. 8–13, p. 12. Also, T. A. Ross (1934) 'The neurotic', *Mental Hygiene*, 10, 86–90.

122. The sociologist Nikolas Rose has been prolific in this area. See, for instance, N. Rose (1985) *The Psychological Complex: Psychology, Politics and Society in England, 1869–1939* (London: Routledge and Kegan Paul); N. Rose (1989) *Governing the Soul: The Shaping of the Private Self* (London: Routledge); N. Rose (1998) *Inventing Our Selves: Psychology, Power and Personhood* (Cambridge: Cambridge University Press); P. Miller and N. Rose (eds) (1986) *The Power of Psychiatry* (Cambridge: Polity Press). See also D. Armstrong (1983) *Political Anatomy of the Body: Medical Knowledge in Britain in the Twentieth Century* (Cambridge: Cambridge University Press); D. Armstrong (2002) *A New history of Identity: A Sociology of Medical Knowledge* (Basingstoke: Palgrave).

123. For example, W. Brown (1935) 'Character and Personality', *Mental Hygiene*, 13, 54–6; R. G. Gordon (1933) 'Habit Formation', *Mental Welfare*, 14, 29–37.

124. W. McDougall, *An Introduction to Social Psychology*, p. 20, 186 and footnote 110.

125. E. Miller (1938) *The Generations: A Study of the Cycle of Parents and Children* (London: Faber and Faber), p. 19.

126. E. Jones (1945) [1933] 'The Unconscious Mind' in C. Burt, E. Jones, E. Miller and W. Moodie (eds) *How the Mind Works*, 2nd edn (London: Allen and Unwin) pp. 61–103, p. 84.

127. E. Miller (1945) [1933] 'How the Mind Works in the Child' in C. Burt *et al. How the Mind Works*, pp. 105–56, p. 109.

128. See H. Crichton Miller (1926) 'Adaptation, Successful and Unsuccessful', *Postgraduate Medical Journal*, 1, 5, 60–3 for an explicit example. But the general claim underlies the bulk of literature associated with the mental hygiene movement.

129. Wills, PP/WDW 1.5 (Journal Record for 1.10.29).

130. Wills, PP/WDW 1.5 (Journal Record for 8.10.29).

131. Wills, PP/WDW 1.5 (Journal Record for 9.10.29).

132. J. Pols (1997) *Managing the Mind*.

133. See, for example, J. R. Lord (1930) 'American Psychiatry'.

134. For an example of such a scheme see C. Hubert Bond (1921) 'The Position of Psychological Medicine in Medical and Allied Services', *Journal of Mental Science*, 67, 404–49, 404. Bond was medical Commissioner of the Board of Control, President of the Medico-Psychological Association, and a member of the Executive Committee of the NCMH.

135. L. Fairfield (1931) 'Crime and Punishment', *Mental Hygiene*, 4, 17–20, p. 18.

136. M. Craig (1933) 'Mental Hygiene in Everyday Life', *Mental Hygiene*, 7, 57–64, p. 63.

137. See S. Sharp and G. Sutherland (1980) 'The Fust Official Psychologist in the Wuurld', *History of Science*, 18, 181–208.

138. M. D. Young (1958) *The Rise of the Meritocracy, 1870–2033: An Essay on Education and Equality* (London: Thames and Hudson).

139. Published as C. Burt (1945) [1933] 'How the Mind Works in society' in C. Burt *et al., How the Mind Works*, pp. 179–333, p. 245.

140. R. D. Gillespie (1931) 'Mental Hygiene as a National Problem', *Mental Hygiene*, 4, 1–9, p. 5.

141. W. R. K. Watson (1926) 'Some Observations on Borderland Cases and Delinquency', CAMW, *Report of a Conference on Mental Welfare Held at the Central*

Hall, Westminster, London, S.W. on Thursday and Friday, December 2ⁿᵈ and 3ʳᵈ, 1926 (London: CAMW), p. 80.

142. W. D. Wills, Unpublished Autobiography, pp. 136–7.

143. B. Bosanquet (2001) [1899] *The Philosophical Theory of the State*, 2nd edn (Kitchener: Batoche Books), pp. 108–9.

144. British Medical Journal (1937) 'Dissociation and Repression: Lecture by Professor McDougall', reported in *British Medical Journal*, 1, 3987, 1169–71.

145. For instance, T. A. Ross (1934) 'The Neurotic', pp. 87–8.

146. J. R. Rees (1929) *The Health of The Mind* (London: Faber & Faber), pp. 19–20, quote at, p. 20.

147. C. Burt (1944) *The Young Delinquent*, 4th edn (London: University of London Press), p. 95.

148. H. V. Dicks, *Fifty Years of the Tavistock Clinic*, pp. 53–4, 67; G. Bléandonu (1994) *Wilfred Bion: His life and works 1897–1979* (London: Free Association Books), p. 43.

149. J. A. Hadfield (1973) [1935] (ed.) 'Introduction', *Psychology and Modern Problems* (Plain View, New York: Books for Libraries Press), p. 15.

150. J. A. Hadfield (ed.) 'Introduction' in *Psychology and Modern Problems*, p. 14.

151. H. Crichton Miller (1922) *The New Psychology and the Parent* (London: Jarrolds), pp. 180–1.

152. C. Burt, *The Young Delinquent*, pp. 465–6.

153. J. R. Rees, *The Health of The Mind*, p. 215.

154. Feversham Committee (1939) *The Voluntary Mental Health Services: The Report of the Feversham Committee* (London: The Committee).

155. C. Burt, 'How the Mind Works in Society', p. 236.

156. C. Burt, 'How the Mind Works in Society', pp. 240–1.

157. C. Burt, 'How the Mind works in Society', p. 245

158. H. Crichton Miller (1973) [1935] 'Educational Ideals and the Destinies of Peoples' in J. A. Hadfield (ed.) *Psychology and Modern Problems*, pp. 131–58, p. 151.

159. This has, in fact, been a perennial concern for classical theorists of democracy. J. S. Mill, for example, advocated the allocation of more votes for educated people. See D. Coates (1990) 'Traditions of Thought and the Rise of Social Science in the United Kingdom' in J. Anderson and M. Ricci (eds) *Society and Social Science: A Reader* (Milton Keynes: The Open University), pp. 239–95, p. 272.

160. MSS.463/box 51: S. C. Brown (1935) 'Objectives and Methods in Social Case Work' (Script of lecture given at Birmingham University), p. 5.

161. S. C. Brown, 'Objectives and Methods in Social Case Work', pp. 5–6.

162. S. C. Brown, 'Objectives and Methods in Social Case Work', pp. 6–7, 12–13.

163. MSS.378/APSW/P17/3:15, S. C. Brown and B. McFie, Memo to Feversham Committee (1938) 'The Functions of the Social Worker in the Mental Health Field'.

164. S. C. Brown, 'Objectives and Methods in Social Case Work', p. 12.

165. Wills, PP/WDW 1.4 Examination papers, University of Birmingham, Social Study Diploma, 1927–8, (PETT Archives).

166. G. Orwell (1989) [1933] *Down and Out in Paris and London* (Harmondsworth: Penguin), Chapters 10–16.

167. G. Orwell, *Down and Out in Paris and London*, p. 50.

168. G. Orwell, *Down and Out in Paris and London*, pp. 74–81, 117–22.

169. Wills, PP/WDW 1.5 (Journal Record for 14.4.30.). The signifying letter for the surname has been changed from the original.
170. M. Cosens (1932) *Psychiatric Social Work and the Family* (London: APSW) p. 9.
171. MSS.463/box 51: S. C. Brown (1935) 'Research into the Causes of Individual and Social Maladjustment and its Application to Social Case Work' (script of lecture given at Birmingham University) p. 20.
172. E. Miller (1939) 'Obsessional and Compulsive States in Childhood' in R. G. Gordon (ed.) *A Survey of Child Psychiatry* (London: Oxford University Press) pp. 106–14, p. 113.
173. S. C. Brown, 'Research into the Causes of Individual and Social Maladjustment, and its Application to Social Case Work', p. 20.
174. M. Cosens, *Psychiatric Social Work and the Family*, p. 10.
175. S. Tuke (1813) *Description of the Retreat: An Institution near York for Insane Persons of the Society of* Friends (York: Alexander), pp. 172–3.
176. These were the Tavistock Clinic and the Home and School Council. The significance of the second of these for our tale will become apparent in the next chapter.
177. M. Payne (1929) *Oliver Untwisted* (London: Edward Arnold and Co.), pp. 95–6.
178. M. Cosens, *Psychiatric Social Work and the Family*, p. 8. See also, for instance, Ian Suttie (1960) [1935] *The Origins of Love and Hate* (Harmondsworth: Penguin).
179. CAMW (1937) *Mental Welfare*, 18, 2, 'News and Notes', 52.
180. W. D. Wills, Unpublished Autobiography, (PETT Archives), p. 183.
181. SA/Q/HM 31.8. Records of Hawkspur Camp for Men, PETT archives.
182. Utopia Britannica (2009) 'Q Camps – The Emotional Vortex': http://www.utopia-britannica.org.uk/pages/Qcamps.htm (accessed 5 November 2009).
183. M. Cranston (1968) Introduction to Jean-Jacques Rousseau, *The Social Contract* (Harmondsworth: Penguin), pp. 18–19.
184. PP/WDW, W. D. Wills Archive Research File. For a fuller description see the extract from the original, Q Camps Committee (1935) *Draft Memorandum on Proposed "Q" Camps (for Offenders Against the Law and Others Socially Inadequate) Under the Management of Grith Fyrd (Pioneer Communities)*: http://archive.pettrust.org.uk/survey-saq2-2memoranda.htm (accessed 15 September 2012).
185. SA/Q/HM 12.1.1 Correspondence Marjorie Franklin/David Wills, 5 March 1936.
186. WEF/D/1 (Institute of Education Archives, University of London), HSC Executive Committee Meeting, 15 July 1930.
187. WEF/D/1 Executive Committee Meeting, 13 December 1933.
188. WEF/D/1 Executive Committee Meeting, 12 May 1930.
189. K. J. Brehony (2004) 'A New Education for a New Era: The Contribution of the Conferences of the New Education Fellowship to the Disciplinary Field of Education 1921–1938', *Paedagogica Historica*, 40, 733–755, p. 734. On 'lay' approaches in psychology see M. Thomson (2006) *Psychological Subjects: Identity, Culture and Health in Twentieth-Century Britain* (Oxford: Oxford University Press).
190. P. Gordon (1983) 'The Writings of Edmond Holmes: A Reassessment and Bibliography', *History of Education*, 12, 15–24, p. 16.
191. SA/Q/HM11.1, Camp Chief Report, April 1936.
192. G. Orwell (1962) [1937] *The Road to Wigan Pier* (Harmondsworth: Penguin), p. 152.
193. On the term 'bullshit' regarding Marxists see G. Cohen (2000) *Karl Marx's Theory of History* (Oxford: Oxford University Press), pp. xxv–vii. Regarding bullshit and academic theorising see G. Cohen (2002) 'Deeper into Bullshit' in S. Buss

and L. Overton (eds) *Contours of Agency, Essays on Themes from Harry Frankfurt* (Cambridge, MA: MIT Press), pp. 321–39.

194. G. Orwell, *The Road to Wigan Pier*, p. 152.

195. WEF/D/1 Minutes of Council meeting 24 February 1930; Executive Committee meetings 1 December 1930, 15 July 1930 and 28 January 1932.

196. W. Boyd and W. Rawson (1965) *The Story of the New Education* (London: Heineman), p. 84.

197. R. Dalal, R. Herzberger, A. Mathur (*c.* 2001) 'Rishi Valley: The First Forty Years' (Krishnamurti Foundation): http://www.arvindguptatoys.com [home-page], p. 13. See also, I. Smith (1999) *The Transparent Mind: A Journey with Krishnamurti* (Ojai, CA: Edwin House), pp. 24–5.

198. See HSC Executive Committee Minutes for the 1930s. He also briefly served on its 'child study' subcommittee.

199. On Glaister's background see M. Pines (1999) 'Forgotten Pioneers: The Unwritten History of the Therapeutic Community Movement', *Therapeutic Communities*, 20, 23–42.

200. W. Trotter (1990) [1916] 'Herd instinct and its bearing on the psychology of civilized man' in *Instincts of the Herd in Peace and War* (London: T Fisher and Unwin), pp. 11–41.

201. D. Prynn (1983) 'The Woodcraft Folk and the Labour Movement 1925–70', *Journal of Contemporary History*, 18, 1, 79–95, p. 80; J. L. Finlay (1970) 'John Hargrave, the Greenshirts and Social Credit', *Journal of Contemporary History*, 5, 1, 53–71, p. 55.

202. M. E. Franklin (1973) 'P.E.T. Trust: Glimpses of a Future Rooted in the Past' in H. Klare and D. Wills (eds) *Studies in Environment Therapy Vol. 2* (Toddington: PETT), pp. 55–62, p. 59.

203. W. D. Wills, Unpublished Autobiography, p. 167.

204. SA/Q/HM 12.1.1 Correspondence Marjorie Franklin/David Wills, 10 May 1936.

205. SA/Q/HM 11.1, Camp Chief Report, April 1936.

206. M. A. Payne (1929) *Oliver Untwisted* (London: Edward Arnold and Co.), pp. 64–5

207. M. A. Payne, *Oliver Untwisted*, p. 4.

208. M. A. Payne, *Oliver Untwisted*, p. 14.

209. M. A. Payne, *Oliver Untwisted*, p. 13.

210. W. D. Wills (1967) [1941] *The Hawkspur Experiment: An Informal Account of the Training of Wayward Adolescents*, 2nd edn (London: George Allen and Unwin), pp. 31–2.

211. W. D. Wills, *The Hawkspur Experiment*, p. 38.

212. M. A. Payne, *Oliver Untwisted*, pp. 1–3, p. 9.

213. M. A. Payne, *Oliver Untwisted*, p. 9.

214. M. A. Payne, *Oliver Untwisted*, p. 6.

215. W. D. Wills, Unpublished Autobiography, p. 167.

216. W. D. Wills, Unpublished Autobiography, p. 168.

217. M. Bridgeland (1971) *Pioneer Work with Maladjusted Children: A Study of the Development of Therapeutic Education* (London: Staple Press), pp. 263–4.

218. This organization was founded in 1931 as 'The Association for the Scientific Treatment of Criminals'. It changed its name in 1932. In 1951 it became 'The Institute for the Study and Treatment of Delinquency'.

219. M. A. Payne, *Oliver Untwisted*, p. 33.

220. W. D. Wills, *The Hawkspur Experiment*, p. 52.

221. M. A. Payne, *Oliver Untwisted*, p. 14.

222. W. D. Wills (1979) 'The Moral Perspective' in P. Righton (ed.) *Studies in Environment Therapy Vol. 3* (Toddington: PETT), pp. 25–35, p. 27.
223. W. D. Wills (1966) [1943] 'Internal Government of the Camp, Its Growth and Changes' in M. E. Franklin (ed.) *Q Camp: An Experiment in Group Living With Maladjusted and Anti-Social Young Men*, pp. 24–8, p. 27.
224. For example, Wills, *The Hawkspur Camp*, p. 25.
225. W. D. Wills, 'The Moral Perspective', p. 30; M. E. Franklin (1966) [1943] 'Summary of the Methods Used' in M. E. Franklin (ed.) *Q Camp: An Experiment in Group Living*, pp. 13–22, pp. 16–17.
226. W. D. Wills, Unpublished Autobiography, p. 176.
227. SA/Q/HM 31.8, letter 3 November, 1936.
228. SA/Q/HM 31.8, letter 7 November, 1936.
229. N. Ellison (1994) *Egalitarian Thought and Labour Politics: Retreating Visions* (London: Routledge), pp. xi, 24–7.
230. J. M. Winter (1970) 'R H Tawney's Early Political Thought', *Past and Present*, 47, 71–96.
231. N. Ellison, *Egalitarian Thought and Labour Politics*, p. 113.
232. HMSO, *Report of the Royal Commission on Lunacy and Mental Disorder* (Cmd. 2700, 1926).
233. HMSO, *Report of the Royal Commission on Lunacy and Mental Disorder*, paras 40–50.
234. C. Unsworth (1987) *The Politics of Mental Health Legislation* (Oxford: Clarendon Press), pp. 169–70.
235. D. Odlum (1931) 'The Meaning of the Mental Treatment Act, 1930', *Mental Hygiene Bulletin*, 3, pp. 8–12, p. 11–12.
236. M. E. Franklin, 'Summary of the Methods Used', p. 21.
237. M. A. Payne, *Oliver Untwisted*, pp. 36–38. Payne mentions that 'some eminent Psychologists' were consulted about the disorder. Certainly, J. A. Hadfield would have been one of them, and most likely others also associated with the Tavistock Clinic. Their advice was that this was a stage that must be lived through. The children needed to 'untwist' before they could move on (p. 70).
238. M. A. Payne, *Oliver Untwisted*, pp. 84–90.
239. W. D. Wills, *The Hawkspur Experiment*, p. 59.
240. W. D. Wills, 'The Moral Perspective', p. 29.
241. SA/Q/HM 31.23.1
242. R. B. Friedman (1990) 'On the Concept of Authority in Political Philosophy' in J. Raz (ed.) *Authority* (New York: New York University Press), pp. 56–91.
243. SA/Q/HM 31.33.1 and 2.
244. SA/Q/HM 31.30.1; SA/Q/HM 31.33.1
245. A. T. Barron (1966) [1943] 'Practical Work at the Camp' in M. E. Franklin (ed.) *Q Camp: An Experiment in Group Living*, pp. 32–9, p. 36.
246. A. T. Barron, 'Practical Work at the Camp', p. 36.
247. Q Camps Committee (1936) 'Memorandum on Proposed Q Camp for Offenders against the Law and others Socially Inadequate' cited in M. E. Franklin (ed.) *Q Camp: An Experiment in Group Living*, p. 68.
248. T .C. Bodsworth (1966) [1943] 'Daily Life at Hawkspur Camp' in M. E. Franklin (ed.) *Q Camp: An Experiment in Group Living*, pp. 29–32, p. 32.
249. A. T. Barron, 'Practical Work at the Camp', pp. 34–38, quote at p. 38.
250. A. T. Barron, 'Practical Work at the Camp', p. 38; Wills, 'The Moral Perspective', p. 29.
251. SA/Q/HM 11.1

252. SA/Q/HM 11.1
253. SA/Q/HM 31.14.1 and 2.
254. W. D. Wills, *The Hawkspur Experiment*, pp. 64–5.
255. SA/Q/HM 11.1 Camp Chief Reports.
256. M. Thomson, *Psychological Subjects*, p. 118.
257. M. Thomson, *Psychological Subjects*, p. 124.
258. M. Bridgeland, *Pioneer Work with Maladjusted Children*, p. 89–91.
259. C. Burt (1944) *The Young Delinquent*, 4th edn (London: University of London Press), pp. 520–22.
260. SA/Q/HM 31.23.1
261. The Medical Superintendent (1924) 'The Manor Institution, Epsom: Some Comments on its First Two Years', *Studies in Mental Inefficiency*, 5, 2, 25–32.
262. The Medical Superintendent, 'The Manor Institution, Epsom', p. 32.
263. This reference to nineteenth-century moral therapy is taken from Robert Castel, 'Moral Treatment: Mental Therapy and Social Control in the Nineteenth Century', in S. Cohen and A. Scull (eds) *Social Control and the State: Historical and Comparative Essays* (Oxford: Martin Robertson), p. 257.
264. The Medical Superintendent, 'The Manor Institution, Epsom', p. 29.
265. The Medical Superintendent, 'The Manor Institution, Epsom', p. 28.
266. The Medical Superintendent, 'The Manor Institution, Epsom', p. 29.
267. F. Kafka (1933), 'In the Penal Settlement' in *Metamorphosis and Other Stories* (London: Secker and Warburg Ltd). The short story was first published in German in 1919.
268. The Superintendent claimed that, although 'absconders were naturally pretty numerous', the Manor experienced relatively few. These escapes, he lamented, were the result of connivance by relatives followed by their concealment of the child. Resort to court proceedings, he complained, was fraught with difficulty as it often resulted in defendants providing 'malicious statements and lies' to local papers (see p. 32). Under the then current Lunacy Acts, an escapee from a mental hospital, if not recaptured within 6 months, became legally released from the orders of certification. There was no similar clause under the Mental Deficiency Act.
269. Feversham Committee (1939) *The Voluntary Mental Health Services: the Report of the Feversham Committee* (London: The Committee), para. 85.
270. Feversham Committee, *The Voluntary Mental Health Services*, para. 85.
271. Feversham Committee, *The Voluntary Mental Health Services*, Recommendations, paras xxviii, 89 and 361.
272. SA/Q/HM 12.2.1 Correspondence David Wills/Marjorie Franklin, 13 July 1936.
273. G. W. F. Hegel (1977) [1807] *Phenomenology of Spirit*, trans. A. V. Miller, (Oxford: Clarendon Press), p. 2.
274. SA/Q/HM: 31.31.2.
275. A notable earlier contribution was made by the Tavistock psychotherapist Ian Suttie with *The Origins of Love and Hate* in 1935.
276. B. Mayhew (2006) 'Between Love and Aggression: The Politics of John Bowlby', *History of the Human Sciences*, 19, 4, 19–35.
277. J. Bowlby (1987) 'A Historical Perspective on Child Guidance', *The Child Guidance Trust News Letter*, 3, 1–2.
278. E. F. M. Durbin and J. Bowlby (1939) *Personal Aggressiveness and War*, pp. 51–126.
279. J. Holmes (1993) *John Bowlby and Attachment Theory* (London: Routledge); F. C. P. Van der Horst, (2011) John Bowlby: From Psychoanalysis to Ethology: Unravelling the Roots of Attachment Theory (Chichester: Wiley-Blackwell).

280. E. F. M. Durbin and J. Bowlby, *Personal Aggressiveness and War*, p. 16, 40, 41.
281. E. F. M. Durbin and J. Bowlby, *Personal Aggressiveness and War*, p. 41.
282. E. F. M. Durbin and J. Bowlby, *Personal Aggressiveness and War*, pp. 41–3, quote at p. 43.
283. E. F. M. Durbin and J. Bowlby, *Personal Aggressiveness and War*, p. 42.
284. E. F. M. Durbin and J. Bowlby, *Personal Aggressiveness and War*, p. 42.
285. E. F. M. Durbin and J. Bowlby, *Personal Aggressiveness and War*, p. 7, 62.
286. E. F. M. Durbin and J. Bowlby, *Personal Aggressiveness and War*, p. 43.
287. E. F. M. Durbin and J. Bowlby, *Personal Aggressiveness and War*, p. 44. See also S. Freud (1994) [1930] *Civilization and its Discontents* (New York: Dover Publications), pp. 38–9.
288. E. F. M. Durbin and J. Bowlby, *Personal Aggressiveness and War*, p. 45.
289. J. Nuttall (2003) '"Psychological Socialist"; "Militant Moderate": Evan Durbin and the Politics of Synthesis', *Labour History Review*, 68, 2, 235–52, p. 242
290. E. F. M. Durbin and J. Bowlby, *Personal Aggressiveness and War*, p. 62.
291. E. F. M. Durbin and J. Bowlby, *Personal Aggressiveness and War*, p. 49.
292. J. Nuttall, '"Psychological Socialist"; "Militant Moderate"', p. 237, 242.
293. It seems that Tawney originally scoffed at Bowlby and Durbin's thesis, but quickly had second thoughts. See Nuttall p. 241.
294. J. Nuttall, 'Psychological Socialist'; 'Militant Moderate', p. 245; D. Paul (1984) 'Eugenics and the Left' *Journal of the History of Ideas*, 45, 4, 567–90.
295. J. Bowlby, D. W. Winnicott and E. Miller (1939) 'Evacuation of Small Children' (letter to the editor) *BMJ* 2, 4119, 1202–3.
296. NAMH published a booklet in the 1950s, written by John Bowlby and based on his maternal deprivation theory, J. Bowlby (n.d.) *Can I Leave My Baby* (London: NAMH).
297. SA/Q/HM 31.38.1 and 2. (PETT Archives).
298. Feversham Committee, *The Voluntary Mental Health Services*.
299. D. Anderson and I. Anderson (1982) 'The Development of the Voluntary Movement in Mental Health' in *The Mental Health Year Book 1981/82* (London: MIND), pp. 427–42, p. 431–2; S. Strong (2000) *Community Care in the Making: A History of MACA 1879–2000* (London: MACA), pp. 33–5.
300. R. Thomas (1945) *Children Without Homes: How can they be compensated for loss of family life?* (London: PNC), p. 8.
301. Visits to any particular area could be prolonged, covering anything from a week to three or even six months. See R. Thomas, *Children Without Homes* pp. 8–9.
302. The wartime issues of *Mental Health*, produced jointly by CGC, CAMW and NCMH, advertised these publications.
303. PNC (1944) *The Care of Children Brought Up Away From Their Homes* (London: PNC).
304. PNC, *The Care of Children Brought Up Away From Their Homes*, p. 2.
305. PNC, *The Care of Children Brought Up Away From Their Homes*, p. 2.
306. Bradford Education Committee (1942) *Report on Juvenile Delinquency* (Bradford: Bradford City Council), p. 29.
307. Bradford Education Committee, *Report on Juvenile Delinquency*, p. 28, 29.
308. *Manchester Guardian* (1943) 'Child Delinquency: Psychologist's Views' (22 February 1943); *Bradford Telegraph and Argus* (1943) 'Causes of Juvenile Delinquency' (15 February 1943).
309. Bradford Education Committee, *Report on Juvenile Delinquency*, p. 27.
310. Bradford Education Committee, *Report on Juvenile Delinquency*, p. 29.

311. Bradford Education Committee, *Report on Juvenile Delinquency*, pp. 29–30.
312. Bradford Education Committee, *Report on Juvenile Delinquency*, pp. 8–9, 28.
313. Bradford Education Committee, *Report on Juvenile Delinquency*, p. 8–9.
314. SA/Q 20 Q Camps Committee, Minutes and Agendas, 25 June 1940.
315. HMSO (1946) *Report of the Care of Children Committee (Curtis Committee)*, Cmd. 6922.
316. R. Thomas, *Children Without Homes*.
317. Thomas trained as a psychoanalyst with Anna Freud during the war and subsequently became director of Training at the Hampstead Child-Therapy Clinic from 1950–1976. See E. E. Model (1983) 'Ruth Thomas 1902–1983: An Appreciation', *Journal of Child Psychotherapy*, 9, 1, 5–6.
318. R. Thomas, *Children Without Homes*, p. 7.
319. R. Thomas *Children Without Homes*, p. 12.
320. R. Thomas, *Children Without Homes*, p. 24.
321. R. Thomas; *Children Without Homes*, p. 11
322. It is notable that despite the claimed focus on what the *child* finds satisfactory the Curtis Committee did not interview a single child from the Homes that they investigated.
323. R. Thomas, *Children Without Homes*, throughout.
324. R. Thomas, *Children Without Homes*, p. 25.
325. R. Thomas, *Children Without Homes* pp. 48, 59, 61–2.
326. R. Thomas, *Children Without Homes*, p. 24.
327. R. Thomas, *Children Without Homes*, pp. 39–43, 58, 101–6.
328. R. Thomas, *Children Without Homes*, p. 42.
329. R. Thomas, *Children Without Homes*, pp. 12, 45, 59–60.
330. R. Thomas, *Children Without Homes*, p. 41 on nursery care and throughout Chapter two for all ages of children.
331. R. Thomas, *Children Without Homes* p. 113.
332. R. Thomas, *Children Without Homes*, p. 63, 82.
333. R. Thomas, *Children Without Homes*, p. 43.
334. R. Thomas, *Children Without Homes*, pp. 49–51, quote at p. 51.
335. R. Thomas, *Children Without Homes*, p. 70.
336. HMSO (1946) *Report of the Care of Children Committee*, paras 508–10.
337. SA/Q 21 Hon. Secretary's Memoranda, Reports, etc. Letter from Margery Franklin to Maurice and Veronica, June 1940.
338. L. G. Fildes (1944) 'Hostels for Children in need of Psychiatric Attention', *Mental Health* 5,2, 31–2, p. 32.
339. PNC, *Second Annual Report 1.4.44–31.3.46*, pp. 17, 27.
340. D.W. Winnicott (1984) [1947] 'Residential Management for Difficult Children', *Human Relations*, reprinted in C. Winnicott, R. Shepherd and M. Davis (eds) *Deprivation and Delinquency* (London: Tavistock), pp. 54–72, p. 65.
341. C. H. W. Tangye (1941) 'Some Observations on the Effect of Evacuation Upon Mentally Defective Children', *Mental Health* 2, 3, 75–8, p. 75. Mr Tangye was Head Master of Lewisham Bridge L.C.C. Elder Boys (M.D.) School. In April 1940 he was a speaker at a one-day conference of the Association of Mental Health Workers (one of the constituent members of the MHEC) dedicated to 'Evacuation and Mental Health Work', *Mental Health* (1940) 1, 2, 61.
342. C. H. W. Tangye, 'Some Observations of the Effect of Evacuation', p. 75.
343. C. H. W. Tangye, 'Some Observations of the Effect of Evacuation', p. 78.
344. H. Barker (1942) 'A Special School Evacuation Unit: Some Observations on its Value', *Mental Health*, 3, 2, 37–9, p. 37.

345. H. Barker, 'A Special School Evacuation Unit', p. 38.
346. R. Thomas, *Children Without Homes*, p. 51–2.
347. W.A.G. Francis (1941) 'Enuresis Record', *Mental Health*, 2, 78.
348. SA/Q/HM 31.42.2.
349. For example, T. Harrison (2000) *Bion, Rickman, Foulkes and the Northfield Experiments: Advancing on a Different Front* (London: Jessica Kingsley); P. Campling and R. Haigh (eds) (1999) *Therapeutic Communities: Past, Present and Future* (London: Jessica Kingsley).
350. E. Shorter (1997) *A History of Psychiatry: From The Era of The Asylum to the Age of Prozac* (New York: John Wiley and Sons); B. Shephard (2000) *A War of Nerves: Soldiers and Psychiatrists 1914–1994* (London: Jonathan Cape).
351. E. Shorter, *A History of Psychiatry*, p. 238. (1999) *Therapeutic Communities: Past, Present and Future* (London: Jessica Kingsley).
352. B. Shephard, *A War of Nerves*, pp. 268, 335.
353. E. Shorter, *A History of Psychiatry*, p. viii.
354. E. Shorter, *A History of Psychiatry*, p. vii.
355. E. Shorter, *A History of Psychiatry*, Chapters 5 and 6.
356. Rees was a psychiatric consultant to the army from 1939.
357. J. Rickman, letter to S. H. Foulkes cited in T. Harrison (1996) 'Battle fields, Social fields, and Northfield', *Therapeutic Communities* 17, 3, 145–8, p. 146.
358. Brigadier L. Bootle-Wilbraham (1946) 'Civil Resettlement of Ex-Prisoners of War', *Mental Health*, 6, 2, 39–42, p. 39. See also T. Harrison (2000) *Bion, Rickman, Foulkes and the Northfield Experiments*, p. 193 and pp. 266–7. Harrison shows that Northfield provided a model for psychiatrists developing the CRS.
359. A. T. M. Wilson, M. Doyle and J. Kelnar (1947) 'Group Techniques in a Transitional Community', *Lancet*, 1, 735–8.
360. SA/Q/HM 31. 23.1.
361. A. T. M. Wilson *et al.*, 'Group Techniques in a Transitional Community', p. 738.
362. A. T. M. Wilson *et al.*, 'Group Techniques in a Transitional Community', pp. 735–737. Quotes at p. 737 and p. 736 respectively.
363. T. F. Main (1946) 'The Hospital as a Therapeutic Institution', *Bulletin of the Menninger Clinic*, 10, 69–70.
364. A. T. M. Wilson *et al.*, 'Group Techniques in a Transitional Community', p. 736; Brigadier L. Bootle-Wilbraham, 'Civil Resettlement of Ex-Prisoners of War', p. 40.
365. T. F. Main, 'The Hospital as a Therapeutic Institution', p. 70.
366. T. F. Main, 'The Hospital as a Therapeutic Institution', p. 66.
367. T. F. Main, 'The Hospital as a Therapeutic Institution', p. 66.
368. T. F. Main, 'The Hospital as a Therapeutic Institution', p. 67.
369. A. T. M. Wilson *et al.*, 'Group Techniques in a Transitional Community', p. 736.
370. SA/Q/HM 31.25.1.
371. *The Guardian* (2011) 'In Pictures: V.E. Day, 1945: "Cheers"' (Picture by Keystone), http://www.guardian.co.uk/gall/0,,1476330,00.html (accessed 19 August 2011).
372. See also J. R. Rees (1945) *The Shaping of Psychiatry by War* (London: Chapman and Hall Ltd).
373. B. Shephard (2000) *A War of Nerves: Soldiers and Psychiatrists 1914–1994* (London: Jonathan Cape), pp. 171–2.
374. British Medical Bulletin (1949) 'Notes on Contributors', *British Medical Bulletin*, 6, 3, 222–3.

375. BMJ (1945) 'Medical News', *BMJ*, 2, 791.
376. K. Soddy (n.d.) *Some Lessons of Wartime Psychiatry* (London: NAMH). (First published as a two-part article in NAMH's journal under the same title.) See *Mental Health* (1946) 6, 230–5 and *Mental Health* (1946) 6, 3, 66–70.
377. K. Soddy, *Some Lessons of Wartime Psychiatry*, p. 16.
378. K. Soddy, *Some Lessons of Wartime Psychiatry*, p. 12.
379. K. Soddy, *Some Lessons of Wartime Psychiatry*, p. 12.
380. K. Soddy, *Some Lessons of Wartime Psychiatry*, p. 13.
381. K. Soddy, *Some Lessons of Wartime Psychiatry*, pp. 13–14.
382. K. Soddy, *Some Lessons of Wartime Psychiatry*, p. 13.
383. K. Soddy, *Some Lessons of Wartime Psychiatry*, p. 14.
384. M Cosens (1932) *Psychiatric Social Work and the Family* (London: APSW), p. 12.
385. K. Soddy (1950) 'Mental Health', *International Health Bulletin of the League of Red Cross Societies*, 2, 2, 8–13, p. 13 (references to 'maturity' and 'vigour' of individuals, parents and the community throughout).
386. D. R. MacCalman (1949) 'Sweet are the Uses of Adversity', *British Journal of Psychiatric Social Work*, 2, 87–94, p. 90.
387. D. R. MacCalman, 'Sweet are the Uses of Adversity', p. 92, 94.
388. Soddy's attitude to the various categories of people deemed 'mentally deficient' through to 'dull' will be obvious. He described what he called the 'psychologically unstable' in *Some Lessons of Wartime Psychiatry* as 'psychologically inferior' (p. 8).
389. K. Soddy, *Some Lessons of Wartime Psychiatry*, p. 15.
390. K. Soddy, *Some Lessons of Wartime Psychiatry*, p. 7.
391. K. Soddy, *Some Lessons of Wartime Psychiatry*, p. 9.
392. K. Soddy, 'Mental Health', p. 11, 13.
393. K. Soddy, 'Mental Health', p. 11.
394. PP/WDW, W. D. Wills Archive Research File, PETT archives.
395. T. Morris (1983) 'Crime and The Welfare State' in P. Bean and S. MacPherson (eds) *Approaches to Welfare* (London: Routledge and Kegan Paul), pp. 166–81, p. 166.
396. B. Kirman (1946) 'Left Turn', *Lancet*, 248, 808.
397. R. H. Ahrenfeldt (1948) 'The Paternal State', *BMJ*, 1, 515–6, p. 515.
398. J. C. Flugel (1921) *The Psychoanalytic Study of the Family*, (London, Hogarth Press).
399. For example, ICMH (1948) *International Congress on Mental Health, London, 1948* [Four volumes], (London: H.K. Lewis & Co. Ltd).
400. R. H. Ahrenfeldt, 'The Paternal State', p. 516.
401. For example, I. E. P. Menzies (1949) 'Factors Affecting Family Breakdown in Urban Communities: A preliminary study leading to the establishment of two pilot Family Discussion Bureaux', *Human Relations* 2, 4, 363–73, p. 364.
402. H. V. Dicks (1954) 'Strains Within the Family' in NAMH, *Strain and Stress in Modern Living: Special Opportunities and Responsibilities of Public Authorities* (London: NAMH), 28–37, p. 35.
403. R. Holman (1973) 'Family Deprivation', *British Journal of Social Work* 3, 431–46, p. 434.
404. R. Holman, 'Family Deprivation', p. 436. Holman's judgement, however, seems to have temporarily failed him some decades later regarding Iain Duncan Smith's policy intentions for people living in poverty. See, for example, R. (Bob) Holman, 'I Thought I Knew Iain Duncan Smith', *The Guardian*,

12 December 2010, http://www.guardian.co.uk/commentisfree/2010/nov/12/ iain-duncan-smith-punishing-the-poor (accessed 1 September 2011).

405. For Miller's rendition, see Chapter 3.

406. D. R. MacCalman (1948) Speech addressing 'Aggression in Relation to Family Life', ICMH, *International Congress*, 2, 50–55.

407. H. V. Dicks, 'Strains Within the Family', p. 31.

408. H. V. Dicks, 'Strains Within the Family', p. 32.

409. H. V. Dicks, 'Strains Within the Family', p. 35.

410. *The Free Lance-Star*, 21 May 1951, http://news.google.com/newspapers? nid=1298&dat=19510521&id=c4kTAAAAIBAJ&sjid=6YoDAAAAIBAJ&pg=6243, 1237465 (accessed 19 April 2010). Torrie alluded to this view over a decade later in a review of Richard Hauser's *The Homosexual Society*. Here, he agreed with Hauser's claim that all wasn't well in the prevailing cultural relationship of men and women, with the latter accepted as the second-class partner. However, this observation was hitched by both Hauser and Torrie to the idea that homosexuality was one of the results—it was a 'distressing social infection'. See A. Torrie (1963) 'Review of Hauser R., The Homosexual Society' *Mental Health*, 22, p. 29. Torrie acted as medical director of the Northern Office of NAMH from 1952 to 1956. He remained a member of NAMH's Advisory Council until 1970.

411. PNC (1944) *After-Care of Service Casualties: Report for the Quarter Ending March 25th, 1944*, p. 1; PNC (1944) 'News and Notes' *Mental Health* 5, 1, 12–16, pp. 12–13.

412. PNC (1944) 'News and Notes', p. 13.

413. E.M. Goldberg (1957) 'The Psychiatric Social Worker in the Community', *British Journal of Psychiatric Social Work* 4, 2, 4–15, p. 4. See also, NAMH (1948) 'News and Notes', *Mental Health*, 7, 3, 76–9, p. 76.

414. NAMH (1949) *Annual Report 1948–1949*, p. 11–12.

415. T. A. Ratcliffe (1951) 'Community Mental Health in Practice' in NAMH, *Proceedings of a Conference on Mental Health held at St. Pancras Town Hall, London, N.W.1, 12–13 March 1951*, pp. 11–23, pp. 11–12.

416. T. A. Ratcliffe, 'Community Mental Health in Practice', pp. 12–13.

417. E. M. Goldberg, 'The Psychiatric Social Worker in the Community', pp. 12–13.

418. T. A. Ratcliffe, 'Community Mental Health in Practice', p. 14.

419. NAMH (1956) *Memorandum Prepared for the Working Party on Social Workers*, paras 23–7.

420. NAMH, *Memorandum for Working Party on Social Workers*, para. 24.

421. NAMH, *Memorandum for Working Party on Social Workers*, para. 26.

422. K. Soddy, (1954) 'Community Care of Psychiatric Patients – A Review', appendix to NAMH and APSW, *Memorandum on Rehabilitation of Psychiatric Patients: Evidence to the Committee of Enquiry into Existing Services for the Rehabilitation of the Disabled*, p. 1.

423. NAMH, *Memorandum for Working Party on Social Workers*, para. 56.

424. N.K. Hunnybun and L. Jacobs (1946) *Interviews With Parents in a Child Guidance Clinic* (London: APSW), p. 4.

425. J. Bowlby (1949) 'The Study and Reduction of Group Tensions in the Family', *Human Relations* 2, 2, 123–8, p. 123.

426. For example, see comments of R. Lucas cited in NAMH (1953) *Tenth Child Guidance Inter-Clinic Conference: Report on Conference and Some Selected Clinic Surveys* (NAMH: London), pp. 28–30, p. 29.

427. D. Hunter (1955) 'An Approach To Psychotherapeutic Work With Children and Parents', *11ᵗʰ Inter-Clinic Conference For Staffs of Child Guidance Clinics: The Family Approach to Child Guidance-Therapeutic Techniques*, (London: NAMH), pp. 11–26, p. 12.

428. J. H. Kahn (*c.* 1957) *Child Guidance* (NAMH leaflet) p. 3.

429. T. A. Ratcliffe, 'Community Mental Health in Practice', p. 14.

430. For example, Dicks, 'Strains within the family', p. 28; Bowlby, 'Group Tensions in the Family', p. 123, 124.

431. K. Soddy, 'Community Care of Psychiatric Patients – A Review', p. 3. This percentage remained constant after civilians were incorporated into the scheme. More than 60 percent of people diagnosed as neurotic were employed. The percentage of those unemployed was given as thirty-five percent of those whose circumstances were known, but this included housewives, children and students.

432. K. Soddy, 'Community Care of Psychiatric Patients – A Review', p. 3, pp. 4–5.

433. K. Soddy, *Some Lessons of Wartime Psychiatry*, p. 19.

434. K. Soddy, 'Community Care of Psychiatric Patients – A Review', p. 2.

435. T. A. Ratcliffe and E. V. Jones (1956) 'Intensive Casework in a Community Setting', *Case Conference*, 2, 10, 17–23, p. 18.

436. NAMH, *Memorandum Prepared for the Working Party on Social Workers*, para. 64.

437. MRC, MSS.378/APSW/P16/5:33, 'The Psychiatric Social Worker as "Duly Authorised Officer"': report of the discussion at APSW General Meeting, 27 September 1947.

438. L. Fildes (1946) 'The Care of the Homeless child' in NAMH, *Report of the Proceedings of a Conference on Mental Health, 14 and 15 November 1946*, pp. 58–62, p. 62.

439. M. Capes (1949), cited in 'Report of the Eighth Inter-Clinic Child Guidance Conference: London, Saturday, December 3, 1949', *Mental Health*, 9, 3 (supplement), p. viii.

440. MSS.378/APSW/P21/4:1a, ' "Westhope" Society for Social Education' (n.d. July 1949?).

441. MSS.378/APSW/P21/4:4 Westhope Manor School: Correspondence May 1949–December 1949.

442. MSS.378/APSW/P21/4:3 Westhope Manor School: Correspondence.

443. MSS.378/APSW/P21/4:6 Westhope Manor School: Correspondence.

444. S. K. Brindley and G. H. Pettingale (1963) 'Swalcliffe Park School', *Mental Health*, 22, 3, 112–14.

445. NAMH, Minutes of Council meeting 31 October 1952, para. 7.

446. R. Balbernie (1966) *Residential Work with Children* (Oxford: Pergamon).

447. M. Bridgeland (1971) *Pioneer Work With Maladjusted Children: A Study of the Development of Therapeutic Education* (London: Staple Press), p. 277.

448. H. C. Gunzburg (1950) 'The Colony and the High-Grade Mental Defective', *Mental Health* 9, 87–92, p. 87.

449. H. C. Gunzburg, 'The Colony', p. 88.

450. H. C. Gunzburg, 'The Colony', p. 87.

451. H. C. Gunzburg, 'The Colony', p. 87.

452. H. C. Gunzburg, 'The Colony', p. 91.

453. H. C. Gunzburg, 'The Colony', p. 89.

454. H. C. Gunzburg, 'The Colony', p. 87, 88.

455. H. C. Gunzburg, 'The Colony', p. 91.

456. H. C. Gunzburg, 'The Colony', p. 88.

457. W. D. Wills (1967) [1941] 2nd edn. *The Hawkspur Experiment: an Informal Account of the Training of Wayward Adolescents* (London: George Allen and Unwin), p. 109.
458. W. D. Wills (1945) *The Barns Experiment* (George Allen and Unwin), p. 95.
459. W. D. Wills, *The Barns Experiment*, p. 24.
460. *Reynolds News*, 16 April 1950.
461. *Reynolds News*, 16 April 1950.
462. *The Observer*, 6 January 2002, 'Civil Liberties are a Communist front: the MI5 letter.' Denis Pritt was one of the most notorious.
463. P. Foot (2005) *The Vote: How it was Won and How it was Undermined* (London: Viking), p. 331.
464. K. Marx (1963) [1845] *The Holy Family* in T. Bottomore and M. Rubel (eds) *Karl Marx: Selected Writings in Sociology and Social Philosophy* (Harmondsworth: Pelican), p. 225.
465. HMSO (1957) *Royal Commission on The Law Relating to Mental Illness and Mental Deficiency 1954–57, Minutes of Evidence* (London: HMSO), cmnd. 169, p. 793; *Reynolds News*, 25 October 1953.
466. U DCL 631.4, Letter from F. Haskell (assistant secretary NCCL) to D. Edwards (Gen. Sec. Newcastle and District Trades Council) 22 February 1957; U DCL 631.6, NCCL publicity letter for a conference on Mental Deficiency Laws and Administration, Broad Street Birmingham November 1956.
467. U DCL 631.6, NCCL publicity letter for a conference on Mental Deficiency Laws, November 1956.
468. For instance, N. Korman and H. Glennerster (1990) *Hospital Closure: A Political and Economic Study* (Milton Keynes: Open University Press), p. 12.
469. *Daily Worker*, 2 March 1950.
470. B. H. Kirman (1952) 'The Law and Mental Deficiency', *Nursing Times* 19 January, p. 63.
471. J. S. Cookson (1945) 'Supervision of Mental Defectives in the Community', *BMJ*, 1, 90–1, p. 90.
472. M. E. Cripps (1958) 'Proposals Concerning Mental Deficiency', *British Journal of Psychiatric Social Work*, 4, 4, 24–26, p. 25.
473. Kirman, 'The Law and Mental Deficiency', p. 63.
474. NCCL (1951) *50,000 Outside the Law: An Examination of the Treatment of those Certified as Mental Defectives* (London: NCCL), p. 27.
475. U DCL/24.2, Letter from Medical Practitioners Union Gen. Sec. Dr. Bruce Cardew to Gen. Sec. NCCL, Elizabeth Allen, 6 December 1954.
476. L. T. Hilliard (1954) 'Resettling Mental Defectives: Psychological and Social Aspects' *BMJ*, 1, 1372–1374, p. 1374.
477. D. Stafford-Clark (1952) *Psychiatry Today* (Harmondsworth: Pelican), p. 89.
478. D. Stafford-Clark, *Psychiatry Today*, p. 89.
479. R. F. Tredgold (1953) Review of D. Stafford-Clark, *Psychiatry Today* in *Mental Health* 13, 1, 37.
480. An example is M. B. Davies (1944) *Hygiene and Health Education For Training Colleges*, 3rd edn (London: Longmans Green & Co.) One passage reads 'Defectives, too, are often sexually uncontrolled, lacking as they do the imagination to evaluate the social stigma that results from lack of control. The women tend to propagate their wastrel class by having illegitimate children; the men are of the class that commits emotionally disturbing and terrifying sex assaults of varying kinds' (p. 244).

481. D. Curran and E. Guttmann (1945) *Psychological Medicine: A Short Introductory to Psychiatry* (Edinburgh: E & S Livingstone), pp. 91–2. 'All that is usually required is to take a brief history and to carry out suitable tests which do not usually take much time' (p. 91); 'a question or two about the school record', [a few about the work record] and 'a short conversation about the front page of the news should enable the doctor to form a fair estimate of the patient's intelligence' (p. 224).

482. *The New Statesman and Nation*, 2 April 1955; 16 April 1955; 23 April 1955. (Quotes from 23 April 1955.) Kenneth Robinson was made a Vice-President of NAMH in 1958.

483. NAMH *Annual Report* (1951) 1950–51, pp. 10–11. See also R. F. Tredgold's review of NCCL's booklet *50,000 Outside the Law*, in *Mental Health*, 10, 3, p. 80; NAMH, Minutes of Fifth Annual General Meeting, 9 January 1952.

484. Letter from NAMH General Secretary to W. S. Shepherd MP, 25 April 1951.

485. NAMH, Minutes of the Mental Deficiency Sub-Committee, 6 December 1954.

486. L. T. Hilliard (1951) Review of NCCL (1951) *50,000 Outside the Law: An Examination of the Treatment of those Certified as Mental Defectives* (London: NCCL) in *Health Education Journal*, 9, 4, p. 202.

487. B. Kirman (1952) *This Matter of Mind* (London: Thrift Books).

488. R. Cohen (2004) 'Tredgold, Rodger Francis (1911–1975)', *Oxford Dictionary of National Biography* (Oxford: Oxford University Press) http://www.oxforddnb.com/view/article/65060

489. R. F. Tredgold (1952) 'Editorial', *Mental Health* 11, 3, 102–3, p. 102.

490. B. Kirman, *This Matter of Mind*, p. 51.

491. B. Kirman, *This Matter of Mind*, pp. 55–7.

492. B. Kirman, *This Matter of Mind*, p. 55.

493. See D. Thom (2004) 'Politics and the People, Brian Simon and the Campaign Against Intelligence Tests in British Schools', *History of Education*, 33, 5, 515–29.

494. See D. Thom, 'Politics and the People', pp. 523–4.

495. D. Thom, 'Politics and the People', p. 529.

496. R. F. Tredgold, 'Editorial', p. 103.

497. B. Kirman, *This Matter of Mind*, p. 87.

498. B. Kirman, *This Matter of Mind*, p. 58.

499. B. Kirman, *This Matter of Mind*, p. 59.

500. B. Kirman, *This Matter of Mind*, p. 34, 73.

501. B. Kirman, *This Matter of Mind*, p. 75.

502. B. Kirman, *This Matter of Mind*, p. 82, 83.

503. B. Kirman, *This Matter of Mind*, p. 86.

504. B. Kirman, *This Matter of Mind*, p. 52.

505. B. Kirman, *This Matter of Mind*, p. 83, 92.

506. B. Kirman, *This Matter of Mind*, pp. 84–6, p. 89.

507. B. Kirman, *This Matter of Mind*, p. 84, 89.

508. R. F. Tredgold (1950) 'Editorial', *Mental Health*, 10, 1–2, 1.

509. B. Kirman, *This Matter of Mind*, p. 84, 86.

510. NAMH, Minutes of the Mental Deficiency Sub-Committee, 16 March 1950.

511. NAMH (1954) *Annual Report* 1953–54, p.19; NAMH (1955) *Annual Report* 1954–55, p. 10. Hilliard didn't actually manage to attend any of the Mental Deficiency Training Sub-Committee meetings and so the committee decided to send him copies of its minutes so that he could keep abreast of what they were doing.

512. NAMH (1951) *Annual Report 1950*–51, p. 10; NAMH (c.1955) *Mentally Handicapped Children: A Handbook for Parents* (London: NAMH).

513. Foreword by J. Tizard in B. Kirman and J. Bicknell (1975) *Mental Handicap* (London: Churchill Livingstone), p. v; S. Segal (1984) *Society and Mental Handicap: Are We Ineducable?* (Tunbridge Wells: Costello), p. 56.
514. See, for instance, NAMH (1953) *Annual Report, 1952*–1953, p. 5.
515. NAMH (1950) *Annual Report 1949*–1950, p. 14; NAMH (1951) *Annual Report 1950*–1951, p. 13.
516. In 1949 Hester Adrian met them to discuss their research into the employability of people designated 'high-grade mentally deficient'. See NAMH, Minutes of Mental Deficiency Sub-Committee, 9 June 1949.
517. NAMH, Minutes of Mental Deficiency Training Sub-Committees, 1954 through to 1960. Though, if the minutes are anything to go by, he doesn't seem to have said a great deal at these meetings. One of the few comments recorded is his suggestion that the proposed diploma course for teachers of 'mentally deficient' children to be run by NAMH should use the Institute of Education as its examining body. But other members dismissed this idea on the grounds that they didn't think education bodies had the 'proper appreciation and understanding of the needs of mentally deficient adults and children.' See Minutes of Mental Deficiency Training Sub-Committee, 11 November 1958.
518. L. T. Hilliard and B. H. Kirman (1957) *Mental Deficiency* (London: J & A Churchill).
519. R. F. Tredgold and K. Soddy (eds) (1970) *Tredgold's Mental Deficiency*, 10th edn (London, Bailliere). There were also some pretty dubious and condescending interpretations of the meaning of the women's facial expressions.
520. He served on standing committees throughout the 1950s and 1960s, and became a Vice-President in 1966. See NAMH *Annual Reports, 1944–6* through to *1966–7*.
521. A. D. B. Clarke and B. Tizard (eds) (1983) *Child Development and Social Policy: The Life and Work of Jack Tizard* (Leicester: British Psychological Society), pp. 1–3.
522. J. Tizard (1950) 'The Abilities of Adolescent and Adult High Grade Mental Defectives', *Journal of Mental Science* 96, 405, 889–907; J. Tizard and N. O'Connor (1952) 'The Occupational Adaptation of High-Grade Mental Defectives', *Lancet* 260, 620–3; NCCL, *50,000 Outside the Law*, p. 22.
523. *Telegraph and Argus*, 29 March 1951, Letter from Elizabeth Allen, Gen. Sec. NCCL on 'Mental Deficiency Laws'; J. Tizard and N. O'Connor (1956) *The Social Problem of Mental Deficiency* (London: Pergamon Press).
524. J. Tizard (1953) speech on 'Adult Defectives and their Employment' in NAMH, *The Practical Application of Research and Experiment to the Mental Health Field: Proceedings of a Conference held at Royal Victoria Halls, London, W.C.1, 5–6 February 1953* (London: NAMH), pp. 46–56.
525. J. Tizard, 'Adult Defectives and their Employment', p. 48.
526. J. Tizard, 'Adult Defectives and their Employment', p. 50.
527. J. Tizard, 'Adult Defectives and their Employment', p. 50.
528. *New Statesman and Nation* 12 June 1954.
529. M. Craft (1959) 'The Place of the Mental Deficiency Hospital in a Community Care Programme', *Mental Health* 18, 2, 60–4, p. 61.
530. NCCL, *50,000 Outside the Law*, p. 7, 26. The NCCL cited the case of 'Victoria' who's licence was cancelled because she had been to the cinema with a young man who attended the same church. She subsequently spent another four years in a mental deficiency institution.
531. NCCL, *50,000 Outside the Law*, p. 26.

532. NAMH, Minutes of Mental Deficiency Sub-Committee, 24 April 1952, 17 July 1952, 26 February 1953.
533. NAMH, Minutes of Mental Deficiency Sub-Committee, 14 October 1952.
534. NAMH, Minutes of Mental Deficiency Sub-Committee, 14 May 1953.
535. NCCL, *50,000 Outside the Law*, p. 25.
536. NCCL, *50,000 Outside the Law*, p. 20, 22.
537. T. P. Rees (1957) 'Back to Moral Treatment and Community Care', *Journal of Mental Science* 103, 303–313, p. 311.
538. Rees had been a member of the Feversham Committee on the voluntary mental health services, set up in 1936, which led, ultimately, to the formation of NAMH. He was subsequently a Royal Medico-Psychological Association representative on the Council of NAMH from its formation after the war until 1957.
539. NAMH (1958) Annual Report, 1957–8, p. 3.
540. T. P. Rees, 'Back to Moral Treatment and Community Care', pp. 309–10.
541. WHO (1953) *The Community Mental Hospital: Third Report of the Expert Committee on Mental Health*, (Geneva: WHO).
542. T. P. Rees, 'Back to Moral Treatment and Community Care', p. 309. Also cited in T. M. Cuthbert (1957) 'The Mental Hospital as a Therapeutic Community' in NAMH, *The Needs of the Mentally Sick: A Challenge to Youth: Report of a conference held on 17 October 1957* (London: NAMH), pp. 3–14, pp. 13–14.
543. WHO, *The Community Mental Hospital*, pp. 19–20.
544. T. P. Rees (1957) Review of M. Greenblatt, R.H. York and E.L. Brown (1955) *From Custodial to Therapeutic Care in Mental Hospitals* (New York: Russell Sage Foundation), *Mental Health* 16, 3, 113.
545. T. P. Rees, 'Back to Moral Treatment and Community Care', p. 311.
546. R. A. Sanderson (1951) 'The Re-Socialization of the Psychiatric Case', *Mental Health* 10, 4, 87–96, p. 91.
547. D. H. Clark (1996) *The Story of a Mental Hospital: Fulbourn 1858–1983* (London: Process Press) p. 89.
548. T. P. Rees (1957) 'Discussion' in NAMH, *The Needs of the Mentally Sick: A Challenge to Youth*, pp. 55–64, pp. 58–9, 62–3 (quote at p. 62).
549. D. V. Martin (1955) 'Institutionalisation', *Lancet*, 266, 1188–90, p. 1188.
550. D. V. Martin, M. M. Glatt and K. F. Weeks (1954) 'An Experimental Unit for the Treatment of Neurosis', *British Journal of Psychiatry*, 100, 983–9.
551. D. V. Martin *et al*, 'An experimental unit for the treatment of neurosis', p. 983, 985.
552. D. V. Martin *et al*, 'An experimental unit for the treatment of neurosis', p. 985, 989.
553. D. V. Martin *et al*, 'An experimental unit for the treatment of neurosis', p. 989.
554. D. H. Clark, *The Story of a Mental Hospital*, pp. 88–9.
555. He ultimately became Vice-Chairman of NAMH in 1966. See NAMH (1967) Annual Report 1966–67, p. 22.
556. D. H. Clark, *The Story of a Mental Hospital*, p. 161.
557. D. H. Clark (1965) 'The Therapeutic Community – Concept, Practice and Future', *British Journal of Psychiatry* 111, 947–54, p. 949.
558. R. Barton (1959) *Institutional Neurosis*, 2nd edn, (Bristol: John Wright & Sons Ltd).
559. NAMH, Annual Reports from 1961–2 through to 1969–70.
560. R. F. Tredgold (1960) Review of R. Barton, *Institutional Neurosis* in *Mental Health* 19, 1, 32–3.

561. In this it was informed by the discovery in Britain of Marx's early theorization of alienation (in the *Economic and Philosophic Manuscripts*) and later by Gramsci's reworking of Marxist understandings of civil society.

562. S. Hall (1960) Editorial, 'Introducing NLR', *New Left Review* 1, 1, 1–3.

563. R. Samuel (1989) 'Born-again Socialism' in R. Archer, D. Bubeck, H. Glock (eds) *Out of Apathy: Voices of the New Left 30 Years On* (London: Verso), pp. 39–57, p. 41, 44; S. Hall, 'The First New Left' in R. Archer *et al.*, *Out of Apathy*, pp. 13–38, pp. 29–34.

564. John Welshman notes that Titmuss had developed close links with NAMH from the late 1950s. See J. Welshman (1999) 'Rhetoric and Reality: Community Care in England and Wales, 1948–74' in P. Bartlett and D. Wright, *Outside the Walls of the Asylum: The History of Care in the Community 1750–2000* (London: The Athlone Press), pp. 204–26, p. 223.

565. F. Powell, (2001) *The Politics of Social Work* (London: Sage), p. 55.

566. N. Ellison (1994) *Egalitarian Thought and Labour Politics: Retreating Visions* (London: Routledge), p. 128.

567. N. Ellison, *Egalitarian Thought and Labour Politics*, p. 146, 133.

568. C. Unsworth (1987) *The Politics of Mental Health* (Oxford: Clarendon Press), p. 265, 267.

569. C. Unsworth, *The Politics of Mental Health Legislation*, p. 259.

570. C. Unsworth, *The Politics of Mental Health Legislation*, p. 230.

571. NAMH Council of Management Meeting, 12.7.57; NAMH (1957) *Annual Report 1956–7*, pp. 4–5.

572. HMSO (1957) *Royal Commission on the Law Relating to Mental Illness and Mental Deficiency 1954–1957, Report* (London: HMSO) Cmnd.169, para. 87.

573. Royal Commission on the Law Relating to Mental Illness and Mental Deficiency, *Report*, para. 601. See also C. Unsworth, *The Politics of Mental Health Legislation*, p. 289.

574. As is witnessed in the somewhat contradictory evidence given by NAMH representatives in their oral evidence to the Royal Commission.

575. C. Unsworth, *The Politics of Mental Health Legislation*, pp. 239–40.

576. C. Unsworth, *The Politics of Mental Health Legislation*, p. 264.

577. M. Appleby, Foreword to, D. W. Wills, *Reynolds House: A Report of the First Five Years*. NAMH April, 1969.

578. W. D. Wills, Unpublished Autobiography, p. 184.

579. NSMHC (1961) *Mentally Handicapped Children Growing Up*, director Basil Wright, Realist Film Unit (Ipswich: Concord Media).

580. The National Society for Mentally Handicapped Children originated in 1946 as the National Association for Parents of Backward Children. It changed its title in 1955.

581. NAMH publicly welcomed the formation of the National Association of Parents of Backward Children and the emergence of associated local parents' associations. However, these clearly increased pressure on NAMH to review and amend its approach.

582. HMSO (1946) *Report of the Care of Children Committee (Curtis Committee)*, cmd. 6922.

583. A. D. B. Clarke and B. Tizard (eds) (1983) *Child Development and Social Policy: The Life and Work of Jack Tizard* (Leicester: British Psychological Society), p. 4.

584. J. Tizard (1964) *Community Services for the Mentally Handicapped* (London: Oxford University Press), p. 60, 121.

585. J. Tizard, *Community Services for the Mentally Handicapped*, p. 101.
586. J. Tizard, *Community Services for the Mentally Handicapped*, p. 97.
587. J. Tizard, *Community Services for the Mentally Handicapped*, p. 96.
588. J. Tizard, *Community Services for the Mentally Handicapped*, p. 79. See the comparable passage in the Curtis Report, para. 418.
589. J. Tizard, *Community Services for the Mentally Handicapped*, p. 130, pp. 133–4.
590. J. Tizard, *Community Services for the Mentally Handicapped*, pp. 102–3.
591. D. W. Wills, *Reynolds House*, p. 33, 38, 47, 57.
592. E. Shorter (1997) *A History of Psychiatry: From The Era of The Asylum to the Age of Prozac* (New York: John Wiley and Sons), p. 274.
593. E. Goffman (1991) [1961] *Asylums: Essays on the Social Situation of Mental Patients and Other Inmates* (Harmondsworth: Penguin), pp. 134–5.
594. E. Goffman, *Asylums*, pp. 311–12.
595. E. Goffman, *Asylums*, pp. 21–2.
596. R. D. King, N. V. Raynes and J. Tizard (1971) *Patterns of Residential Care: Sociological Studies in Institutions for Handicapped Children* (London: Routledge and Kegan Paul).
597. N. V. Raynes and R. D. King (1974) [1967] 'Residential Care for the Mentally Retarded' in D. M. Boswell and J. M. Wingrove (eds) (1974) *The Handicapped Person in the Community* (London: Tavistock and OUP), pp. 299–306, p. 299.
598. R. D. King *et al.*, *Patterns of Residential Care*, pp. 115–16.
599. A. Kushlick (1972) 'The Need for Residential Care' in V. Shennan (ed.) *Subnormality in the 70s: Action for the Retarded* (London: NCMHC and WFMH), pp. 13–26. A. D. B. Clarke and B. Tizard (eds) *Child Development and Social Policy: The Life and Work of Jack Tizard*, p. 4.
600. R. Barton (1973) 'The Institutional Mind and the Subnormal Mind' in H.C. Gunzburg (ed.) *Advances in the Care of the Mentally Handicapped* (London: British Society for the Study of the Mental Subnormality), pp. 13–20.
601. D. H. Clark (1974) *Social Therapy in Psychiatry* (Harmondsworth: Penguin), pp. 12, 42–3.
602. For instance, Douglas Bennett a psychiatrist closely associated with NAMH favourably reviewed Goffman's later book *Stigma* in 1964, see 'Social Attitudes and Mental Disorder', *Mental Health* 23, 6, 241–2.
603. G. Cohen (1964) 'Hospitals and People', *Mental Health* 23, 6, 236–8, p. 236.
604. N. Rose (1986) 'Law, Rights and Psychiatry' in P. Miller, N. Rose, *The Power of Psychiatry* (Cambridge: Polity Press) pp. 177–213, p. 179, fn. 6; E. Shorter, *A History of Psychiatry*, pp. 274–5; P. Sedgwick (1982) *Psycho Politics* (London: Pluto Press), Chapter 2.
605. R. D. Laing (1960) *The Divided Self* (London: Tavistock Publications).
606. P. Sedgwick, *Psychopolitics*, Chapter 3; C. Unsworth, *The Politics of mental Health*, p. 345, N. Crossley (2006) *Contesting Psychiatry: Social Movements in Mental Health* (London, Routledge), Chapter 5.
607. S. Hall (1964) 'Growing Up Absurd?' in NAMH, *Adolescence: Report of the 20th Child Guidance Inter-Clinic Conference* 1964 (London: NAMH), pp. 8–18.
608. J. H. Kahn, 'Foreword' in *Adolescence: Report of the 20th Child Guidance Inter-Clinic Conference*, p. 3.
609. S. Hall, 'Growing Up Absurd?', p. 9, 10.
610. S. Hall, 'Growing Up Absurd?', p. 8.
611. S. Hall, 'Growing Up Absurd?', p. 12.
612. S. Hall, 'Growing Up Absurd?', p. 9, 12.

613. S. Hall, 'Growing Up Absurd?', pp. 12–15, quote at p. 15.
614. S. Hall, 'Growing Up Absurd?', pp. 12–13, 14, 16.
615. S. Hall, 'Growing Up Absurd?', p. 17.
616. S. Hall, 'Growing Up Absurd?', p. 16.
617. S. Hall, 'Growing Up Absurd?', p. 16.
618. W. D. Wills, *Reynolds House*, pp. 74–5.
619. C. L. C. Burns (1959) 'Conference "Summing Up" ' in NAMH, *Truancy or School Phobia? Being the Proceedings of the 15th Inter-Clinic Conference* (London: NAMH), pp. 26–32, p. 26.
620. J. L. Green (1959) 'Truancy – Or School Phobia' in NAMH, *Truancy or School Phobia?*, 8–16, p. 11.
621. J. L. Green, 'Truancy or School Phobia', p. 11, 18, 29.
622. J. L. Green, 'Truancy – Or School Phobia', p. 13. See also Burns, 'Conference "Summing Up"', p. 29 for similar assertions.
623. C. L. C. Burns, 'Conference "Summing Up"', p. 29.
624. C. L. C. Burns, 'Conference "Summing Up"', p. 29.
625. C. L. C. Burns, 'Conference "Summing Up"', p. 27.
626. J. L. Green, 'Truancy – Or School Phobia', p. 15.
627. Comments cited in NAMH (1946) *Report of the Proceedings of a Conference on Mental Health Held at St Pancras Town Hall, London, NW1, 14–15 November 1946* (London: NAMH) pp. 100–1.
628. Comments cited in NAMH (1953) *The Practical Application of Research and Experiment to the Mental Health Field: Proceedings of a Conference held at Royal Victoria Halls, London, W.C.1, 5–6 February 1953* (London: NAMH), p. 94.
629. E. M. Goldberg (1949) 'Comment' on F. E. Waldron, 'The Meaning of the Word "Social" in Psychiatric Social Work', *British Journal of Psychiatric Social Work* 1, 3, 18–21, p. 20.
630. C. Beedell (1957) 'The Psychopathology of Inter-Clinic Conferences', *British Journal of Psychiatric Social Work* 4, 1, 24–8.
631. C. Beedell, 'The Psychopathology of Inter-Clinic Conferences', p. 25.
632. C. Beedell, 'The Psychopathology of Inter-Clinic Conferences', p. 28.
633. R. Clough (2001) 'Christopher Beedell' (Obituary), *Guardian*, 6 September 2001. http://www.guardian.co.uk/news/2001/sep/06/guardianobituaries.socialsciences (accessed 30 November 2011).
634. D. H. Clark (1956) 'Functions of the Mental Hospital', *Lancet* 268, 1005–9, p. 1008.
635. E. Lewis (1964) review of D.V. Martin, *Adventure in Psychiatry: Social Change in a Mental Hospital*, *Mental Health* 23, 1, 26.
636. D. H. Clark, *The Story of a Mental Hospital*, p. 128.
637. D. H. Clark, *The Story of a Mental Hospital*, p. 142.
638. See NAMH, Annual Reports.
639. M. James (1965) 'Summing Up' in NAMH, *Child Guidance: Function and Social Role: Report of the 21st Child Guidance Inter-Clinic Conference, 1965* (London: NAMH), pp. 51–2.
640. B. Butler (1971) 'The Changing Role of the Social Worker' in J. D. Sutherland (ed.) *Towards Community Mental Health* (London: Tavistock Publications), pp. 77–95, p. 84. (Papers delivered to the Psychotherapy Section of the Royal Medico-Psychological Association between 1966 and 1967.)
641. The reports of the Inter-Clinic Child Guidance Conferences show this clearly. See, for example, NAMH (1967) *Child Guidance From Within: Reactions to New Pressures. Papers Given at the 23rd Child Guidance Inter-Clinic Conference 1967* (London: NAMH); NAMH (1966) *Mental Illness in the Family: Its Effect on the*

Child. Proceedings of the Twenty-Second Child Guidance Inter-Clinic Conference, 1966 (London: NAMH).

642. H. M. Holden (1965) 'The Ethic of Child Guidance: Can Child Psychiatry Be Justified?', NAMH, *Child Guidance: Function and Social Role*, pp. 12–25.

643. W. D. Wills, *Reynolds House*, p. 60.

644. D. Cooper, Introduction to M. Foucault, (1989) [1967] *Madness and Civilization: A History of Insanity in the Age of Reason*, trans. Richard Howard, (London: Routledge), pp. vii–ix, p. viii.

645. M. Foucault, *Madness and Civilization*, p. 276.

646. M. Foucault, *Madness and Civilization*, p. 254.

647. E. Goffman, *Asylums*, p. 309, pp. 335–6.

648. D. Cooper (1972) [1971] *The Death of the Family* (Harmondsworth: Penguin), pp. 5–6.

649. D. Cooper, *The Death of the Family*, p. 6.

650. NAMH Council of Management Minutes, 22 November 1966.

651. NAMH (1970) 'Dramatic charade', NAMH, *Annual Report, 1969–1970*, p. 10.

652. See NAMH (1961) *Mental Health*, 20, 1, 15.

653. NAMH (1964) 'The Whole truth About Care of the Mentally Disordered', *Mental Health* 23, 2, 43–9, pp. 45–6.

654. NAMH (1969) 'Consumer Panel' in NAMH, *New Ways with Old Problems: Report of the Annual Conference 1969* (London: NAMH), pp. 20–38.

655. NAMH (1969) *Annual Report, 1968–9*, p. 9. A sustained and ongoing battle between NAMH and the Church of scientology may also have partly informed these moves, as well as the later development of NAMH into MIND.

656. For an outline of these scandals and their effects see J. Martin (1984) *Hospitals in Trouble* (London: Basil Blackwell).

657. B. Robb (ed.) (1967) *Sans Everything: A Case to Answer* (London: Thomas Nelson & Sons Ltd).

658. J. Martin, *Hospitals in Trouble*, p. 4.

659. See Chapter 6.

660. NAMH (1968) *Annual Report, 1967–68*, p. 13.

661. B. Abel-Smith (1968) in NAMH *What's Wrong with the Mental Health services? Report of the Annual Conference 1968* (London: NAMH), pp. 15–16; P. Townsend (1969) 'New Structures: A Critical Review' in NAMH, *New Ways with Old Problems: Report of the Annual Conference 1969* (London: NAMH), pp. 10–19, p. 19.

662. The journalist C. H. Rolph, closely involved with the NAMH since the 1950s, contributed an article criticizing the Court of Protection for its inability to 'protect the defenceless invalid against physical discomfort, emotional exploitation and deprivation, indifference, exasperation and neglect.' See C. H. Rolph, 'Cruelty in the Old People's Ward' in B. Robb (ed.), *Sans Everything: A Case to Answer*, 3–7, p. 3. This had previously been published by NAMH in its journal *Mental Health*.

663. R. Barton, 'Foreword' to B. Robb (ed.) *Sans Everything: A Case to Answer*, pp. ix–xi.

664. NAMH (1968) 'Scene 2: A Grim Old Workhouse', *Mental Health* Summer, pp. 36–7.

665. NAMH (1968) 'Scene 2: A Grim Old Workhouse', *Mental Health* Summer, p. 37.

666. P. Mittler (1968) 'The State of the Nation', *Mental Health* Summer, 2–3, p. 3.

667. J. Martin, *Hospitals in Trouble*, p. 7. After Ely, the Welsh Hospital Board instituted a survey of all long-stay hospitals. This report was described as giving an 'extremely depressing picture' and showed that the conditions at Ely weren't untypical. See Martin, *Hospitals in Trouble*, p. 68.

668. NAMH, 'Consumer Panel', Charles Hannam, pp. 30–1, p. 31 and Jean Slee-Smith, pp. 31–4, pp. 32–3.
669. 'Consumer Panel', Jean Slee-Smith, pp. 32–4.
670. 'Consumer Panel', Charles Hannam, p. 30.
671. 'Consumer Panel', Clare Marc Wallace, pp. 24–6, Mr G, pp. 26–8, Diana Williams, pp. 28–9.
672. 'Consumer Panel', Mr G, p. 26.
673. 'Consumer Panel', Mr G, p. 26, 27.
674. 'Consumer Panel', Clare Marc Wallace, p. 24, 25.
675. 'Consumer Panel', Clare Marc Wallace, p. 25, Diana Williams, p. 28
676. 'Consumer Panel', J. H. Kahn, Chairman's introduction to consumer panel session, p. 20.
677. W. D. Wills (1968) Letter to the editor, *British Journal of Criminology*, 8, 448.
678. A Member of the Executive Committee (1969) 'The Schools Action Union', *The Marxist*, no. 10, reprinted at Encyclopedia of Anti-Revisionism On-Line, accessed at http://www.marxists.org/history/erol/uk.firstwave/sau.htm (accessed 10 September 2012). 'Smash the Dictatorship of the Head' was an often used slogan. See G. Stevenson, 'Anatomy of Decline – The Young Communist League 1967–86', accessed at http://www.grahamstevenson.me.uk/ [home page] (accessed 10 September 2012).
679. PP/WDW, W. D. Wills Archive Research File.
680. K. Jones (1993) *Asylums and After: A Revised History of the Mental Health Services: From the Early 18th Century to the 1990s* (London: The Athlone Press); C. Unsworth (1987) *The Politics of Mental Health* (Oxford: Clarendon Press).
681. K. Jones, *Asylums and After*, p. 203; P. Sedgwick (1982) *Psycho Politics* (London: Pluto Press), pp. 213–14; C. Unsworth, *The Politics of Mental Health Legislation*, p. 4, see also pp. 23–4.
682. The phrase is from K. Jones, *Asylums and After*, title to Chapter 6. See also V. Coppock and J. Hopton (2000) *Critical Perspectives on Mental Health* (London: Taylor and Francis), pp. 140–1.
683. K. Jones, *Asylums and After*, p. 203.
684. C. Unsworth, *The Politics of Mental Health Legislation*, p. 4, 21; A. Clare (1981) 'Can The Law Reform Psychiatry?', *MIND OUT*, 48, 17.
685. K. Jones, *Asylums and After*. See also A. Clare (1980) *Psychiatry in Dissent: Controversial Issues in Thought and Practice*, 2nd edn (London: Tavistock), Chapter 8.
686. P. Bean (1986) *Mental Disorder and Legal Control* (Cambridge: Cambridge University Press), pp. 180–2.
687. N. Rose (1985) 'Unreasonable Rights: Mental Illness and the Limits of the Law', *Journal of Law and Society*, 12, 2, 199–218. The argument here was subsequently revised and extended. See N. Rose (1986) 'Law, Rights and Psychiatry' in P. Miller, N. Rose (eds) *The Power of Psychiatry* (Cambridge: Polity Press). My argument refers to both versions.
688. N. Rose, 'Law, Rights and Psychiatry', pp. 200–1.
689. N. Rose, 'Law, Rights and Psychiatry', pp. 200–2.
690. N. Rose, 'Unreasonable Rights', p. 204.
691. M. Foucault (1980) [Interview originally published 1977] 'Truth and Power' in C. Gordon (ed.) *Power/Knowledge* (Harlow: The Harvester Press), pp. 109–33, pp. 118–9.
692. M. Foucault (1991) [1971] 'Nietzsche, Genealogy, History' in P. Rabinow (ed.) *The Foucault Reader* (Harmondsworth: Pengiun), pp. 76–100, p. 78, 89.

693. M. Foucault, 'Nietzsche, Genealogy, History', pp. 87–8.

694. I. Berlin (2002) *Freedom and its Betrayal: Six Enemies of Human Liberty* (Princeton: Princeton University Press), Chapter 2 (quotes at p. 31, 40, 41).

695. M. Foucault, 'Truth and Power', p. 133.

696. M. Foucault, 'Truth and Power', p. 119.

697. M. Foucault (1977) [1975] *Discipline and Punish: The Birth of the Prison*, trans. Alan Sheridan (Harmondsworth: Penguin), p. 27.

698. M. Foucault, *Discipline and Punish*, p. 27.

699. M. Foucault (1991) [1969] 'What is an Author?' in P. Rabinow (ed.) *The Foucault Reader*, pp. 101–20, p. 118.

700. M. Foucault, *Discipline and Punish*, p. 194.

701. M. Foucault, 'Truth and Power', p. 121.

702. D. Philpott (2010) 'Sovereignty', *The Stanford Encyclopedia of Philosophy*, Summer 2010 Edition, E. N. Zalta (ed.), accessed at http://plato.stanford.edu/archives/sum2010/entries/sovereignty/ (accessed 10 January 2011).

703. MIND (1971) *The MIND Manifesto*, p. 1.

704. MIND (1974) MIND Report no. 12, *Psychotherapy: Do we need more 'talking treatment'?* p. 8. (Emphasis in the original).

705. MIND (1973) 'MIND Matters', *MIND OUT*, 4, 2.

706. T. Smythe (1984) 'Mental Patients and Civil Liberties' in P. Wallington (ed.) *Civil Liberties, 1984* (Oxford: Robertson), pp. 309–23, pp. 310–11.

707. MIND (1974) Report No. 13, *Co-ordination or Chaos?: The Run-down of the Psychiatric Hospitals* (London: MIND), pp. 20–1, p. 26. MIND cited a passage from Circular 22/59 of 7 August 1959 which read 'the Minister...hereby directs...that...arrangements shall be made by every local health authority for the purpose of the prevention of mental disorder and...the care of persons suffering from mental disorder and the after care of such persons' (p. 21).

708. L. O. Gostin (1975) *A Human Condition: The Mental Health Act from 1959 to 1975: Observations, Analysis and Proposals for Reform*, Volume 1 (London: MIND), p. 13.

709. L. O. Gostin, *A Human Condition*, pp. 13–14.

710. MIND (c1975) *The State of MIND* (pamphlet), p. 1.

711. MIND (1972) *A Right to Love?* (London: MIND), reprinted in MIND (1972) *MIND and Mental Health*, 14–17.

712. For example, MIND (1979) 'Sex Education for the Mentally Handicapped', *MIND OUT*, 35, 7; MIND (1982) *Getting Together: Sexual and Social Expression for Mentally Handicapped People* (London: MIND); MIND (1981) 'We Work for Mentally Handicapped People Too', *MIND In Action '81* (publicity newsletter), p. 3.

713. MIND (1978) Editorial, 'The Issues of Mental Handicap', *MIND OUT*, 28, 2; L. Knight (1978) 'Better Services in Wessex', *MIND OUT*, 28, 6–8, p. 6.

714. A. Wertheimer (1980) 'Researching into Mental Handicap', *MIND OUT*, 40, 16–18, p. 17.

715. MIND (1974) Report No. 13, *Co-ordination or Chaos?: The Run-down of the Psychiatric Hospitals* (London: MIND), pp. 14–15; MIND (1978) Editorial, 'The Issues of Mental Handicap', *MIND OUT*, 28, 2.

716. MIND (1981) 'Wessex Research Unit Threatened with Closure', *MIND OUT*, 45, 4–5; MIND (1981) 'Threat to Wessex Research Unit Postponed', *MIND OUT*, 46, 4.

717. MIND (1976) *MIND Campaign 1976: Help a Healthy Mind Leave Hospital* (pamphlet); MIND (1977) *MIND 1976 Campaign, Home From Hospital: Progress and Results* (London: MIND), 2.

718. MIND (1976), Occasional Paper 5 - *Effective Community Care and the Mentally Ill: Back up Resources for GPs, Hostel Staff, Day Centre Staff and Others Caring for the Mentally Ill in the Community* (London: MIND).
719. MIND, Report No. 13, *Co-ordination or Chaos?*, p. 7. (Emphasis in the original.)
720. MIND, Report No. 13, *Co-ordination or Chaos?*, p. 7.
721. MIND (1976) *MIND Report: 'Room to Let': A Report on Nine Social Services Lodging Schemes* (London: MIND); MIND (1975) 'Life Begins at Sixty Plus' *MIND OUT*, 12, 3; MIND (1975) 'Wanted! More Landladies', *MIND OUT*, 12, 4–5; MIND (1975) 'Stepping Stones', *MIND OUT*, 12, 5.
722. A. Wertheimer (1975) 'Co-ordination or Chaos – The Rundown of Psychiatric Hospitals', *Royal Society of Health Journal*, July 1975 (MIND reprint), pp. 2–3.
723. Cited in speech by E. Morgan to Camden Association for Mental Health, 20 March 1973. (Edith Morgan private papers).
724. NAMH (1972) *Starting and Running a Group Home* (London: NAMH), p. 27.
725. NAMH, *Starting and Running a Group Home*, p. 23.
726. MIND, *Getting Together*, p. 8.
727. MIND, *Getting Together*, p. 8.
728. Reported in M. Manning (1975) 'Do Staff Really Need Qualifications?', *Community Care* 1 October 1975, p. 7. Ann Shearer described her experience of L'Arche communities as coming closest to the residential ideal. The first of these had been set up by a French Canadian Jean Vanier in 1964. By 1979 there was an international network of 50 branches.
729. A. Tyne (1979) 'Who's Consulted?' *MIND OUT*, 32, 8.
730. MIND (1979) *Annual Report* 1978–1979, p. 14; J. Melville (1980) 'Nobody Labels Us Here', *MIND OUT*, 40, 14–16.
731. MIND (1978) *Introducing the Junction Road Project* (news sheet).
732. W. David Wills (1979) 'The Moral Perspective', *Studies in Environment Therapy* 3, 25–35, p. 25.
733. W. David Wills, 'The Moral Perspective', p. 29.
734. W. David Wills, 'The Moral Perspective', p. 30.
735. We might note here that this historical contention need not imply (as Rose appears to suggest) a functionalist explanation. It does not necessarily follow that as 'the rights of the individual' and 'disciplinary mechanisms for the management of individuals' became manifest amongst, what he calls, the same broad ranging 'social, political and intellectual forces', they therefore have a necessarily functional fit with each other.
736. P. Sedgwick, *Psycho Politics*, p. 217.
737. T. Campbell (1983) *The Left and Rights* (London: Routledge and Kegan Paul).
738. P. Fennell (1986) 'Law and Psychiatry: The Legal Constitution of the Psychiatric System', *Journal of Law and Society*, 13, 35–65, p. 58
739. R. Cooter, 'The Ethical Body', in R. Cooter and J. Pickstone (eds) *Medicine in the Twentieth Century* (Amsterdam: Harwood Academic Publishers), pp. 451–85, p. 464; A. Rogers and D. Pilgrim (1996) *Mental Health Policy in Britain: A Critical Introduction* (London: MacMillan), pp. 78–9.
740. For an example of where Rose rules out these approaches and also provides a description of the notion of governmentality see P. Miller and N. Rose (1988) 'The Tavistock Programme: The Government of Subjectivity and Social Life', *Sociology*, 22, 171–92, pp. 173–4.
741. P. Fennell; 'Law and Psychiatry', p. 59.

742. N. Rose (1985) 'Unreasonable Rights', p. 212.
743. N. Rose, 'Law, Rights and Psychiatry', p. 211.
744. N. Rose (1989) *Governing the Soul: The Shaping of the Private Self* (London: Routledge), p. 254.
745. R. Cooter (2007) 'After Death/After-Life: The Social History of Medicine in Post-Modernity', *Social History of Medicine*, 20, 441–64, p. 449.
746. N. Rose (1998) *Inventing Our Selves* (Cambridge: Cambridge University Press), p. 16 (my emphasis).
747. N. Rose, *Governing the Soul*, pp. 246–7.
748. N. Rose, *Governing the Soul*, p. 11.
749. N. Rose, 'Law, Rights and Psychiatry', pp. 199–200.
750. N. Rose, 'Unreasonable Rights', p. 203.
751. N. Rose, 'Law, Rights and Psychiatry', p. 212.
752. N. Rose, 'Law, Rights and Psychiatry', p. 212.
753. N. Rose (1985) *The Psychological Complex: Psychology, Politics and Society in England, 1869–1939* (London: Routledge and Kegan Paul), p. 27.
754. N. Rose (1986) 'Psychiatry: The Discipline of Mental Health' in P. Miller, N. Rose (eds) *The Power of Psychiatry*, pp. 43–84, p. 74, 77.
755. MIND's evidence on Corporal Punishment in Schools, 1 February 1978.
756. D. Jolley (2003) 'John Anthony Whitehead: Formerly Consultant Psychiatrist, Brighton Health District', *Psychiatric Bulletin*, 27, 478.
757. Charles Clark was a prominent publisher and copyright expert who had taken an interest in mental health for some years.
758. MIND's evidence on Corporal Punishment in Schools, p. 1.
759. MIND's evidence on Corporal Punishment in Schools, p. 1.
760. MIND's evidence on Corporal Punishment in Schools, p. 1.
761. MIND's evidence on Corporal Punishment in Schools, p. 2.
762. MIND's evidence on Corporal Punishment in Schools, p. 2.
763. MIND's evidence on Corporal Punishment in Schools, p. 2.
764. MIND's evidence on Corporal Punishment in Schools, p. 2.
765. NAMH (1968) Annual Report 1967–68, p. 15; NAMH (1969) Annual Report 1968–9, p. 12.
766. N. K. Manning (1976), 'What Happened to the Therapeutic Community?' in K. Jones and S. Baldwin (eds), *The Year Book of Social Policy in Britain, 1975* (London: Routledge and Kegan Paul). Chapter 9, pp. 141–54, p. 152.
767. MIND (c1975) *MIND's Training and Staff Development Consultancy Service* (London: MIND); MIND (1975) 'MIND', *MIND OUT*, 9, 15.
768. MIND (1979) 'Proposed Temporary Closure of Henderson Hospital, Sutton – Mind's Response to the Consultative Paper Issued by the Merton, Sutton and Wandsworth Area Health Authority', pp. 1–2; MIND (1979) 'Unique Mental Health Service Threatened with Closure: this highlights health authorities' irresponsible actions, warns MIND Director', *MIND OUT*, 36, 5.
769. T. Smythe (1991) 'Health Warning', *The Raven: Anarchist Quarterly*, 4, 3, 216–219, p. 217. A report to MIND's Executive Committee on 20 September 1974 said that it was highly debatable whether Duncroft and Springhead Park's 'containing functions' were appropriate for delegation to a voluntary organization.
770. R. Stansfield (1977) 'A Special Kind of School', *MIND OUT*, 23, 5–7, p. 5. This was an acknowledgement of influence made by MIND about its approach to child care in general; see D. Winn (1975) 'Editorial', *MIND OUT*, 9, 3.

771. The development of Fairhaven also, however, expressed some of the tensions and conflicts inherent in the connection of what I have called the partial inversion of the Family, with wider elements of social critique.
772. D. Winn (1975) 'Editorial', *MIND OUT*, 9, 3.
773. MIND, Council of Management Minutes, 7.10.77.
774. MIND, *MIND Manifesto*, p. 1.
775. MIND, *MIND Manifesto*, pp. 2–3.
776. T. Smythe (1979) 'The Personal Social Services in the Next Decade', address to the Social Services Conference, 1979, pp. 16–17.
777. D. Winn (1974) 'Editorial', *MIND OUT*, 5, 3.
778. The initiative was publicized in various newspapers, as well as the magazine *Time Out* and the BBC's *Woman's Hour*.
779. MIND, Council of Management Minutes, 4 October 1974.
780. D. Winn (1974) 'Editorial', *MIND OUT*, 7, 2.
781. MIND (1974) 'No Place for Children', *MIND OUT*, 7, 15.
782. MIND (1974) 'Letters to the Editor', *MIND OUT*, 8, 4.
783. MIND, (1974) 'Letters to the Editor', *MIND OUT*, 8, 5.
784. D. Winn (1974) 'Editorial', *MIND OUT*, 7, 2.
785. T. Smythe and D. Winn (1975) 'A Problem Shared', *MIND OUT*, 10, 2–3, p. 2.
786. T. Smythe and D. Winn, 'A Problem Shared', p. 2.
787. T. Smythe and D. Winn, 'A Problem Shared', p. 3.
788. D. Graeber (2006) 'Beyond Power/Knowledge, an Exploration of the Relation of Power, Ignorance and Stupidity', p. 4. Paper presented to the London School of Economics, 25 May 2006, accessed at htt://www2.lse.ac.uk/PublicEvents/pdf20060525-Graeber.pdf (accessed 22 July 2009).
789. N. Rose (1996) 'Identity, Genealogy, History' in, S. Hall and P. du Gay (eds) *Questions of Cultural Identity* (London: Sage), pp. 128–50. See also P. Miller and N. Rose (1994) 'On Therapeutic Authority: Psychoanalytic Expertise under Advanced Liberalism', *History of the Human Sciences*, 7, 3, 29–64.
790. R. Cooter (2007) 'After Death/After-'Life': The Social History of Medicine in Post-Modernity', *Social History of Medicine*, 20, 441–64.

Bibliography

Archival Sources

Planned Environment Therapy Trust Archive (PETT), Church Lane, Toddington, Cheltenham.

W. D. Wills, Personal and Professional Papers, PP/WDW

PP/WDW 1.4, W. D. Wills, Examination papers, University of Birmingham, Social Study Diploma, 1927–8.
PP/WDW 1.5, W. D. Wills, Personal Journal of New York, September 1929 to May 1930.
PP/WDW, W. D. Wills, Unpublished Autobiography.
PP/WDW, W. D. Wills Archive Research File.

Archives of the Q Committee and of the Hawkspur Camp, SA/Q

SA/Q 20 Q Camps Committee, Minutes and Agendas, 25 June 1940.
SA/Q 21 Hon. Secretary's Memoranda, Reports, etc.
SA/Q/HM: Records of Hawkspur Camp for Men:
SA/Q/HM: 11, Camp Chief Reports, April 1936–March 1940.
SA/Q/HM: 12, Correspondence:
SA/Q/HM 12.1.1 Correspondence Marjorie Franklin/David Wills, 5 March 1936
SA/Q/HM 12.1.1 Correspondence Marjorie Franklin/David Wills, 10 May 1936
SA/Q/HM 12.2.1 Correspondence David Wills/Marjorie Franklin, 13 July 1936.
SA/Q/HM: 31, Members Files:
SA/Q/HM: 31.8
SA/Q/HM: 31.8
SA/Q/HM: 31.14.1 and 2
SA/Q/HM: 31.23.1
SA/Q/HM: 31.25.1
SA/Q/HM: 31.30.1
SA/Q/HM: 31.31. 1 and 2
SA/Q/HM: 31.33.1 and 2
SA/Q/HM: 31.38.1 and 2
SA/Q/HM: 31.42.2.

MIND Archives, Stratford

NCMH Annual Report, *Tenth Report of the National Council for Mental Hygiene 1932.*
PNC Annual Reports 1944–1946.
NAMH Annual Reports 1947–1985.
NAMH, Minutes of Council meeting 31 October 1952.
NAMH Council of Management Meeting, 12 July 1957.
NAMH Council of Management Minutes, 22 November 1966.
MIND, Council of Management Minutes, 4 October 1974.
MIND, Council of Management Minutes, 7 October 1977.

NAMH, Minutes of Fifth Annual General Meeting, 9 January 1952.

NAMH, Minutes of Mental Deficiency Sub-Committee, 9 June 1949.

NAMH, Minutes of the Mental Deficiency Sub-Committee, 16 March 1950.

NAMH, Minutes of Mental Deficiency Sub-Committee, 24 April 1952.

NAMH, Minutes of Mental Deficiency Sub-Committee, 17 July 1952.

NAMH, Minutes of Mental Deficiency Sub-Committee, 14 October 1952.

NAMH, Minutes of Mental Deficiency Sub-Committee, 26 February 1953.

NAMH, Minutes of Mental Deficiency Sub-Committee, 14 May 1953.

NAMH, Minutes of the Mental Deficiency Sub-Committee, 6 December 1954.

NAMH, Minutes of Mental Deficiency Training Sub-Committees 1954–1960.

Appleby, M. (1969) Foreword to D. W. Wills, *Reynolds House: A Report of the First Five Years*. NAMH April, 1969.

Letter from NAMH General Secretary to W. S. Shepherd MP, 25 April 1951.

NAMH and APSW (1954) *Memorandum on Rehabilitation of Psychiatric Patients: Evidence to the Committee of Enquiry into Existing Services for the Rehabilitation of the Disabled*.

NAMH (1956) *Memorandum Prepared for the Working Party on Social Workers*.

NCMH (1930) 'First International Congress on Mental Hygiene', *News Bulletin* (London: NCMH) 8–13.

MIND (1974) MIND Report no.12, *Psychotherapy: Do we need more 'talking treatment'?*

MIND (1974) Report No.13, *Co-ordination or Chaos?: The Run-down of the Psychiatric Hospitals* (London: MIND).

MIND (c1975) *MIND's Training and Staff Development Consultancy Service* (London: MIND).

MIND (c1975) *The State of MIND* (pamphlet).

MIND (1976) *MIND Campaign 1976: Help a Healthy Mind Leave Hospital* (pamphlet).

MIND (1976) *MIND Report: 'Room to Let': A Report on Nine Social Services Lodging Schemes* (London: MIND).

MIND (1976), Occasional Paper 5 – *Effective Community Care and the Mentally Ill: Back up Resources for GPs, Hostel Staff, Day Centre Staff and Others Caring for the Mentally Ill in the Community* (London: MIND).

MIND (1977) *MIND 1976 Campaign, Home From Hospital: Progress and Results* (London: MIND).

MIND (1978) *Introducing the Junction Road Project* (news sheet).

MIND's evidence on Corporal Punishment in Schools (1978) (evidence to the Secretary of State for Education and Science on corporal punishment in Schools, 1 February 1978).

MIND (1979) 'Proposed Temporary Closure of Henderson Hospital, Sutton – MIND's Response to the Consultative Paper Issued by the Merton, Sutton and Wandsworth Area Health Authority'.

MIND (1981) 'We Work for Mentally Handicapped People Too', *MIND In Action '81* (publicity newsletter).

PNC (1944) *After-Care of Service Casualties: Report for the Quarter Ending March 25th, 1944*.

Soddy, K. (1954) 'Community Care of Psychiatric Patients – A Review', appendix to NAMH and APSW, *Memorandum on Rehabilitation of Psychiatric Patients: Evidence to the Committee of Enquiry into Existing Services for the Rehabilitation of the Disabled*.

Smythe, T. (1979) 'The Personal Social Services in the Next Decade', address to the Social Services Conference, 1979.

Wills, D. W. (1969) *Reynolds House: A Report of the First Five Years*. NAMH April, 1969.

Modern Records Centre (MRC) University of Warwick

Archives of the Association of Psychiatric Social Workers

MSS.378/APSW/ P16/5:33, 'The Psychiatric Social Worker as "Duly Authorised Officer"': report of the discussion at APSW General Meeting, 27 September 1947.

MSS.378/APSW/P17/3:7, CGC (1936) Response to questionnaire circulated by the Feversham Committee.

MSS.378/APSW/P17/3:15, S. Clement Brown and B. McFie, Memo to the Feversham Committee (1938) 'The Functions of the Social Worker in the Mental Health Field'.

MSS.378/APSW/P21/4:1a, ' "Westhope" Society for Social Education' (n.d. July 1949?).

MSS.378/APSW/P21/4:1–9, Westhope Manor School: Correspondence May 1949–December 1949.

Archives of the National Institute for Social Work (NISW) and Related Organizations

MRC, MSS.463/box 51, S. Clement Brown (1935) 'Objectives and Methods in Social Case Work' (script of lecture given at Birmingham University).

MRC, MSS.463/box 51, S. Clement Brown (1935) 'Research into the Causes of Individual and Social Maladjustment and its Application to Social Case Work' (script of lecture given at Birmingham University).

Institute of Education Archives, University of London

Records of the World Education Fellowship

Records of the Home and School Council of Great Britain

WEF/D/1, vol. 1, Home and School Council Executive Committee Meetings 1929–1950.

Hull History Centre (Hull University Archives)

Archives of Liberty (formerly the National Council For Civil Liberties)

U DCL/24.2, Filing Case: 'No.24' 1911–1960.

U DCL 631 Filing Case: 'Mental Health' 1956–1979.

Private Papers of Edith Morgan O.B.E.

Edith Morgan, Speech to Camden Association for Mental Health, 20 March 1973.

Wellcome Library, Wellcome Collection, London

Robina Addis Collection, PP/ADD

Official Publications

Feversham Committee (1939) *The Voluntary Mental Health Services: The Report of the Feversham Committee* (London: The Committee).

HMSO, *Report of the Royal Commission on Lunacy and Mental Disorder*, Cmd 2700, 1926.

Bradford Education Committee (1942) *Report on Juvenile Delinquency* (Bradford: Bradford City Council).

HMSO (1946) *Report of the Care of Children Committee (Curtis Committee)*, Cmd 6922.

HMSO (1957) *Royal Commission on The Law Relating to Mental Illness and Mental Deficiency 1954–57, Report* (London: HMSO), Cmd 169.
HMSO (1957) *Royal Commission on The Law Relating to Mental Illness and Mental Deficiency 1954–57, Minutes of Evidence* (London: HMSO), Cmd 169.

Newspapers

Manchester Guardian; Bradford Telegraph and Argus; Guardian; Free Lance-Star; Reynolds News; Observer; Daily Worker; The New Statesman and Nation; Telegraph and Argus

Books and Articles

A Member of the Executive Committee (1969) 'The Schools Action Union', *The Marxist*, no. 10, reprinted in Encyclopedia of Anti-Revisionism On-Line, http://www.marxists.org/history/erol/uk.firstwave/sau.htm (accessed 10 September 2012).
Abel-Smith, B. (1968) in NAMH, *What's Wrong with the Mental Health services? Report of the Annual Conference 1968* (London: NAMH).
Ahrenfeldt, R. H. (1948) 'The Paternal State', *BMJ*, 1, 515–6.
Anderson, D., Anderson, I. (1982) 'The Development of the Voluntary Movement in Mental Health' in *The Mental Health Year Book 1981/82* (London: MIND), 427–42.
Anderson, J., Ricci, M. (eds) (1990) *Society and Social Science: A Reader* (Milton Keynes: The Open University).
Anon (1940) 'Evacuation and Mental Health Work', *Mental Health*, 1, 2.
Archer, R., Bubeck, D., Glock, H. (eds) (1989) *Out of Apathy: Voices of the New Left 30 Years On* (London: Verso).
Armstrong, D. (1983) *Political Anatomy of the Body: Medical Knowledge in Britain in the Twentieth Century* (Cambridge: Cambridge University Press).
Armstrong, D. (2002) *A New history of Identity: A Sociology of Medical Knowledge* (Basingstoke: Palgrave).
Balbernie, R. (1966) *Residential Work with Children* (Oxford: Pergamon).
Barker, H. (1942) 'A Special School Evacuation Unit: Some Observations on its Value', *Mental Health*, 3, 2, 37–9.
Barron, A. T. (1966) [1943] 'Practical Work at the Camp' in M. E. Franklin (ed.) Q *Camp: An Experiment in Group Living*, pp. 32–9.
Bartlett, P., Wright, D. (1999) *Outside the Walls of the Asylum: The History of Care in the Community 1750–2000* (London: The Athlone Press).
Barton, R. (1959) *Institutional Neurosis*, 2nd edn, (Bristol: John Wright & Sons Ltd).
Barton, R. (1973) 'The Institutional Mind and the Subnormal Mind' in H.C. Gunzburg (ed.) *Advances in the Care of the Mentally Handicapped* (London: British Society for the Study of the Mental Subnormality), pp. 13–20.
Barton, R. 'Foreword' to B. Robb (ed.) (1967) *Sans Everything: A Case to Answer* (London: Thomas Nelson & Sons Ltd).
Bean, P. (1986) *Mental Disorder and Legal Control* (Cambridge: Cambridge University Press).
Bean, P., MacPherson, S. (eds) (1983) *Approaches to Welfare* (London: Routledge and Kegan Paul).
Beedell, C. (1957) 'The Psychopathology of Inter-Clinic Conferences', *British Journal of Psychiatric Social Work* 4, 1, 24–8.

Bennett, D. (1964) 'Social Attitudes and Mental Disorder', *Mental Health* 23, 241–2.

Berlin, I. (2002) *Freedom and its Betrayal: Six Enemies of Human Liberty* (Princeton: Princeton University Press).

Bléandonu, G. (1994) *Wilfred Bion: His life and works 1897–1979* (London: Free Association Books).

BMJ (1937) 'Dissociation and Repression: Lecture by Professor McDougall', *British Medical Journal*, 1, 3987, 1169–71.

BMJ (1945) 'Medical News', *British Medical Journal*, 2, 791.

Bodsworth, T. C. (1966) [1943] 'Daily Life at Hawkspur Camp' in M. E. Franklin (ed.) *Q Camp: An Experiment in Group Living*, pp. 29–32.

Bond, C. H. (1921) 'The Position of Psychological Medicine in Medical and Allied Services', *Journal of Mental Science*, 67, 279, 404–49.

Bootle-Wilbraham, Brigadier, L. (1946) 'Civil Resettlement of Ex-Prisoners of War', *Mental Health*, 6, 2, 39–42.

Bosanquet, B. (1968) [1895] 'Character in its Bearing on Social Causation' in B. Bosanquet (ed.) *Aspects of the Social Problem*, pp. 103–17.

Bosanquet, B. (1968) [1895] 'Socialism and Natural Selection' in B. Bosanquet (ed.) *Aspects of the Social Problem* (London: MacMillan), pp. 289–307.

Bosanquet, B. (1968) [1895] 'The Principle of Private Property' in *Aspects of the Social Problem* (London: MacMillan), pp. 308–18.

Bosanquet, B. (1968) [1895] 'The Reality of the General Will' in B. Bosanquet (ed.) *Aspects of the Social Problem*, pp. 319–32.

Bosanquet, B. (ed.) (1968) [1895] *Aspects of the Social Problem* (London: MacMillan).

Bosanquet, B. (2001) [1899] *The Philosophical Theory of the State*, 2nd edn (Kitchener: Batoche Books).

Bosanquet, H. (1906) *The Family* (London: MacMillan).

Bosanquet, H. (1973) [1914] *Social Work in London 1869–1912*, (Brighton: The Harvester Press).

Boswell, D. M., Wingrove, J. M. (1974) *The Handicapped Person in the Community* (London: Tavistock and Oxford University Press).

Bowlby, J. (1949) 'The Study and Reduction of Group Tensions in the Family', *Human Relations* 2, 2, 123–8.

Bowlby, J. (n.d.) *Can I Leave My Baby?* (London: NAMH).

Bowlby, J. (1987) 'A Historical Perspective on Child Guidance', *The Child Guidance Trust News Letter*, 3, 1–2.

Bowlby, J., Winnicott, D. W. and Miller, E. (1939) 'Evacuation of Small Children' (letter to the editor) *BMJ* 2, 4119, 1202–3.

Boyd, W. and Rawson, W. (1965) *The Story of the New Education* (London: Heineman).

Brehony, K. J. (2004) 'A New Education for a New Era: The Contribution of the Conferences of the New Education Fellowship to the Disciplinary Field of Education 1921–1938', *Paedagogica Historica*, 40, 5, 733–55.

Bridgeland, M. (1971) *Pioneer Work with Maladjusted Children: A Study of the Development of Therapeutic Education* (London: Staple Press).

Brindley, S. K., Pettingale, G. H. (1963) 'Swalcliffe Park School', *Mental Health*, 22, 3, 112–14.

British Medical Bulletin (1949) 'Notes on Contributors', *British Medical Bulletin*, 6, 3, 222–3.

Brown, W. (1935) 'Character and Personality', *Mental Hygiene*, 13, 54–6.

Burns, C.L.C. (1959) 'Conference "Summing Up"' in NAMH, *Truancy or School Phobia? Being the Proceedings of the 15th Inter-Clinic Conference* (London: NAMH), pp. 26–32.

Burt, C. (1944) *The Young Delinquent*, 4th edn (London: University of London Press).

Burt, C. (1945) [1933] 'How the Mind Works in society' in C. Burt, E. Jones, E. Miller, W. Moodie (eds) *How the Mind Works* (London: Allen and Unwin), pp. 179–333.

Burt, C. (1952) Introduction to W. McDougall, *Psychology: The Study of Behaviour*, 2nd edn (London: Oxford University Press).

Burt, C., Jones, E., Miller, E., Moodie, W. (1945) [1933] *How the Mind Works*, 2nd edn (London: Allen and Unwin).

Butler, B. (1971) 'The Changing Role of the Social Worker' in J. D. Sutherland (ed.) *Towards Community Mental Health* (London: Tavistock Publications), pp. 77–95.

Campbell, T. (1983) *The Left and Rights* (London: Routledge and Kegan Paul).

Campling, P., Haigh, R. (eds) (1999) *Therapeutic Communities: Past, Present and Future* (London: Jessica Kingsley).

CAMW (1926) *Report of a Conference on Mental Welfare Held at the Central Hall, Westminster, London, S.W. on Thursday and Friday, December 2nd and 3rd, 1926* (London: CAMW).

CAMW (1937) 'News and Notes', *Mental Welfare*, 18, 2, 52.

Castel, R. (1983) 'Moral Treatment: Mental Therapy and Social Control in the Nineteenth Century' in S. Cohen, A. Scull (eds) *Social Control and the State: Historical and Comparative Essays* (Oxford: Martin Robertson).

Clare, A. (1980) *Psychiatry in Dissent: Controversial Issues in Thought and Practice*, 2nd edn (London: Tavistock).

Clare, A. (1981) 'Can The Law Reform Psychiatry?', *MIND OUT*, 48, 17.

Clark, D. H. (1956) 'Functions of the Mental Hospital', *Lancet*, 268, 1005–9.

Clark, D. H. (1965) 'The Therapeutic Community – Concept, Practice and Future', *British Journal of Psychiatry*, 111, 947–54, p. 949.

Clark, D. H. (1974) *Social Therapy in Psychiatry* (Harmondsworth: Penguin).

Clark, D. H. (1996) *The Story of a Mental Hospital: Fulbourn 1858–1983* (London: Process Press).

Clarke, A.D.B., Tizard, B. (eds) (1983) *Child Development and Social Policy: The Life and Work of Jack Tizard* (Leicester: British Psychological Society).

Clough, R. (2001) 'Christopher Beedell' (Obituary), *Guardian*, 6 September 2001, http://www.guardian.co.uk/news/2001/sep/06/guardianobituaries.socialsciences (accessed 12 October 2010).

Coates, D. (1990) 'Traditions of Thought and the Rise of Social Science in the United Kingdom' in J. Anderson, M. Ricci (eds), *Society and Social Science: A Reader* (Milton Keynes: The Open University), pp. 239–95.

Cohen, G. (1964) 'Hospitals and People', *Mental Health*, 23, 6, 236–8.

Cohen, G. (2000) *Karl Marx's Theory of History* (Oxford: Oxford University Press).

Cohen, G. (2002) 'Deeper into Bullshit' in S. Buss and L. Overton *Contours of Agency, Essays on Themes from Harry Frankfurt* (Cambridge Massachusetts: MIT Press), pp. 321–39.

Cohen, R. (2004) 'Tredgold, Rodger Francis (1911–1975)', *Oxford Dictionary of National Biography* (Oxford University Press), accessed at http://www.oxforddnb.com/view/article/65060 (accessed 21 August 2012).

Collini, S. (1976) 'Hobhouse, Bosanquet and the State: Philosophical Idealism and Political Argument in England 1880–1918', *Past and Present*, 72, 1, 86–111.

Cookson, J. S. (1945) 'Supervision of Mental Defectives in the Community', *BMJ*, 1, 90–1.

Cooper, D. (1972) [1971] *The Death of the Family* (Harmondsworth: Penguin).

Cooper, D. (1989) [1967] Introduction to M. Foucault, *Madness and Civilization: A History of Insanity in the Age of Reason*, trans. Richard Howard (London: Routledge).

Cooter, R. (1992) (ed.) *In The Name of The Child* (London: Routledge).

Cooter, R. (2007) 'After Death/After-Life: The Social History of Medicine in Post-Modernity', *Social History of Medicine*, 20, 441–64.

Cooter, R. 'The Ethical Body', in R. Cooter, J. Pickstone (eds) *Medicine in the Twentieth Century* (Amsterdam: Harwood Academic Publishers), pp. 451–485.

Cooter, R. and Pickstone, J. (eds) *Medicine in the Twentieth Century* (Amsterdam: Harwood Academic Publishers).

Coppock, V. and Hopton, J. (2000) *Critical Perspectives on Mental Health* (London: Taylor and Francis).

Cosens, M. (1932) *Psychiatric Social Work and the Family* (London: APSW).

Craft, M. (1959) 'The Place of the Mental Deficiency Hospital in a Community Care Programme', *Mental Health*, 18, 2, 60–4.

Craig, M. (1933) 'Mental Hygiene in Everyday Life', *Mental Hygiene*, 7, 57–64.

Cranston, M. (1968) Introduction to J. J. Rousseau (1993) *The Social Contract* (Harmondsworth: Penguin).

Cripps, M. E. (1958) 'Proposals Concerning Mental Deficiency', *British Journal of Mental Deficiency*, 4, 4, 24–6.

Crossley, N. (2006) *Contesting Psychiatry: Social Movements in Mental Health* (London, Routledge).

Curran, D. and Guttmann, E. (1945) *Psychological Medicine: A Short Introductory to Psychiatry* (Edinburgh: E & S Livingstone).

D. R. MacCalman (1948) Speech addressing 'Aggression in Relation to Family Life', ICMH, *International Congress*, 2, 50–5.

D. Winn (1974) 'Editorial', *MIND OUT*, 7, 2.

Dain, N. (1980) *Clifford Beers: Advocate for the Insane* (Pittsburg: Pittsburg University Press).

Dalal, R., Herzberger, R., Mathur, A. (c. 2001) 'Rishi Valley: The First Forty Years' (Krishnamurti Foundation), accessed at http//:www.arvindguptatoys.com/ (see 'Books in English').

Davies, M. B. (1944) *Hygiene and Health Education For Training Colleges*, 3rd edn (London: Longmans Green & Co.)

Den Otter, S. M. (1996) *British Idealism and Social Explanation: A Study in Late Victorian Thought* (Oxford: Clarendon Press).

Dendy, H. (1968) [1895] 'The Industrial Residuum' in B. Bosanquet (ed.) *Aspects of the Social Problem* (London: MacMillan), pp. 82–102.

Dicks, H. V. (1954) 'Strains Within the Family' in NAMH, *Strain and Stress in Modern Living: Special Opportunities and Responsibilities of Public Authorities* (London: NAMH), pp. 28–37.

Dicks, H. V. (1970) *Fifty Years of the Tavistock Clinic* (London: Routledge and Kegan Paul).

Digby, A. (1985) 'Moral Treatment at the Retreat, 1796–1846' in W.F. Bynum, R. Porter and M. Shepherd (eds), *The Anatomy of Madness: Essays in the History of Psychiatry, Volume 2* (London: Tavistock Publications), pp. 52–72.

Digby, A. (1985) *Madness, Morality and Medicine: A Study of The York Retreat 1796–1914* (Cambridge: Cambridge University Press).

Dionne Jr, E. J. (1997) 'Authority, Community, and a Lost Voice', *The Responsive Community*, 7, 4, 67–9.

Durbin, E. F. M. and Bowlby, J. (1939) *Personal Aggressiveness and War* (London: Kegan Paul, Trench and Trubner).

Ellison, N. (1994) *Egalitarian Thought and Labour Politics: Retreating Visions* (London: Routledge).

Etymonline (2008) 'Alienation', http://www.etymonline.com [home page] (accessed 1 August 2008).

Fairfield, L. (1931) 'Crime and Punishment', *Mental Hygiene*, 4, 17–20.

Fennell, P. (1986) 'Law and Psychiatry: The Legal Constitution of the Psychiatric System', *Journal of Law and Society*, 13, 1, 35–65.

Fildes, L. G. (1944) 'Hostels for Children in need of Psychiatric Attention', *Mental Health*, 5, 2, 31–2.

Fildes, L. G. (1946) 'The Care of the Homeless child' in NAMH, *Report of the Proceedings of a Conference on Mental Health, 14 and 15 November 1946*, pp. 58–62.

Finlay, J. L. (1970) 'John Hargrave, the Greenshirts and Social Credit', *Journal of Contemporary History*, 5, 1, 53–71.

Flugel, J. C. (1921) *The Psychoanalytic Study of the Family*, (London: Hogarth Press).

Foot, P. (2005) *The Vote: How it was Won and How it was Undermined* (London: Viking).

Foucault, M. (1977) [1975] *Discipline and Punish: The Birth of the Prison*, trans. Alan Sheridan (Harmondsworth: Penguin).

Foucault, M. (1980) [Interview originally published 1977] 'Truth and Power' in C. Gordon (ed.) *Power/Knowledge* (Harlow: The Harvester Press), pp. 109–33.

Foucault, M. (1989) [1967] *Madness and Civilization: A History of Insanity in the Age of Reason*, trans. Richard Howard (London: Routledge).

Foucault, M. (1991) [1969] 'What is an Author?' in P. Rabinow (ed.) *The Foucault Reader* (Harmondsworth: Penguin).

Foucault, M. (1991) [1971] 'Nietzsche, Genealogy, History' in P. Rabinow (ed.) *The Foucault Reader* (Harmondsworth: Pengiun), pp. 76–100.

Foucault, M. (2006) *History of Madness*, trans. Jonathan Murphy and Jean Khalfa, (London: Routledge).

Francis, W. A. G. (1941) 'Enuresis Record', *Mental Health*, 2, 3.

Franklin, M. E. (1966) [1943] 'Summary of the Methods Used' in M. E. Franklin (ed.) *Q Camp: An Experiment in Group Living*, pp. 13–22.

Franklin, M. E. (1973) 'P.E.T. Trust: Glimpses of a Future Rooted in the Past' in H. Klare and D. Wills (eds) *Studies in Environment Therapy Vol. 2* (Toddington: PETT).

Friedman, R. B. (1990) 'On the Concept of Authority in Political Philosophy' in J. Raz (ed.) *Authority* (New York: New York University Press), pp. 56–91.

Gillespie, R. D. (1931) 'Mental Hygiene as a National Problem', *Mental Hygiene*, 4, 1–9.

Goffman, E. (1991) [1961] *Asylums: Essays on the Social Situation of Mental Patients and Other Inmates* (Harmondsworth: Penguin).

Goldberg, E. M. (1949) 'Comment' on F. E. Waldron, 'The Meaning of the Word "Social" in Psychiatric Social Work', *British Journal of Psychiatric Social Work* 1, 3, 18–21.

Goldberg, E. M. (1957) 'The Psychiatric Social Worker in the Community', *British Journal of Psychiatric Social Work* 4, 2, 4–15.

Good, J. A. (2006) *A Search for Unity in Diversity: The 'Permanent Hegelian Deposit' in the Philosophy of John Dewey* (Oxford: Lexington Books).

Gordon, C. (ed.) (1980) *Power/Knowledge: Selected Interviews and Other Writings* (Harlow: The Harvester Press).

Gordon, P. (1983) 'The Writings of Edmond Holmes: A Reassessment and Bibliography', *History of Education*, 12, 1, 15–24.

Gordon, R. G. (1933) 'Habit Formation', *Mental Welfare* 14, 2, 29–37.

Gostin, L. O. (1975) *A Human Condition: The Mental Health Act from 1959 to 1975: Observations, Analysis and Proposals for Reform*, Volume 1 (London: MIND).

Graeber, D. (2006) 'Beyond Power/Knowledge, an Exploration of the Relation of Power, Ignorance and Stupidity', p. 4. Paper presented to the London School of Economics, 25 May 2006, accessed at http://libcom.org/files/20060525-Graeber.pdf (accessed 12 December 2009).

Green, J. L. (1959) 'Truancy – Or School Phobia' in NAMH, *Truancy or School Phobia? Being the Proceedings of the 15th Inter-Clinic Conference* (London: NAMH), pp. 8–16.

Guardian (2011) 'In Pictures: V.E. Day, 1945: "Cheers"' (Picture by Keystone), accessed at: http://www.guardian.co.uk/gall/0,,1476330,00.html (accessed 19 August 2011).

Gunzburg, H. C. (1950) 'The Colony and the High-Grade Mental Defective', *Mental Health* 9, 87–92.

Gunzburg, H. C. (ed.) (1973) *Advances in the Care of the Mentally Handicapped* (London: British Society for the Study of the Mental Subnormality).

Gutting, G. (1989) *Michel Foucault's Archaeology of Scientific Reason* (Cambridge: Cambridge University Press).

Hadfield, J. A. (1973) [1935] (ed.) 'Introduction', *Psychology and Modern Problems* (Plain View, NY: Books for Libraries Press).

Hall S., du Gay, P. (eds) *Questions of Cultural Identity* (London: Sage).

Hall, G. S. (1885) 'The New Psychology', *Andover Review*, 120–35, 239–48, accessed at http://psychclassics.yorku.ca/Hall/newpsych.htm (accessed 10 December 2012).

Hall, S. (1960) Editorial, 'Introducing NLR', *New Left Review* 1, 1, 1–3.

Hall, S. (1964) 'Growing Up Absurd?' in NAMH, *Adolescence: Report of the 20th Child Guidance Inter-Clinic Conference* 1964 (London: NAMH), pp. 8–18.

Hall, S. (1989) 'The First New Left' in R. Archer, D. Bubeck, H. Glock (eds) *Out of Apathy: Voices of the New Left 30 Years On* (London: Verso), pp. 13–38.

Hargrove, A. (1965) *Serving the Mentally Handicapped* (London: NAMH).

Harrison, T. (1996) 'Battle fields, Social fields, and Northfield', *Therapeutic Communities* 17, 3, 145–8.

Harrison, T. (2000) *Bion, Rickman, Foulkes and the Northfield Experiments: Advancing on a Different Front* (London: Jessica Kingsley).

Hayes, S. (2007) 'Rabbits and Rebels: The Medicalisation of Maladjusted Children in Mid-Twentieth-Century Britain' in M. Jackson (ed.) *Health and the Modern Home* (Abingdon: Routledge), pp. 128–52.

Hegel, G. W. F. (1977) [1807] *Phenomenology of Spirit*, trans. A. V. Miller, (Oxford: Clarendon Press).

Hendrick, H. (1994) *Child Welfare: England 1872–1989* (London: Routledge).

Hilliard, L. T. (1951) 'Review of NCCL, *50,000 Outside the Law: An Examination of the Treatment of those Certified as Mental Defectives* (London: NCCL)', *Health Education Journal*, 9, 4, 202.

Hilliard, L. T. (1954) 'Resettling Mental Defectives: Psychological and Social Aspects' *BMJ*, 1, 1372–4.

Hilliard, L. T., Kirman, B. H. (1957) *Mental Deficiency* (London: J & A Churchill).

Holden, H. M. (1965) 'The Ethic of Child Guidance: Can Child Psychiatry Be Justified?', *Child Guidance: Function and Social Role: Report of the 21st Child Guidance Inter-Clinic Conference, 1965* (London: NAMH), pp. 12–25.

Holman, R. (1973) 'Family Deprivation', *British Journal of Social Work*, 3, 4, 431–46.

Holman, R. (Bob) (2010) 'I Thought I Knew Iain Duncan Smith', *Guardian*, 12 December 2010, accessed at http://www.guardian.co.uk/commentisfree/2010/nov/12/iain-duncan-smith-punishing-the-poor (accessed 1 September 2011).

Holmes, J. (1993) *John Bowlby and Attachment Theory* (London: Routledge).

Hunnybun, N. K., Jacobs, L. (1946) *Interviews With Parents in a Child Guidance Clinic* (London: APSW).

Hunter, D. (1955) 'An Approach To Psychotherapeutic Work With Children and Parents', *11th Inter-Clinic Conference For Staffs of Child Guidance Clinics: The Family Approach to Child Guidance-Therapeutic Techniques*, (London: NAMH), pp. 11–26.

ICMH (1948) *International Congress on Mental Health, London, 1948* [four volumes], (London: H.K. Lewis & Co. Ltd).

James, M. (1965) 'Summing Up' in NAMH, *Child Guidance: Function and Social Role: Report of the 21st Child Guidance Inter-Clinic Conference, 1965* (London: NAMH), pp. 51–2.

Jolley, D. (2003) 'John Anthony Whitehead: Formerly Consultant Psychiatrist, Brighton Health District', *Psychiatric Bulletin* 27, 478.

Jones, E. (1945) [1933] 2nd edn, 'The Unconscious Mind' in C. Burt, E. Jones, E. Miller, W. Moodie, *How the Mind Works* (London: Allen and Unwin) pp. 61–103.

Jones, G. (1986) *Social Hygiene* (London: Croom Helm).

Jones, K. (1993) *Asylums and After: A Revised History of the Mental Health Services: From the Early 18th Century to the 1990s* (London: The Athlone Press).

Jones, K., Baldwin, S. (eds) (1976), *The Year Book of Social Policy in Britain, 1975* (London: Routledge and Kegan Paul).

Jones, K. W. (1999) *Taming the Troublesome Child: American Families, Child Guidance, and the Limits of Psychiatric Authority* (London: Harvard University Press).

Kafka, F. (1933), 'In the Penal Settlement' in *Metamorphosis and Other Stories* (London: Secker and Warburg Ltd).

Kahn, J. H. (1964) 'Foreword' to NAMH, *Adolescence: Report of the 20th Child Guidance Inter-Clinic Conference*, p. 3.

Kahn, J. H. (c1957) *Child Guidance* (NAMH leaflet).

King, R. D., Raynes, N. V., Tizard, J. (1971) *Patterns of Residential Care: Sociological Studies in Institutions for Handicapped Children* (London: Routledge and Kegan Paul).

Kirman, B. H. (1946) 'Left Turn' (letter to the editor) *Lancet*, 248, 808.

Kirman, B. H. (1952) 'The Law and Mental Deficiency', *Nursing Times*, 19 January 1963.

Kirman, B. H. (1952) *This Matter of Mind* (London, Thrift Books).

Kirman, B. H., Bicknell, J. (1975) *Mental Handicap* (London: Churchill Livingstone).

Knight, L. (1978) 'Better Services in Wessex', *MIND OUT*, 28, 6–8.

Korman, N., Glennerster, H. (1990) *Hospital Closure: A Political and Economic Study* (Milton Keynes: Open University Press).

Kushlick, A. (1972) 'The Need for Residential Care', in V. Shennan (ed.) *Subnormality in the 70s: Action for the Retarded* (London: NCMHC and WFMH), pp. 13–26.

Laing, R. D. (1960) *The Divided Self* (London: Tavistock Publications).

Lewis, E. (1964) review of D.V. Martin, *Adventure in Psychiatry: Social Change in a Mental Hospital*, *Mental Health*, 23, 1, 26.

Lewis, J. (1995) *The Voluntary Sector, The State and Social Work in Britain: The Charity Organisation Society/Family Welfare Association since 1869* (Aldershot: Edward Elgar).

Linebaugh, P. (2003) *The London Hanged: Crime and Civil Society in the Eighteenth Century* (London: Verso).

Lord, J. R. (1930) 'American Psychiatry and its Practical Bearings on the Application of Recent Local Government and Mental Treatment Legislation, Including a

Description of the Author's Participation in the First International Congress on Mental Hygiene, Washington, D.C., May 5–10, 1930', *Journal of Mental Science*, 76, 456–95.

Lucas, R. (1953) Comments at, NAMH, *Tenth Child Guidance Inter-Clinic Conference: Report on Conference and Some Selected Clinic Surveys* (London: NAMH), pp. 28–30.

Lukes, S. (1979) 'Power and Authority' in T. B. Bottomore, R. Nisbet (eds) *A History of Sociological Analysis* (London: Heinemann).

MacCalman, D. R. (1949) 'Sweet are the Uses of Adversity', *British Journal of Psychiatric Social Work*, 2, 87–94.

Main, T. F. (1946) 'The Hospital as a Therapeutic Institution', *Bulletin of the Menninger Clinic*, 10, 66–70.

Manning, M. (1975) 'Do Staff Really Need Qualifications?', *Community Care*, 1 October 1975, 7.

Manning, N. K. 'What Happened to the Therapeutic Community?' in K. Jones, S. Baldwin (eds), *The Year Book of Social Policy in Britain, 1975* (London: Routledge and Kegan Paul, 1976).

Martin, D. V. (1955) 'Institutionalization', *Lancet*, 266, 1188–90.

Martin, D. V. (1962) *Adventure in Psychiatry: Social Change in a Mental Hospital* (Oxford: Cassirer).

Martin, D. V., Glatt M. M., K. F. Weeks (1954) 'An Experimental Unit for the Treatment of Neurosis', *British Journal of Psychiatry*, 100, 983–9.

Martin, J. (1984) *Hospitals in Trouble* (London: Basil Blackwell).

Marx, K. (1963) [1845] *The Holy Family* in T. Bottomore, M. Rubel (eds) *Karl Marx: Selected Writings in Sociology and Social Philosophy* (Harmondsworth: Pelican).

Mayhew, B. (2006) 'Between Love and Aggression: The Politics of John Bowlby', *History of the Human Sciences*, 19, 4, 19–35.

McBriar, A. M. (1987) *An Edwardian Mixed Doubles, The Bosanquets Versus the Webbs: A study in British Social Policy 1890-1929* (Oxford: Clarendon Press).

McDougall, W. (1952) *Psychology: The Study of Behaviour*, 2nd edn (London: Oxford University Press).

Melville, J. (1980) 'Nobody Labels Us Here', *MIND OUT*, 40, 14–16.

Menzies, I. E. P. (1949) 'Factors Affecting Family Breakdown in Urban Communities: A preliminary study leading to the establishment of two pilot Family Discussion Bureaux', *Human Relations*, 2, 4, 363–73.

Miller P., Rose, N. (eds) (1986) *The Power of Psychiatry* (Cambridge: Polity Press).

Miller, E. (1938) *The Generations: A Study of the Cycle of Parents and Children* (London: Faber and Faber).

Miller, E. (1939) 'Obsessional and Compulsive States in Childhood' in R. G. Gordon (ed.) *A Survey of Child Psychiatry* (London: Oxford University Press).

Miller, E. (1945) [1933], 'How the Mind Works in the Child' in C. Burt, E. Jones, E. Miller, W. Moodie, *How the Mind Works* (London: Allen and Unwin), pp. 105–56.

Miller, H. C. (1922) *The New Psychology and the Parent* (London: Jarrolds).

Miller, H. C. (1926) 'Adaptation, Successful and Unsuccessful', *Postgraduate Medical Journal*, 1, 5, 60–3.

Miller, H. C. (1973) [1935] 'Educational Ideals and the Destinies of Peoples', in J. A. Hadfield, *Psychology and Modern Problems* , 131–58.

Miller, P., Rose, N. (1988) 'The Tavistock Programme: The Government of Subjectivity and Social Life', *Sociology*, 22, 2, 171–92.

Miller, P., Rose, N. (1994) 'On Therapeutic Authority: Psychoanalytical Expertise under Advanced Liberalism', *History of the Human Sciences*, 7, 3, 29–64.

MIND (1971) *The MIND Manifesto* (London: MIND) (pamphlet).

MIND (1972) *A Right to Love?* (London: MIND), reprinted in MIND (1972) *MIND and Mental Health*, 14–17.

MIND (1973) 'MIND Matters', *MIND OUT*, 4, 2.

MIND (1974) 'Letters to the Editor', *MIND OUT*, 8, 4–5.

MIND (1974) 'No Place for Children', *MIND OUT*, 7, 15.

MIND (1975) 'Life Begins at Sixty Plus' *MIND OUT*, 12, 3.

MIND (1975) 'MIND', *MIND OUT*, 9, 15.

MIND (1975) 'Stepping Stones', *MIND OUT* 12, 5.

MIND (1975) 'Wanted! More Landladies', *MIND OUT*, 12, 4–5.

MIND (1978) Editorial, 'The Issues of Mental Handicap', *MIND OUT*, 28, 2.

MIND (1978) Editorial, 'The Issues of Mental Handicap', *MIND OUT*, 28, 2.

MIND (1979) 'Sex Education for the Mentally Handicapped', *MIND OUT*, 35, 7.

MIND (1979) 'Unique Mental Health Service Threatened with Closure: This Highlights Health Authorities' Irresponsible Actions, Warns MIND Director', *MIND OUT*, 36, 5.

MIND (1981) 'Threat to Wessex Research Unit Postponed', *MIND OUT*, 46, 4.

MIND (1981) 'Wessex Research Unit Threatened with Closure', *MIND OUT*, 45, 4–5.

MIND (1982) *Getting Together: Sexual and Social Expression for Mentally Handicapped People* (London: MIND).

MIND (1982) *The Mental Health Year Book 1981/82* (London: MIND).

Mittler, P. (1968) 'The State of the Nation', *Mental Health*, Summer, 2–3.

Model, E. E. (1983) 'Ruth Thomas 1902–1983: An Appreciation', *Journal of Child Psychotherapy*, 9, 1, 5–6.

Morris, T. (1983) 'Crime and The Welfare State' in P. Bean, S. MacPherson (eds) *Approaches to Welfare* (London: Routledge and Kegan Paul), pp. 166–81.

NAMH (1946) *Report of the Proceedings of a Conference on Mental Health, 14 and 15 November 1946* (London: NAMH).

NAMH (1946) *Report of the Proceedings of a Conference on Mental Health Held at St Pancras Town Hall, London, NW1, 14–15 November 1946* (London: NAMH).

NAMH (1948) 'News and Notes', *Mental Health*, 7, 3, 76–9.

NAMH (1949) 'Report of the Eighth Inter-Clinic Child Guidance Conference: London, December 3, 1949', supplement to *Mental Health*, 9, 3 (supplement).

NAMH (1951) *Proceedings of a Conference on Mental Health held at St. Pancras Town Hall, London, N.W.1, 12–13 March 1951.*

——*Tenth Child Guidance Inter-Clinic Conference: Report on Conference and Some Selected Clinic Surveys* (London: NAMH).

NAMH (1953) *The Practical Application of Research and Experiment to the Mental Health Field: Proceedings of a Conference held at Royal Victoria Halls, London, W.C.1, 5–6 February 1953* (London: NAMH).

NAMH (1954) *Strain and Stress in Modern Living: Special Opportunities and Responsibilities of Public Authorities* (London: NAMH), pp. 28–37.

NAMH (1955) *11ᵗʰ Inter-Clinic Conference For Staffs of Child Guidance Clinics: The Family Approach to Child Guidance-Therapeutic Techniques*, (London: NAMH).

NAMH (1957) *The Needs of the Mentally Sick: A Challenge to Youth: Report of a conference held on 17 October 1957* (London: NAMH).

NAMH (1959) *Truancy or School Phobia? Being the Proceedings of the 15ᵗʰ Inter-Clinic Conference* (London: NAMH).

NAMH (1961) 'Everybody's Business', *Mental Health*, 20, 1, 15.

NAMH (1964) 'The Whole truth About Care of the Mentally Disordered', *Mental Health* 23, 2, 43–49.

NAMH (1964) *Adolescence: Report of the 20th Child Guidance Inter-Clinic Conference 1964* (London: NAMH).

NAMH (1965) *Child Guidance: Function and Social Role: Report of the 21st Child Guidance Inter-Clinic Conference, 1965* (London: NAMH).

NAMH (1966) *Mental Illness in the Family: Its Effect on the Child. Proceedings of the Twenty-Second Child Guidance Inter-Clinic Conference, 1966,* (London: NAMH).

NAMH (1967) *Child Guidance From Within: Reactions to New Pressures. Papers Given at the 23rd Child Guidance Inter-Clinic Conference 1967,* (London: NAMH).

NAMH (1968) 'Scene 2: A Grim Old Workhouse', *Mental Health*, Summer, 36–7.

NAMH (1969) 'Consumer Panel' in NAMH, *New Ways with Old Problems: Report of the Annual Conference 1969* (London: NAMH) pp. 20–38.

NAMH (1972) *Starting and Running a Group Home* (London: NAMH).

NAMH (c1955) *Mentally Handicapped Children: A Handbook for Parents* (London: NAMH).

NAMH and APSW (1954) *Memorandum on Rehabilitation of Psychiatric Patients: Evidence to the Committee of Enquiry into Existing Services for the Rehabilitation of the Disabled.*

NCCL (1951) *50,000 Outside the Law: An Examination of the Treatment of those Certified as Mental Defectives* (London: NCCL).

Nisbet, R. (1966) *The Sociological Tradition* (London: Basic Books).

NSMHC (1961) *Mentally Handicapped Children Growing Up*, director Basil Wright, Realist Film Unit (Ipswich: Concord Media).

Nuttall, J. (2003) ' "Psychological Socialist"; "Militant Moderate": Evan Durbin and the Politics of Synthesis', *Labour History Review*, 68, 2, 235–52.

Odlum, D. (1931) 'The Meaning of the Mental Treatment Act, 1930', *Mental Hygiene Bulletin*, 3, 8–12.

Orwell, G. (1962) [1937] *The Road to Wigan Pier* (Harmondsworth: Penguin).

Orwell, G. (1989) [1933] *Down and Out in Paris and London* (Harmondsworth: Penguin).

Paul, D. (1984) 'Eugenics and the Left' *Journal of the History of Ideas*, 45, 4, 567–90.

Payne, M. (1929) *Oliver Untwisted* (London: Edward Arnold and Co.).

Philpott, D. (2010) 'Sovereignty', in E. N. Zalta (ed.) *The Stanford Encyclopedia of Philosophy*, Summer 2010 Edition, accessed at http://plato.stanford.edu/archives/sum2010/entries/sovereignty/ (accessed 10 January 2011).

Pines, M. (1999) 'Forgotten Pioneers: The Unwritten History of the Therapeutic Community Movement', *Therapeutic Communities*, 20, 1, 23–42.

PNC (1944) 'News and Notes' *Mental Health*, 5, 1, 12–16.

PNC (1944) *The Care of Children Brought Up Away From Their Homes* (London: PNC).

Pols, J. (1997) *Managing the Mind: The Culture of American Mental Hygiene, 1910–1950,* PhD Dissertation, University of Pennsylvania.

Pols, J. (2010) ' "Beyond the Clinical Frontiers": The American Mental Hygiene Movement, 1910–1945' in V. Roelcke, P. Weindling, L. Westwood (eds) *International Relations in Psychiatry: Britain, Germany and the United States to World War Two* (Rochester: University of Rochester Press), pp. 111–33.

Porter, R. (1987) *Mind Forg'd Manacles: A History of Madness in England from the Restoration to the Regency* (London: Athlone Press).

Porter, R. (1996) *A Social History of Madness: Stories of the Insane* (London: Phoenix).

Powell, F. (2001) *The Politics of Social Work* (London: Sage).

Prynn, D. (1983) 'The Woodcraft Folk and the Labour Movement 1925–70', *Journal of Contemporary History*, 18, 1, 79–95.

Q Camps Committee (1935) [Extract of] *Draft Memorandum on Proposed "Q" Camps (for Offenders Against the Law and Others Socially Inadequate) Under the Management of Grith Fyrd (Pioneer Communities)*, accessed at http://archive.pettrust.org.uk/survey-saq2-2memoranda.htm (accessed 15 September 2012).

Q Camps Committee (1936) 'Memorandum on Proposed Q Camp for Offenders against the Law and others Socially Inadequate' cited in M. E. Franklin (ed.) *Q Camp: An Experiment in Group Living*, p. 68.

Rabinow, P. (ed.) (1991) *The Foucault Reader* (Harmondsworth: Pengiun).

Randall, Jr, J. H. (1966) 'Idealistic Social Philosophy and Bernard Bosanquet', *Philosophy and Phenomenological Research*, 26, 4, 473–502.

Ratcliffe, T. A. (1951) 'Community Mental Health in Practice' in NAMH, *Proceedings of a Conference on Mental Health held at St. Pancras Town Hall, London, N.W.1, 12–13 March 1951*, pp. 11–23.

Ratcliffe, T. A., Jones, E. V. (1956) 'Intensive Casework in a Community Setting', *Case Conference*, 2, 10, 17–23.

Raynes, N. V., King, R. D. (1974) [1967] 'Residential Care for the Mentally Retarded' in, D. M. Boswell and J. M. Wingrove (1974) *The Handicapped Person in the Community* (London: Tavistock and Oxford University Pres), pp. 299–306.

Rees, J. R. (1929) *The Health of The Mind* (London: Faber & Faber).

Rees, J. R. (1945) *The Shaping of Psychiatry by War* (London: Chapman and Hall Ltd).

Rees, T. P. (1957) 'Back to Moral Treatment and Community Care', *Journal of Mental Science*, 103, 303–13.

Rees, T. P. (1957) 'Discussion' in NAMH, *The Needs of the Mentally Sick: A Challenge to Youth: Report of a conference held on 17 October 1957* (London: NAMH), pp. 55–64, pp. 58–9, 62–3.

Rees, T. P. (1957) Review of M. Greenblatt, R.H. York and E.L. Brown (1955) *From Custodial to Therapeutic Care in Mental Hospitals* (New York: Russell Sage Foundation), *Mental Health*, 16, 3, 113.

Robb, B. (ed.) (1967) *Sans Everything: A Case to Answer* (London: Thomas Nelson & Sons Ltd).

Roelcke, V., Weindling, P., Westwood, L. (eds) (2010) *International Relations in Psychiatry: Britain, Germany and the United States to World War Two* (Rochester: University of Rochester Press)

Rogers, A., Pilgrim, D. (1996) *Mental Health Policy in Britain: A Critical Introduction* (London: MacMillan).

Rolph, C. H. (1967) 'Cruelty in the Old People's Ward' in B. Robb (ed.), *Sans Everything: A Case to Answer* (London: Thomas Nelson & Sons Ltd), pp. 3–7.

Rolph, S. (2002) *Reclaiming the Past: The Role of Local Mencap Societies in the Development of Community Care in East Anglia, 1946–1980* (Milton Keynes: Open University).

Rose, N. (1985) 'Unreasonable Rights: Mental Illness and the Limits of the Law', *Journal of Law and Society*, 12, 2, 199–218.

Rose, N. (1985) *The Psychological Complex: Psychology, Politics and Society in England, 1869–1939* (London: Routledge and Kegan Paul).

Rose, N. (1986) 'Law, Rights and Psychiatry' in P. Miller, N. Rose (eds), *The Power of Psychiatry* (Cambridge: Polity Press), pp. 177–213.

Rose, N. (1986) 'Psychiatry: The Discipline of Mental Health' in N. Rose, P. Miller (eds) *The Power of Psychiatry* (Cambridge: Polity Press), pp. 43–84.

Rose, N. (1989) *Governing the Soul: The Shaping of the Private Self* (London: Routledge).

Rose, N. (1996) 'Identity, Genealogy, History' in S. Hall, P. du Gay (eds) *Questions of Cultural Identity* (London: Sage), pp. 128–150.

Rose, N. (1998) *Inventing Our Selves: Psychology, Power and Personhood* (Cambridge: Cambridge University Press).

Ross, T. A. (1934) 'The Neurotic', *Mental Hygiene*, 10, 86–90.

Rousseau, J. J. (1993) [1762] *The Social Contract*, trans. C. Cranston (Harmondsworth: Penguin).

Samuel, R. (1989) 'Born-again Socialism' in R. Archer, D. Bubeck, H. Glock (eds) *Out of Apathy: Voices of the New Left 30 Years On* (London: Verso), pp. 39–57.

Sanderson, R. A. (1951) 'The Re-Socialization of the Psychiatric Case', *Mental Health*, 10, 4, 87–96.

Schochet, G. J. (1988) *The Authoritarian Family and Political Attitudes in Seventeenth-Century England* (Oxford: Blackwell).

Scott, Rt. Hon. Lord Justice, (1926) Chairman's introduction to discussion on, 'The Proper Care of Defectives Outside Institutions' in Central Association for Mental Welfare, *Report of a Conference on Mental Welfare Held in the Central Hall, Westminster, London, SW on Thursday and Friday, December the 2nd and 3rd, 1926* (London: CAMW), pp. 17–23.

Scott, Rt. Hon. Lord Justice, (1937) 'Miss Evelyn Fox, CBE: A Tribute', *Mental Welfare*, 18, 2, 48.

Scull, A. T. (1979) *Museums of Madness: The Social Organization of Madness in 19th Century England* (London: Allen Lane).

Sedgwick, P. (1982) *Psycho Politics* (London: Pluto Press).

Segal, S. (1984) *Society and Mental Handicap: Are We Ineducable?* (Tunbridge Wells: Costello).

Sharp, S., Sutherland, G. (1980) 'The Fust Official Psychologist in the Wuurld', *History of Science*, 18, 181–208.

Shennan, V. (ed.) (1972) *Subnormality in the 70s: Action for the Retarded* (London: NCMHC and WFMH).

Shephard, B. (2000) *A War of Nerves: Soldiers and Psychiatrists 1914–1994* (London: Jonathan Cape).

Shorter, E. (1997) *A History of Psychiatry: From The Era of The Asylum to the Age of Prozac* (New York: John Wiley and Sons).

Smith, I. (1999) *The Transparent Mind: A Journey with Krishnamurti* (Ojai, CA: Edwin House).

Smythe, T. (1984) 'Mental Patients and Civil Liberties' in P. Wallington (ed.) *Civil Liberties, 1984* (Oxford: Robertson), pp. 309–23.

Smythe, T. (1991) 'Health Warning', *The Raven: Anarchist Quarterly*, 4, 3, 216–19.

Smythe, T., Winn, D. (1975) 'A Problem Shared', *MIND OUT*, 10, 2–3.

Soddy, K. (1946) 'Some Lessons of Wartime Psychiatry, I', *Mental Health*, 6, 2, 30–5.

Soddy, K. (1946) 'Some Lessons of Wartime Psychiatry, II', *Mental Health*, 6, 3, 66–70.

Soddy, K. (1950) 'Mental Health', *International Health Bulletin of the League of Red Cross Societies*, 2, 2, 8–13.

Soddy, K. (n.d.) *Some Lessons of Wartime Psychiatry* (London: NAMH).

Stafford-Clark, D. (1952) *Psychiatry Today* (Harmondsworth: Pelican).

Stansfield, R. (1977) 'A Special Kind of School', *MIND OUT*, 23, 5–7.

Steadman Jones, G. (1971) *Outcast London: A Study in the Relationship Between Classes in Victorian Society* (Oxford: Clarendon Press).

Stevenson, G. (2012) 'Anatomy of Decline – The Young Communist League 1967–86', http://www.grahamstevenson.me.uk/ [home page] (accessed 10 September 2012).

Strong, S. (2000) *Community Care in the Making: A History of MACA 1879–2000* (London: MACA).

Sutherland, J. D. (ed.) (1970) *Towards Community Mental Health* (London: Tavistock Publications).

Suttie, I. D. (1960) [1935] *The Origins of Love and Hate* (Harmondsworth: Penguin).

Swart, K. W. (1962) ' "Individualism" in the Mid-Nineteenth Century (1826–1860)', *Journal of the History of Ideas*, 23, 1, 77–90.

T.M. Cuthbert, T. M. (1957) 'The Mental Hospital as a Therapeutic Community' in NAMH, *The Needs of the Mentally Sick: A Challenge to Youth: Report of a conference held on 17 October 1957* (London: NAMH), pp. 3–14.

Tangye, C. H. W. (1941) 'Some Observations on the Effect of Evacuation Upon Mentally Defective Children', *Mental Health*, 2, 3, 75–8.

The Free Lance-Star (1951) 'British Called Immature Because Men Boss Women', accessed at http://news.google.com/newspapers?nid=1298&dat=1951 0521&id=c4kTAAAAIBAJ&sjid=6YoDAAAAIBAJ&pg=6243,1237465 (accessed 12 December 2012).

The Medical Superintendent (1924) 'The Manor Institution, Epsom: Some Comments on its First Two Years', *Studies in Mental Inefficiency*, 5, 2, 25–32.

Thom, D. (2004) 'Politics and the People, Brian Simon and the Campaign Against Intelligence Tests in British Schools', *History of Education*, 33, 5, 515–29.

Thomas, R. (1945) *Children Without Homes: How can they be compensated for loss of family life?* (London: PNC).

Thomas, R. R. (2004) 'Fox, Dame Evelyn Emily Marion (1874–1955)', *Oxford Dictionary of National Biography* (Oxford University Press), accessed at http://www.oxforddnb.com/view/article/33231 (accessed 3 September 2011).

Thomson, C. (1922) 'A National Council for Mental Hygiene' (letter), *BMJ*, 1, 3196, 538.

Thomson, M. (1996) 'Family, Community and State: The Micro-politics of Mental Deficiency' in D. Wright and A. Digby (eds) *From Idiocy to Mental Deficiency: Historical Perspectives on People with Learning Disabilities* (London: Routledge).

Thomson, M. (1998) *The Problem of Mental Deficiency: Eugenics, Democracy and Social Policy in Britain c. 1870–1959* (Oxford: Clarendon Press).

Thomson, M. (2006) *Psychological Subjects: Identity, Culture and Health in Twentieth-Century Britain* (Oxford: Oxford University Press).

Thomson, M. (2010) 'Mental Hygiene in Britain During the First Half of the Twentieth Century: The Limits of International Influence' in V. Roelcke, P. Weindling, L. Westwood (eds) *International Relations in Psychiatry: Britain, Germany and the United States to World War Two* (Rochester: University of Rochester Press), pp. 134–55.

Thomson, M. 'Constituting Citizenship: Mental Deficiency, Mental Health and Human Rights in Inter-War Britain' in C. Lawrence, A. Meyer (eds) *Regenerating England: Science, Medicine and Culture in Inter-War Britain* (Amsterdam: Clio Medica), pp. 231–50.

Thornton, A. (2005) *Reading History Sideways, The Fallacy and Enduring Impact of the Developmental Paradigm on Family Life* (Chicago: University of Chicago Press).

Tizard, J. (1950) 'The Abilities of Adolescent and Adult High Grade Mental Defectives', *Journal of Mental Science*, 96, 405, 889–907.

Tizard, J. (1953) speech on 'Adult Defectives and their Employment' in NAMH, *The Practical Application of Research and Experiment to the Mental Health Field: Proceedings of a Conference held at Royal Victoria Halls, London, W.C.1, 5–6 February 1953* (London: NAMH), pp. 46–56.

Tizard, J. (1964) *Community Services for the Mentally Handicapped* (London: Oxford University Press).

Tizard, J., O'Connor, N. (1952) 'The Occupational Adaptation of High-Grade Mental Defectives', *Lancet*, 260, 620–23.

Tizard, J., O'Connor, N. (1956) *The Social Problem of Mental Deficiency* (London: Pergamon Press).

Tizard, J. (1975) 'Foreword' in B. Kirman and J. Bicknell (eds) *Mental Handicap* (London: Churchill Livingstone).

Townsend, P. (1969) 'New Structures: A Critical Review' in NAMH, *New Ways with Old Problems: Report of the Annual Conference 1969* (London: NAMH), pp. 10–19.

Tredgold, A. F. (1924) Review of, A. Wohlgemuth, *A Critical Examination of Psycho-Analysis* (London: Allen & Unwin, 1924) in *Studies in Mental Inefficiency*, 5, 3, 67.

Tredgold, R. F. (1950) 'Editorial', *Mental Health*, 10, 1–2.

Tredgold, R. F. (1951) Review of NCCL, *50,000 Outside the Law: An Examination of the Treatment of those Certified as Mental Defectives* (London: NCCL), in *Mental Health*, 10, 3, 80.

Tredgold, R. F. (1952) 'Editorial', *Mental Health*, 11, 3, 102–3.

Tredgold, R. F. (1953) Review of D. Stafford-Clark, *Psychiatry Today* in *Mental Health*, 13, 1, 37.

Tredgold, R. F. (1960) Review of R. Barton, *Institutional Neurosis* in *Mental Health*, 19, 1, 32–3.

Tredgold, R. F., Soddy, K. (eds) (1970) *Tredgold's Mental Deficiency*, 10th edn (London, Bailliere).

Trotter, W. (1990) [1916] 'Herd instinct and its bearing on the psychology of civilized man' in *Instincts of the Herd in Peace and War* (London: T Fisher and Unwin).

Tuke, S. (1813) *Description of the Retreat: An Institution near York for Insane Persons of the Society of* Friends (York: Alexander).

Tyne, A. (1979) 'Who's Consulted?' *MIND OUT*, 32, 8.

Unsworth, C. (1987) *The Politics of Mental Health* (Oxford: Clarendon Press).

Urwin, C., Sharland, E. (1992) 'From Bodies to Minds in Childcare Literature: Advice to Parents in Interwar Britain' in R. Cooter (ed.) *In The Name of The Child* (London : Routledge), pp. 174–99.

Utopia Britannica (2009) 'Q Camps – The Emotional Vortex', accessed at http://www.utopia-britannica.org.uk/pages/Qcamps.htm (accessed 5 November 2009).

Van der Horst, F.C.P. (2011) *John Bowlby: From Psychoanalysis to Ethology: Unravelling the Roots of Attachment Theory* (Chichester: Wiley-Blackwell).

Walford, E. (1878) *Old and New London: Volume 4* (Thornbury and Walford), pp. 14–26, (apparently citing an 1875 article in the *Builder*), British History Online, accessed at http://www.british-history.ac.uk/report.aspx?compid= 45179 (accessed 5 January 2009).

Wallington, P. (ed.) *Civil Liberties, 1984* (Oxford: Robertson).

Watson, W. R. K. (1926) 'Some Observations on Borderland Cases and Delinquency', CAMW, *Report of a Conference on Mental Welfare Held at the Central Hall, Westminster, London, S.W. on Thursday and Friday, December 2nd and 3rd, 1926* (London: CAMW).

Welshman, J. (1999) 'Rhetoric and Reality: Community Care in England and Wales, 1948–74' in P. Bartlett and D. Wright, *Outside the Walls of the Asylum: The History of Care in the Community 1750–2000* (London: The Athlone Press), pp. 204–26.

Wertheimer, A. (1975) 'Co-ordination or Chaos – The Rundown of Psychiatric Hospitals', *Royal Society of Health Journal*, July 1975 (MIND reprint).

Wertheimer, A. (1980) 'Researching into Mental Handicap', *MIND OUT*, 40, 16–18.

WHO (World Health Organization) (1953) *The Community Mental Hospital: Third Report of the Expert Committee on Mental Health*, (Geneva: WHO).

Williams, R. (1988) *Keywords: A Vocabulary of Culture and Society* (London: Fontana).

Wills, W. D. (1945) *The Barns Experiment* (London: George Allen and Unwin).

Wills, W. D. (1966) [1943] 'Internal Government of the Camp, Its Growth and Changes' in M. E. Franklin (ed.) *Q Camp: An Experiment in Group Living With Maladjusted and Anti-Social Young Men*, pp. 24–8.

Wills, W. D. (1967) [1941] *The Hawkspur Experiment: An Informal Account of the Training of Wayward Adolescents*, 2nd edn (London: George Allen and Unwin).

Wills, W. D. (1968) Letter to the editor, *British Journal of Criminology*, 8, 448.

Wills, W. D. (1979) 'The Moral Perspective' in P. Righton (ed.) *Studies in Environment Therapy Vol. 3* (Toddington: PETT), pp. 25–35.

Wilson, A.T.M., Doyle, M., Kelnar, J. (1947) 'Group Techniques in a Transitional Community', *Lancet*, 1, 735–738.

Winn, D. (1975) 'Editorial', *MIND OUT*, 9, 3.

Winnicott, C., Shepherd, R., Davis M. (eds) (1984) *Deprivation and Delinquency* (London: Tavistock).

Winnicott, D. W. (1984) [1947] 'Residential Management for Difficult Children', *Human Relations*, reprinted in C. Winnicott, R. Shepherd, M. Davis (eds) *Deprivation and Delinquency* (London: Tavistock), pp. 54–72.

Winter, J. M. (1970) 'R H Tawney's Early Political Thought', *Past and Present*, 47, 71–96.

Young, M. D. (1958) *The Rise of the Meritocracy, 1870–2033: An Essay on Education and Equality* (London: Thames and Hudson).

Index

50,000 Outside the Law, 124, 125, 130
  see also National Council for Civil
    Liberties (NCCL)
A Human Condition, 181, 183
Abel-Smith, Brian, 146, 167
Adrian, Hester, 131 fn. 515, 147,
  167
Ahrenfeldt, R.H. 108–9
'Alan' 90–1
'Alasdair', 68
alienation
  and anti-psychiatry, 163–4
  associated with 'egoism', 'idleness' and
    'individualism', 9
  conflict of 'tradition' and 'modernity',
    9–10, 16–17
  'the Family', and 4, 6
  'mental alienation' at the Retreat, 3
  and mental hospital care, 197
  and power/knowledge, 174–6, 188
  separation and estrangement, 9
  and '*The Social Contract*', 14
  and the New Left, 143–6, 155
Allen, Elizabeth, Ch. 6, 121, 124
Appleby, Mary, 149, 164
Association of Psychiatric Social Workers
    (APSW), 30, 48, 91, 105–6, 114, 116,
    117, 132
Association of Therapeutic
    Communities, 193
authority
  authoritarianism, 46–7, 62, 92, 102,
    105, 108, 113, 116, 127, 132, 135,
    141, 144–5, 149, 158, 159, 161,
    168
  importance regarding Charity
    Organisation Society (COS), 8, 10,
    11–15, 16–17, 18–19, 20–2, 22–4,
    26, 27, 54
  and 'The Family', 10, 16–17, 18–19,
    20–1, 26, 27, 33, 34, 36, 40–1,
    47–52, 54, 59–60, 70–1, 78, 94,

99–100, 111,112–3, 139, 143–4,
  164,191, 194–5, 198
  and 'the idle', 8, 16–17, 33
  and critiques of institutional care,
    94–6, 101–2, 118–20, 138–9, 141,
    168, 180, 197, 180, 197
  and *Madness and Civilization*'s analysis
    of Moral Treatment, 3–6, 10, 47–8,
    49–50
  and mental deficiency (later termed
    'mental handicap'), 26, 80–2, 97,
    118–20, 135, 184–5
  importance to mental hygiene
    movement, 2, 27, 48–52, 58,
    59–60, 92–3, 139, 159, 160,
    179–80, 191
  importance to MIND's activity, 191–3,
    194–6, 197, 198
  importance to moral treatment, 2–6
  and Robert Nisbet, R., 9–10
  'paradox of self-government', 12–15
  parental authority and children's
    development, 6, 48, 54, 86–8, 128,
    191–3
  patriarchy, 4, 18–19, 139–40
  'power/knowledge', 176–9, 190, 200
  self-government, 3, Ch. 2 *passim*,
    40–4, 54, 60, 62, 66–7, 70–73, 85,
    88–90, 142–3, 143–4, 186–7
  and the State 8, 11–15, 21–2, 41,
    88–9, 108, 110, 143–4, 148–9, 164,
    198
  'teenage culture', 156–7
  The Social Contract, 14

Balbernie, Richard, 117
Barron, Arthur (Bunny), 74–6, 185
Barton, Russell, 142, 153, 154, 167, 183,
  192
Beedell, Christopher, 159–160
Beers, Clifford, 31–2
Berlin, Isaiah, 175
Boorman, John, 155

Bosanquet, Bernard, 11–15, 16–17, 20–22, 23–5, 104, 127
  and the mental hygiene movement, 34, 35, 37, 40, 41, 42, 43, 48, 54, 64, 70–1, 74–5, 88, 164
  and the 'New Psychology', 32–33
Bosanquet, Helen, 11, 16–17, 18–19, 20, 21, 22–4, 25
  and the mental hygiene movement, 34, 35, 42, 48, 51, 54, 64, 70, 74–5, 86, 88, 89, 164
Bowlby, John, 85–90, 94, 102, 104, 106, 113, 114, 151, 192
Braddock, Bessie, 147
Bradford City Council, 92
Brecknock Community Centre, 185
Brehony, Kevin, Ch, 4, 55
British Idealism, 11, 56, 69
British Psycho-analytical Society, 65
Brooklands Study, 150–2, 154, 183
Brown, Sybil Clement, 44–5, 49, 63, 159, 180
Burt, Cyril, 38–9, 41, 42, 43–4, 75, 79, 118, 126, 127, 132
Butler, Barbara, 161
Buttle Trust, 149

Cambridgeshire Mental Welfare Association, 161
Campaign for Nuclear Disarmament (CND), 146, 181
Campaign for the Mentally Handicapped (CMH), 183, 185
Capes, Mary, 117
Care of Children Committee (Curtis Committee) ,94, 96, 97, 150, 151, 152
Carroll, Denis, 53, 100
Castel, Robert, 7, 80
Central Association for Mental Welfare (CAMW), 28–9, 30, 34, 35, 39, 55, 80, 90, 91, 149
Charity Organisation Society (COS), 8–10, 11, 16–17, 21, 23–6, 33, 34, 48, 54, 64, 109, 110, 127, 143
  and the Bosanquets, 11–15, 16–17, 18–19, 20–27
Child Guidance Council (CGC), 30, 55, 91
Children Without Homes, 94–6, 97–8

Civil Resettlement Service (CRS), 99, 100, 101–2
Clark, Charles, 92, 194
Clark, D.H., 141–2, 154, 161
Claybury Mental Hospital, 141
'Clive', 84–5
Cohen, Gerda, 154
Cole, G.D.H., 132, 145
Committee of 100, 146, 181
Communist Party of Great Britain, 121, 127, 132
Contractarianism, 9, 11–14, 16, 18, 19, 21, 43, 53–4
Cooper, David, 154–5, 162–4, 168
Cooter, Roger, 189, 200
Cotswold Community, 117
Craig, Maurice, 37
Crossman, Richard, 167–8
Curran, Desmond, 123–4

Dewey, John, 32, 201
'dialectic, the'
  and 'the Family', 6, 27, 47, 52, 54, 54–5, 55, 85, 104, 111, 138, 143, 164, 173, 175, 179, 200–1
  and Foucault, 6, 174–5, 200–1
  Hegelian, 6, 25, 33, 54, 84, 127
  Marxist, 128, 129
  as relation between parental authority and children's development, 6, 48, 54, 60, 86–7, 112–13, 128
Dicks, H. V., 109, 110
Dockar-Drysdale, Barbara, 117–18
Donzelot, Jacques, 47
Doyle, Martin, 100
  see also Civil Resettlement Service
Dulwich College, 172
Duly Authorising Officers (DAOs)
  debate with APSW, 116
Durbin, Evan 85–90, 113, 192

egalitarianism,
  and the Bosanquets' theorizing, 13–14, 21–2
  and 'the Family', 41–2, 42–5, 46–7, 51–2
  and Hawkspur, 54, 66–7, 68–9, 70–1, 73, 74–6, 78, 84, 142, 144, 186–7
  and notion of 'in authority', 72–3
  associated with individualism, 9

and Leytonstone Poor Law Homes, 62, 67, 70
and 'mental handicap', 184–5
and 'the New Left', 144–5
and 'progressive' education, 56
and socialist values, 68–9, 145, 146
R. H. Tawney's conception compared with Hawkspur approach, 68–9, 142
and *The Social Contract*, 14, 53–4
Ely Hospital Enquiry, 167–8
emotionality
    and 'the crowd', 20–1, 40–3
    'emotional security', 51–2, 60, 66–7, 72, 78, 84, 87, 90, 92–3, 94–6, 98, 101, 104, 111, 113, 114, 119, 144, 151–2, 168, 170, 191
    and 'the Family', 19, 34–6, 48–52, 54, 59–60, 62, 78, 99–100, 112–3, 143–4, 179–80, 191, 194–5
    and individuality, 14–15, 35–6, 39, 42–3, 48–9, 50–2
    and progressive history, 26, 35–6, 138
Ensor, Beatrice, 55, 56
enuresis, 98
Epsom Manor Mental Deficiency Institution, 80–1
Erikson, Erik, 155
Ex-Serviceman's Movement for Peace, 132

Fairhaven hostel, 181, 193
'Family, the'
    role in 'alienation', 4
    and 'the dialectic', 6, 27, 47, 52, 54, 104, 111, 138, 143, 164, 173, 175, 179, 200–1
    role as 'disalienation', 4
    as organizing principle, 3–4, 7, 9–10, 11, 16–17, 18–19, 20, 21–22; Ch. 3, 29, 34, 35–6, 43, 47–8, 49–52, 54, 57, 59–60, 70–1, 78, 90, 99–100, 111, 112–3, 119, 139, 143–4, 150, 152, 163–4, 164–5, 170, 173, 174, 175, 191, 194–5, 198
    and 'mental deficiency'(later termed 'mental handicap'), 26, 98, 119, 135–6, 150, 152, 184–5

and 'paradox of self-government', 13, 15
and paternalism, 5, 112–13, 156, 198
and patriarchy, 4, 18–19, 139–40
and 'power/knowledge', 174–5, 175–6
Fennell, Phil, 188
Ferenczi, Sandor, 65
Feversham Committee, 81–2, 91
Feversham School, 193
Fildes, Lucy, 51, 96–7, 116
Flugel, J. C., 108, 109
Foot, Paul, 121
Forest School,58–9
Foucault, Michel
    and authority, 2–6, 10, 47–8, 49–50, 176–9
    and 'the Family', 2–6, 10, 27, 48, 49–50, 143, 174–5, 200
    History of Madness, 2, fn. 3
    Madness and Civilization, 2–6, 10, 49–50, 163
    and 'power/knowledge', 3, 47, 174–9, 189, 199–200
    and the Retreat, 2–6, 10, 49–50
Fountain Mental Deficiency Hospital, 121, 123, 131, 151
    see also Hilliard, L. T. and Kirman, B.
Fox, Evelyn, 28–9, 30, 34, 149
Francis, W.A.G., 98
Franklin, Marjorie, 53, 54, 56, 59, 64–5, 67, 70, 71, 79, 82–3, 96, 100
Freud, Anna, 94
Friedman, R. B., 73
Fulbourn Mental Hospital, 141, 161

Gaitskell, Hugh, 86
'General Bugger Abouts' (G.B.A.s), 75, 85
'Gerald', 72
Gillespie, R. D., 38–9, 75, 118
Glaister, Dorothy, 146
Glaister, Norman, 57, 58–9, 100, 146
Goffman, Irving, 153–4, 163, 169, 183
Goldberg, E.M., 112, 159, 162
Gostin, Larry, 181, 183
Graeber, David, 199
Green, T. H., 11, 69
Greenwich House, 16, 22
Grith Fyrd, 58–9, 77–8
Gunzburg, H. C., 118–20, 129
Guttmann, Eric, 123–4

Hadfield, J. A., 41
Hall, Stanley G., 32–3
Hall, Stuart, 143, 155–7
Hawkspur Camp, Ch. 4 *passim*, 84–5, 87,
    88, 90–1, 93, 98, 100–1, 102–3, 120,
    142–3, 144, 145, 146, 149, 160,
    185–7
  Camp Council, 61, 68, **71–2, 72–3**, 74,
    75, **76–78**
Hegel, G. W. F., 84
  and Bernard Bosanquet, 11–15, 25, 70
  and Foucault, 6
  and the 'New Psychology', 33
Henderson Hospital, 193
Hilliard, L. T., 123, 125, 130–2, 134
Hoare, Russell, 79
Hobbes, Thomas
  and Bernard Bosanquet, 11–12
  and Freud, 88
  and *Personal Aggressiveness and War*,
    88–9
Holman, Robert (Bob), 109–10
Holmes, Edmond, 56
Home and School Council (HSC), 30, 55,
    57–8, 87
'Home from Hospital' campaign, 183–4
Howard League, 65
Hunter, Dugmore, 114

'idle, the'
  and 'alienation', 9
  and authority, 8
  and Charity Organisation Society
    social casework, 23–4
  associated with 'the mass', 21
  and 'mental deficiency', 25–6
individualism
  associated with 'alienation', 9
  associated with 'egoism' and 'idleness',
    9
  and the 'mass', 20–2, 42–3
  and mental deficiency, 26
Institute for the Scientific Treatment of
    Delinquency (ISTD), 30, 53, and 53,
    65 and fn. 218, 100
'institutionalism', 62, 64, 183
'institutionalization', 118, 120, 140–2,
    148, 164, 176, 182–4, 185, 197
Isaacs, Susan, 87

James, Martin, 161
James, William, 32
'John', 100
Jones, Ernest, 36
'Joseph', 63–4
'Julian', Ch. 4, 73–4, 91
Junction Road Project, 185

Kahn, Jack, 114, 155, 169
Kelnar, John, 100
  see also Civil Resettlement Service
King, Roy, Ch. 7, 154, 167
Kirman, Brian, 108, 121, 122, 123,
    125–30, 131, 132, 145
Krishnamurti, Jiddu, 58
Kushlick, Albert, 154, 183
Kydd, J.C., 8, 10–11

Laing, R.D., 154–5, 162, 163
Lane, Homer, 56, 185–6
Le Play, Frédéric, 22–3, 35
Lennhoff, F.G., 117
Lewis Aubrey, 132
Lewis, Jane, 24
Leytonstone Poor Law Homes, 50, 58,
    59, 60, 62, 66, 67, 69–71, 73, 78, 80,
    101, 102
Linebaugh, Peter, 8
Little Commonwealth, 56, 78
Loch, C.S., 25

MacCalman, D. R., 106–10
McDougall, William, 33, 34, 35, 40
Mailer, Norman, 157
Main, Tom, 100, 101–2
Mannin, Ethel, 120
Martin, D. V., Ch. 7, 141, 166
Martin, John, 167
Marxism
  and 'alienation', 9, 144
  compared with civil libertarianism,
    121–2
  and intelligence testing, 126–7
  Edward Shorter on, 99
  This Matter of Mind, 125–30.
  'maternalism', 64, 102, 112–13
  and Bowlby's theorizing, 90, 192

'medical model'
  critique and disruption of, 95, 102,
    114–15, 122, 133, 139, 153, 154,
    159, 170, 180, 182, 185
Medical Practitioners Union, 122
Medical Research Council Social
  Psychiatry Unit, 132
Mental Deficiency Act, 1913 26, 27,
  28–9, 37, 81 fn. 268, 82
Mental Deficiency Legislation
  Sub-Committee, 131
Mental Deficiency Sub-Committee, 124,
  125, 130, 131, 134, 135
Mental Deficiency Training
  Sub-Committee, 131
mental deficiency, 25–7, 28–9, 30, 33,
  36, 37, 37–9, 43, 80–2, 96–8, 106–7,
  115, 118–20, 120–5, 127, 129,
  130–6, 146–7, 148, 150, 152, 166–7,
  169, 180
  see also mental handicap
mental handicap [also termed
  'subnormality], 149–52, 166, 167–8,
  170, 183, 184–5
  see also mental deficiency
Mental Health Emergency Committee
  (MHEC), 91, 92, 97 fn. 341
mental testing, 37–9, 126–7
Mental Treatment Act, 1930 69–70, 81,
  101, 147
*Mentally Handicapped Children Growing
  Up*, 149–50
Meyer, Adolph, 32, 65
Miller, Emmanuel, 35–6, 49, 110
Miller, Hugh Crichton, 30, 41–2, 43, 44,
  100
MIND, Ch. 8 *passim*
  activity in relation to 'the Family',
    179–80, 184, 191–5, 198
  continuities with mental hygiene
    movement, 180–1, 181–5, 191–3,
    194–5, 197, 199
  and power as totalized, rationalized,
    centralized and popular, 195–9
  rights approach, 173, 181–2, 187
  *see also references under National
    Association for Mental Health
    (NAMH)*
Mittler, Peter, 167

moral treatment
  and anti-psychiatry, 163
  Charity Organisation Society social
    casework as, 7, 8–10, 23–4, 48
  and the mental hygiene movement,
    48, 137–8
  and negation of history, 7
  related to New Left, 143–4
  and Nikolas Rose, 190–1
  and reanimation of history, 7
  at the Retreat, 1–6, 163
  and work, 4, 74–5
Morris, Terence, 107–8

National Association for Mental Health
  (NAMH)
  1969 Annual Conference, 166, 168–9
  1970 Annual Conference, 165–6
  and Russell Barton, 142
  and D.H. Clark, 141
  and support for community therapy,
    117–8, 193–4
  founded, 105
  and Fountain Mental Deficiency
    Hospital, 130–1
  and Martin James, 161
  concerns over mental deficiency
    institutions, 131
  Mental Deficiency Sub-Committee,
    124, 125, 130, 131, 134, 135
  Mental Deficiency Training
    Sub-Committee, 131
  'National Trends and the Mental
    Health Services', 164–5
  and T.P. Rees, 137, 147
  Reynolds House, 149
  adoption of rights approach, 173, 174,
    182
  and Royal Commission on the Law
    Relating to Mental Illness and
    Mental Deficiency, 147, 148, 182
  and *Sans Everything: A Case to Answer*,
    166–7
  Social After-Care Service (SACS),
    111–15, 182
  Kenneth Soddy – Medical Director, 105
  Alfred Torrie – Medical Director, 111
  and Westhope Manor, 117
  see also references under MIND

National Association for Mentally
    Handicapped Children, 150
National Association for the Care of the
    Feebleminded (NACF), 25
National Council for Civil Liberties
    (NCCL), 121–3, 124–5, 130–1,
    133–4, 135, 146, 150, 166, 181
National Council for Mental Hygiene
    (NCMH), 30, 37, 55, 90, 91
National Council of Women, 134
National Health Service Act 1946, 112,
    182
New Education Fellowship (NEF), 55, 56
'New Left', 143–6, 149, 155–7
New York School of Social Work, 1, 7–8,
    10–11, 16, 17–18, 19–20, 22, 29–30,
    32, 34, 36–7,39–40 46, 47
Nisbet, Robert, 9–10
    and 'alienation', 9–10, 144
    and conflict between 'tradition' and
        'modernity', 9–10
Northfield Military Hospital, 98–9, 100,
    101–2, 161

O'Connor, Neil, 131, 132, 133
Order of Woodcraft Chivalry, 58, 59
Orwell, George, 46–7, 57, 62

Parent and Child, 55, 87
Parents National Education Union, 65
paternalism, 5, 105, 112–13, 156, 161,
    182, 198
patriarchy
    and Helen Bosanquet's theorizing,
        18–19
    and 'the Family', 4, 18–19, 139–40
    and Le Play, 22–3
    and 'traditional' social order, 9–10
Payne, Muriel, 50–1, 57–8, 59, 62, 64,
    66, 70–1, 102
Personal Aggressiveness and War, 85–90
Pettit, Walter, 11, 17, 29, 34
Pols, Johannes, 37
Polsky, Ned, 157
Porter, Roy, 3, 32
power
    as centralized, 21, 23, 25, 43, 66, 88,
        108, 135, 143–5, 177, 178, 194–5,
        196, 198, 199

as popular and participatory, 21, 66,
        69, 71, 88, 108, 135, 143–4, 145,
        177, 178, 194–5, 196, 198, 199
    as rationalization, 24–5, 43, 66, 88–9,
        108, 129, 135, 143–4, 145, 155,
        177, 178, 194–5, 196, 198, 199
    as totalized, 25, 43, 66, 88, 108, 129,
        135, 143–4, 145, 155, 177, 178,
        194–5, 196, 198, 199
    see also Foucault, 'power/knowledge'
'power/knowledge', 3, 47,174–9, 187–8,
    189, 190, 199–201
Provisional National Council for Mental
    Health (PNC), 91, 94–6, 96–8,
    101–2, 105, 111
Psychiatric Social Work and the Family,
    48–9, 51
psychiatric social work, 1, 30, 34, 44–5,
    48–9, 51, 55, 63, 86, 91, 105–6, 111,
    112–14, 115–16, 117, 159, 160, 161,
    162, 180
psychoanalysis
    as exemplar of 'the Family', 6, 163
    Brian Kirman's criticism of, 128
    and Norman Mailer, 157
    importance to the mental hygiene
        movement, 34–5, 180–1
    and partial inversion of 'the Family',
        60, 66, 73
    and *Personal Aggressiveness and War*, 86
    Edward Shorter on, 99

Q Camps, 53, 59, 68, 75, 85, 117, 186
Q Camps Committee, 53, 59, 186
Quakers
    the Retreat, 1, 3
    and Order of Woodcraft Chivalry, 58–9
    and 'progressive' communities, 57, 58,
        93
    David Wills, 1
    Woodbrooke College, 7–8, 10–11, 46,
        67–8

Ratcliffe, T. A., 112, 114, 115
Raynes, Norma, 154, 167
Reed, Herbert, 120
Rees, J. R., 42, 100, 108
Rees, T.P., 137–9, 140, 141, 142, 147
Report on Juvenile Delinquency (Bradford
    Education Committee), 92–3

Retreat, the, 1–6, 8, 10, 18, 49–50, 74–5, 163

Reynolds House, 149, 152–3, 157, 162

'Richard', 102–3

Richmond Fellowship, 193

Rickman, John, 100

'Robbie', 61–2

Robinson, Kenneth, Rt Hon., 124, 166–7

Rolph, Sheena, 29 fn. 98

Rose, Nikolas, 47, 174, 177, 179, 184, 187–91, 199–200

Rosen, Maurice, H., 192

Rousseau, Jean-Jacques
  and Isaiah Berlin, 175
  and Bernard Bosanquet, 13–15, 70
  and Helen Bosanquet, 18
  and Hawkspur Camp, 53–4, 68
  and legitimate authority, 14, 89

Royal Commission on Lunacy and Mental Disorder, 69

Royal Commission on the Law Relating to Mental Illness and Mental Deficiency, 146–8, 150, 182–3

Royal Medico-Psychological Association, 137, 161

Russell Cotes Home and School of Recovery, 67–8

Rutter, Cuthbert, 59

Sandison, R. A., 139–40

Sans Everything: A Case to Answer, 166–7

Schochet, Gordon, 11

'school phobia', 158–9

Schools Action Union, 172

Scott, Rt Hon. Lord Justice, 28, 29

Scottish Education Department, 107

Sedgwick, Peter, 187

self-government
  and authority, 3, 11–15, 16–17, 18–19, 20–2, 23, 26, 41–4, 54, 59–60, 62, 66–7, 70–3, 78, 85, 142–3, 143, 178–9,186–7
  and Charity Organisation Society theorizing, 11–15, 16–17, 18–19, 20–2, 23, 26
  and mental deficiency (later termed 'mental handicap'), 25–7, 152
  and moral treatment, 3
  and 'paradox of self-government', 12–15

'power/knowledge', 175–6, 178–9

psychological and political definitions linked, Ch. 2 *passim*, 40–4, Ch. 4 *passim*, 84, 85, 86–90, 108–10, 112–13, 142–3, 143–4, 175–6, 185–7, 195, 198

Shaftsbury, Lord, 25

Shapiro, M.,127

Shearer, Anne, 167, 183, 185

Shephard, Ben, 99

Shepherd, W. S., 124

Shorter, Edward, 99

Simkhovitch, Mary Kingsbury, 16

Simon, Brian, 127

Simon, Joan, 127

'Simon', Ch. 4, 74

Smythe, Tony, (*see also MIND*) 146, 181–2, 192, 193, 194, 195–6, 198–9

Social After-Care Service (SACS), 111–15, 182

socialism
  and the Bosanquets, 21–2
  and J.C. Flugel, 108–9
  and infantilization, 21–2, 108–9
  'Keynesian', 68–9, 89, 143
  New Left, 143, 144–6
  and National Council for Civil Liberties, 121, 122
  and George Orwell, 57
  and *Personal Aggressiveness and War*, 85–6, 88, 89–90
  and power/knowledge, 187–8
  and psychoanalysis, 128
  'qualitative', 68–9, 89–90, 142–3, 145, 146, 149
  and 'rights', 187
  and R. H. Tawney, 68–9, 89, 90 fn. 293, 142–3, 145, 149
  'technocratic', 68–9, 89, 145
  and D.W. Wills, 20, 22, 120

Socialist Medical Association, 122

Soddy, Kenneth, 104–7, 112, 115, 116, 123, 132, 145

Some Lessons of Wartime Psychiatry, 105–7, 115

Spastics Society, 150

Stafford-Clark, David, 123, 125, 129

Steadman Jones, Gareth, 16–17

Sysonby House, 79

Tangye C.H.W., 97
Tavistock Clinic, 30, 41, 42, 55, 58, 70
    fn. 237, 100, 109, 114, 162, 185
Tawney, R. H., 68–9, 89, 90 fn. 293, 142,
    145, 149
The Family, 18–19, 22–3
The Hawkspur Experiment, 63, 100, 103,
    120, 160
Theosophical Fraternity in Education, 56
Theosophy, 56, 57, 58
therapeutic communities
    D.H. Clark, 141–2, 154
    importance regarding 'the Family',
        99–100
    Northfield Military Hospital, 98–100,
        101–2
    Laing and Cooper, 162
    D. V. Martin, 141
    MIND, 193
    T.P. Rees, 137–39, 140, 141, 142
    and Nikolas Rose, 190–1
    Villa 21, 168–9
This Matter of Mind, 125–30
Thom, Deborah, 127
Thomas, D.H.H., 147
Thomas, Ruth (*see also* Children Without
    Homes), 94–5 and 94 fn. 315
Thomson, Mathew, 34, 78
Titmuss, Richard, 146
Tizard, Jack, 131, 132–3, 150–2, 154,
    167, 183
'Tommy', 98
Torrie, Alfred, 111, 117, and 117 fn. 407
Townsend, Peter, 146, 167
Tredgold, R.F., 123, 125–7, 129, 132, 142,
    145
Tuke, Samuel, 4, 5, 6, 8, 9, 10, 13, 18,
    49–50, 74, 163

Unsworth, Clive, 69–70, 81–2, 147,
    148–9

Van Waters, Miriam, 36–7

War Resisters International, 181
Warlingham Park Mental Hospital, 137,
    139–40, 141
Webb, Beatrice, 24
Westhope Manor, 117

Westlake, Ernest, 58
White Lion Street Free School, 194
Whitehead, Anthony, 191–2
Wills, D. W., viii, 1, 5
    and anarchism, 77, 120, 172–3, 186
    Barns Hostel, 93, 100–1,107
    Bicester camp, 93
    and Evelyn Fox, 34, 149
    and Hawkspur Camp, Ch. 4 *passim*,
        84–5, 90–1, 185–7;
    and correspondence with
        ex-Hawkspur members, 84–5,
        90–1, 93, 98, 100–1, 102–3
    MIND – influence on, 193, 194.
    and New York Charity Organization
        Society, 17–18, 19–20, 31
    at New York School of Social Work, 1,
        7–8, 10–11, 16, 17–18, 19–20, 22,
        29–30, 32, 34, 36–7, 39–40 46, 47
    as Quaker, 1
    Reynolds House, 149, 152–3, 157, 162
    and Schools Action Union, 172–3
    and Woodbrooke College, 7–8, 10–11,
        46, 67–8
Wills, Ruth, 53, 59, 61, 84–5, 98, 100,
    101
Wilson, A.T.M., 100
    see also Civil Resettlement Service
Winn, Denise, 198–9
Winnicott, D. W., 97
Winston House, 161
Woodbrooke College, 7–8, 10–11, 46, 68
work
    employment categories fused with
        medical definitions, 38–9
    and Epsom Manor Mental Deficiency
        Institution, 80–1
    and Hawkspur, 74–6, 76–8, 186–7
    mental and moral value of, 4, 19,
        74–6, 186–7
    as register of intellect and status, 38–9
    as social and political disciplining, 8,
        19, 46–7
World Health Organization (WHO), 137,
    138, 141, 142

Young Communist League, 172
Young, Michael, 38